Datalevin: The Definitive Guide to Logical and Intelligent Databases

Huahai Yang

Calero Creek Press

Copyright

Published by Calero Creek Press.

First edition published July 12, 2026.

This edition updated September 2, 2026.

The examples and operational guidance in this guide are provided for educational purposes. Readers should verify behavior against the Datalevin version named in this guide and against their own production requirements.

To Yunyao and Boyan.

Your love, patience, and steady support made this work possible. Every page carries something of the time, trust, and encouragement you have given me.

Contents

Contents

Preface

SQL has been the default language of application databases for half a century. That default is now holding application state back. Datalevin[1] is a database built to replace SQL databases at the center of application systems: it stores data as small facts and queries those facts with Datalog [1].

Modern applications rarely fit neatly into SQL's table-first model. A product may need transactions, graph relationships, documents, full-text search, and vector similarity in the same workflow. The usual response is to split these capabilities across services or stretch a SQL database with extensions. Both responses create costs. Services copy data and push consistency into application code. SQL-centered designs have their own tax: queries live as strings inside application code, composition pushes teams toward ORMs or query builders, dialect differences leak through, and non-relational workloads retreat into extension-specific syntax.

Complex queries expose a deeper structural problem. In conventional row-shaped SQL storage, the values a query cares about are packed inside rows, where cross-condition statistics are hard to maintain, and missing values are often represented with `NULL`. The planner must estimate which conditions are selective and which joins will stay small; bad estimates can turn a reasonable query into a brittle performance problem.

Datalevin replaces the table as the center of gravity with the fact. A **fact** is a statement such as "Alice follows Bob" or "document 42 is in English": a logical assertion with one entity, one attribute, and one value. Together, those three elements form a **triple**. Datalevin combines the basic model with LMDB [2], a fast key-value store, and a careful physical layout. The result can represent relational records, graph relationships, documents, search hits, and vector neighbors as different views over the same durable facts.

To query these facts, Datalevin uses **Datalog**. SQL joined relational algebra to a structured-English surface, part of an era that tried to make computing approachable by making programs read like business prose. That choice was useful for interactive database work, but it is awkward inside application programs: a query becomes a string in one language describing work for another.

Datalog takes a different path. It comes from logic programming and asks which facts must be true together. Datalevin uses a user-friendly form of Datalog and makes it the main application query language: one query model for relational joins, graph traversal, rules, document predicates, full-text search, and vector or embedding similarity.

A query is just data. Even before the syntax is explained, the query reads like a small description of the facts to match:

```
[:find ?person-eid
 :where [?person-eid :person/name "Alice"]]
```

Application code can compose, generate, inspect, and test such data directly.

[1] The name joins LMDB's "Lightning" with "levin", a Middle English word for lightning.

Why Datalevin Can Replace SQL Database

Datalevin is not meant to be one more database beside SQL databases. Its case as the center of application state rests on two practical advantages: better developer ergonomics and better performance on complex application queries.

Developer ergonomics. In my experience, many developers do not miss SQL once they use Datalog in a real application. Here, "developers" includes both human programmers and LLMs that generate or revise application queries. This matters because a language that replaces SQL must support composition, inspection, testing, and generation in normal application code. A fuller example appears in the next section. There are two concrete reasons: Datalog is more **composable** in programs, and its declarative model stays close to application facts.

Datalevin queries are data structures, not strings assembled inside another language. A program can build a query by adding or removing pieces, combining common parts, or supplying parameters without concatenating SQL text or depending on a SQL query builder. The query remains ordinary data that can be inspected, transformed, tested, logged, and generated by tools.

Datalog is truly declarative and it maps directly to the fact model. A query describes the facts that must be true, and the engine finds the matching values. The programmer states what relationships and constraints should hold in the result, not how the engine should implement them. The system takes responsibility for planning the query, choosing an efficient execution order, and handling the database-specific details needed to produce the answer.

Together, composable query data and declarative constraints let a query express the logic of the application domain directly. The programmer can write in terms of people, documents, relationships, events, and constraints, instead of first translating that logic into database-specific jargon. This helps human developers, new team members, and LLMs for the same reason: there is less distance between the question being asked and the query that answers it.

Performance for complex application queries. The performance advantage follows from the same design. Datalevin's fact indexes keep data close to the shapes that Datalog queries ask for. This addresses the planner problem described earlier: the engine can count and sample candidate facts under query constraints, giving the query planner better information and reducing unnecessary intermediate work. Integrated full-text, vector, embedding, and document indexes also avoid shuttling candidate ids between separate systems before applying logical constraints.

The published Datalevin benchmarks make the challenge concrete. In the current Join Order Benchmark (JOB) snapshot, Datalevin completed 113 IMDb-derived join queries in 38.1 seconds. The retained PostgreSQL baseline took 128.2 seconds, and SQLite took at least 821.8 seconds when each of its nine timed-out queries is counted at the 60-second cutoff [3]. Datalevin was 3.37x faster than PostgreSQL and at least 21.59x faster than SQLite. In a separate LDBC Social Network Benchmark scale-factor-1 read-query comparison with Neo4j Community Embedded [4], Datalevin recorded the lower time on 20 of 21 query types. The equal-query geometric mean and summed time favored Datalevin by 4.996x and 7.046x, respectively. A representative OpenRuleBench run adds recursive-rule evidence: Datalevin recorded the lowest latency in all ten tasks in a field that included SQLite, PostgreSQL, XSB, Soufflé, Clara Rules, and O'Doyle Rules [9]. Across the three suites, the results provide evidence for Datalevin's Datalog engine, fact indexes, and optimizer on demanding relational, graph, and recursive-logic workloads. These are workload-specific observations, not universal performance guarantees; Chapters 13 and 21 explain their methodology and limits.

These advantages define Datalevin's purpose: retire SQL as the default database model for day-to-day application state. Application data belongs in a model that can be queried, searched, composed, and evolved without SQL strings, ORM ceremony, extension mini-languages, or cross-system glue code.

The same Datalevin engine can run embedded, as a server, or from the command line. For AI-oriented applications, embedding and text-generation providers are built in, and a bundled MCP server exposes Datalevin to AI agents without a separate adapter. Datalevin is the database for applications that have outgrown SQL as their center of gravity.

A Look in Code

The example below has four parts:

1. Declare two attributes: `:person/name` is a unique identity, and `:person/follows` is a reference to another person.
2. Open a connection.
3. Write three people: Alice follows Bob, and Bob follows Carol.
4. Query for Alice's friend and friend-of-a-friend.

```
(require '[datalevin.core :as d])

(def schema
  {:person/name    {:db/unique :db.unique/identity}
   :person/follows {:db/valueType :db.type/ref}})

(def conn (d/get-conn "/tmp/preface-demo" schema))

(d/transact! conn
  [{:db/id -1
    :person/name "Alice"
    :person/follows -2}
   {:db/id -2
    :person/name "Bob"
    :person/follows -3}
   {:db/id -3
    :person/name "Carol"}])

(d/q '[:find ?friend ?friend-of-friend
       :where
       ;; Start from Alice.
       [?alice-eid :person/name "Alice"]

       ;; Follow Alice -> Bob.
       [?alice-eid :person/follows ?friend-eid]
       [?friend-eid :person/name ?friend]

       ;; Follow Bob -> Carol.
       [?friend-eid :person/follows ?fof-eid]
```

```
        [?fof-eid :person/name ?friend-of-friend]]
      (d/db conn))
;; => #{["Bob" "Carol"]}

(d/close conn)
```

Read the query from top to bottom. It first finds Alice's entity id, follows `:person/follows` to Bob's entity id, reads Bob's name, follows one more `:person/follows` reference to Carol's entity id, and reads Carol's name. The result is the pair `["Bob" "Carol"]`.

The point here is Datalevin's basic shape: data is represented as facts, relationships are ordinary values, and queries are data. The database answers questions by matching facts, following references, and applying logic. Later chapters explain the syntax and APIs in detail.

How to Read the Examples

The print version of this guide shows Clojure example code for its conciseness. The web version at https://datalevin.org also shows Java, Python, and JavaScript examples. You do not need to know Clojure to follow the code listings; read them as EDN data plus a small number of Datalevin function calls. EDN [8] is a data notation similar to JSON. In EDN data, text after a semicolon, such as `;` or `;;`, is considered comment, not part of the data. Appendix B covers EDN. EDN has readers and printers in many languages, and Datalevin exposes EDN helpers or typed builders where useful.

The code examples follow one set of conventions:

- Clojure examples use EDN directly.
- Java examples use the high-level `datalevin` package, especially typed schema and transaction builders. Keyword strings start with a colon, such as `":person/name"` and `":tx-fn"`, the same as Python and JavaScript.
- Python examples import the `connect` factory with `from datalevin import connect`, bind the returned connection as `conn`, and call methods such as `conn.transact`, `conn.query`, and `conn.pull`.
- JavaScript examples import `connect` from `datalevin-node`. Connection methods are asynchronous and are shown with `await`.

Teaching examples use this variable convention: an internal Datalevin entity id ends in `-eid`, such as `?person-eid` or `?order-eid`. An application or domain identifier ends in `-id`, such as `?person-id` or `?order-id`. Variables such as `?name` or `?total` hold ordinary values.

Built for Performance and Flexibility

Datalevin's implementation is built for both performance and flexibility. Storage is performance-critical, so Datalevin uses a close-to-metal native code implementation for the transactional storage layer. The layers above storage need flexibility to support tight integration between the data model and a variety of paradigms and capabilities.

From those needs, a deliberately layered implementation follows. At the bottom, Datalevin uses LMDB as the native storage foundation for its robust transaction mechanism and raw read performance. Above that foundation, Datalevin uses the JVM for its portability, maturity, tooling, and extensive integration ecosystem.

In particular, Datalevin's higher layers use a data-oriented implementation style [5]: schemas, transactions, queries, and plans are represented as plain immutable data structures that can be inspected, transformed, tested, and passed around. This keeps the implementation close to the database's data model and makes the code easier to adapt to evolving requirements. It also reduces complexity by separating data from behavior: facts and queries remain explicit data inside the engine, not state hidden behind a large object hierarchy. Immutable data also helps with concurrency, because it can be shared, snapshotted, compared, and passed between threads with less hidden mutable state.

This layered approach leads to a mixed language implementation. LMDB is written in C. A Java layer packages and exposes the native LMDB operations to managed code. On top, a Clojure layer implements schema, queries, planning, and high-level APIs. The same implementation is exposed through Clojure, Java, Python, and JavaScript/Node.js APIs. Appendix A gives installation details for these language environments and links to the current compatibility matrix.

The result is a pragmatic stack: native code where the storage operations must be fast and obey strict transaction boundaries, and data-oriented managed code where database semantics must evolve. The remaining chapters cover this design in practice.

Why This Guide Exists

SQL is not a law of nature. As a successful compromise from an earlier era of application architecture, it has table-shaped records, string-shaped queries, and a sharp divide between storage, search, logic, and application code. Datalevin is a compact database system designed to replace that compromise for modern applications. It can be embedded in applications written in multiple languages, run as a server, and used from scripts.

This guide exists to make the replacement practical. A serious SQL replacement cannot stop at a nicer query language. It must explain transactions, schema, identity, indexes, performance, deployment, failure modes, and the discipline needed to model application state as facts rather than tables plus glue code.

The first goal is to teach Datalevin as a complete database, not just a query language: how to open it, transact data, read data, design schemas, model relationships, use the key-value layer, query with Datalog, tune storage, and operate the system.

The second goal is to change the modeling reflex. Datalevin works best when you stop treating rows, documents, graphs, search results, and similarity results as separate conceptual universes. They are different projections over durable facts. Once that shift lands, much of the machinery around SQL applications starts to look accidental.

The third goal is to show why this matters now. Large language models and agentic applications are pushing databases into a new role. A useful AI system needs persistent memory, scoped task state, retrieval with permissions and provenance, and a way to integrate new observations into a coherent long-term model. Part VI shows how Datalevin can serve as that memory substrate.

How This Guide Is Organized

Part I builds the foundation. It explains why Datalevin exists, how to get started, how to think in facts rather than containers, how LMDB shapes the storage model, and how attributes, entities, and namespaces work.

Part II covers the core APIs. You will transact data atomically, read with lookup, pull, and entity APIs, learn Datalog fundamentals, rules, recursion, and derived knowledge, and then drop down to the direct key-value API when sorted keys or custom storage structures are the right tool.

Part III is about modeling. It shows how relational, graph, and document patterns map onto Datalevin's fact-first model, and how to make schema choices that can evolve without losing clarity.

Part IV treats indexes as capabilities. You will learn the core EAV and AVE indexes, full-text search, vector and embedding search, and hybrid queries that combine those capabilities.

Part V is about performance and operations. It covers durability, storage tuning, batching, ingestion, query planning and diagnostics, recursive-rule execution, deployment, security, replication, monitoring, and production checklists.

Part VI applies the database to intelligent systems. It follows a path from persistent agent memory to episodic, semantic, and working memory; then to recall and context assembly; finally to apperception, truth maintenance, and stateful AI application patterns.

Part VII is reference material. It covers installation, runtimes, and deployment modes; EDN, the data notation used throughout this guide; schema keys; Datalog built-ins; and the core, key-value, and client APIs.

Who This Guide Is For

This guide is for engineers who build data-intensive applications and are ready to stop treating SQL as inevitable. If tables, SQL strings, ORMs, side indexes, and auxiliary services have made ordinary application data harder to model and change, this guide is written for you. You may be building an embedded application, a backend service, a knowledge graph, a search-heavy product, a durable workflow system, or an AI application that needs persistent memory.

That does not mean every byte in a system belongs in Datalevin. Warehouses, analytics stores, logging stores, object stores, and specialized infrastructure still have their jobs. The target here is SQL's default role as the place where application state, relationships, constraints, search, and derived knowledge are forced to meet.

This guide assumes basic database literacy: transactions, indexes, schemas, and querying. You do not need prior Datalog experience. This guide introduces Datalog from first principles, then returns to it throughout the modeling, search, performance, and intelligent-systems chapters.

How to Read This Guide

If you are new to Datalevin, read Parts I and II in order. They establish the mental model and the APIs that later chapters assume.

If you are designing an application schema, continue into Part III before optimizing anything. Good Datalevin design starts with clear facts, identities, and relationships.

If you are building search or retrieval features, read Parts III and IV together. The most useful retrieval systems combine modeling choices with the right indexes.

If you are operating Datalevin in production, read Part V after the foundations. The operational chapters make more sense once you know how data, indexes, and queries are represented.

If you are building AI systems, resist the urge to jump straight to Part VI. Persistent memory and agent state are only reliable when the underlying data model, transactions, indexes, and query semantics are clear. You should not consider Part VI to be a separate AI appendix; it is the natural consequence of the earlier parts.

For source code, issues, release history, and project documentation, use the Datalevin GitHub repository (https://github.com/datalevin/datalevin).

What You Should Take Away

By the end of this guide, you should be able to:

- move appropriate application state off SQL and into Datalevin's fact-first model;
- model application data as durable facts without losing relational, graph, document, search, or vector expressiveness;
- use Datalog as an application query language for joins, rules, recursive relationships, and exact constraints;
- choose when to use the Datalog layer and when to use the lower-level key-value API directly;
- design schemas and indexes that support correctness, performance, and change over time;
- operate Datalevin in embedded, server, scripting, and production contexts;
- build persistent-memory systems for AI applications without treating a chat transcript as the source of truth.

The larger argument of this guide is this: the application database should be a logical substrate, a place where facts are stored, relationships are traversed, knowledge is derived, evidence is preserved, and intelligent systems can maintain state over time. Datalevin is built around that idea.

Acknowledgments

Datalevin stands on a line of ideas and software that made this project possible. I am grateful to Rich Hickey, whose EDN-based Datalog flavor in Datomic [6] showed how logic queries could become a practical application programming interface; to Nikita Prokopov, whose DataScript [7] code was the starting point for Datalevin; and to Howard Chu, whose LMDB code became the storage foundation on which Datalevin is built [2].

I also want to thank the Datalevin users in the Clojure community. Their questions, bug reports, feature requests, examples, and production experience have shaped the project in ways that are hard to separate from the code itself. Thanks also to everyone who contributed code, wrote about

Datalevin, or supported the work through GitHub Sponsors: Dennis Heihoff, Anders Murphy, Aleksandr Bogdanov, Nils Grünwald, Christophe Grand, Garrett Hopper, Daniel Vingo, Samuel Ludwig, itonomi, Amar Mehta, Ryan Domigan, Lars Rune Nøstdal, Clay Hopperdietzel, Isaac Ballone, and many others I have missed.

Datalevin was originally motivated by needs at Juji, where intelligent applications required persistent memory, fast retrieval, and a more flexible way to represent user and agent state. I am grateful to Michelle Zhou, Amar Mehta, Wenxi Chen, Ransom Williams, Taejun Song, and other colleagues at Juji, whose work made those needs concrete and whose applications kept the project grounded in real systems.

This guide has already benefited from volunteer reviewers in the Datalevin community who read draft chapters, ran examples, questioned unclear explanations, improved phrasing, and made suggestions: Estevo U. C. Castro, Max Rothman, Norbert Wójtowicz, Jeroen van Dijk, Siavash Mohammady, Weidong Cai, Thomas Moerman, Amar Mehta, and Vlad Poh. I am grateful for their contributions. Any remaining mistakes are my own responsibility.

Huahai Yang
July 2026

References

[1] Stefano Ceri, Georg Gottlob, and Letizia Tanca, "What You Always Wanted to Know About Datalog (And Never Dared to Ask)," *IEEE Transactions on Knowledge and Data Engineering* 1(1):146-166, 1989. URL: https://hdl.handle.net/11311/665510.

[2] Symas, "Lightning Memory-Mapped Database (LMDB)." URL: https://www.symas.com/symas-embedded-database-lmdb.

[3] Datalevin project, "Join Order Benchmark," September 1, 2026 Datalevin run with August 23, 2026 PostgreSQL and SQLite baselines, methodology, and artifacts. URL: https://github.com/datalevin/datalevin/blob/master/benchmarks/JOB-bench/README.md#results.

[4] Datalevin project, "LDBC SNB Benchmark," August 23, 2026 comparison using the build intended for Datalevin 1.1.0, methodology, and implementation, commit `0ac42023`. URL: https://github.com/datalevin/datalevin/blob/0ac42023b330ceaf478e0a31b046f5a251f50dc2/benchmarks/LDBC-SNB-bench/README.md#current-local-datalevinneo4j-comparison.

[5] Yehonathan Sharvit, *Data-Oriented Programming*, Manning, 2022.

[6] Rich Hickey, "Datomic." URL: https://www.datomic.com/.

[7] Nikita Prokopov, "DataScript." URL: https://github.com/tonsky/datascript.

[8] EDN format project, "Extensible Data Notation." URL: https://github.com/edn-format/edn.

[9] Datalevin project, "Portable OpenRuleBench Tasks," August 26, 2026 representative run, methodology, and results. URL: https://github.com/datalevin/datalevin/tree/master/benchmarks/openrulebench.

PART I

Foundations: A Fact-First Database

CHAPTER 1

Why Datalevin

This chapter makes the preface's claim concrete: Datalevin is built to replace SQL's default role at the center of application state, not to sit beside SQL as one more specialized database.

The problem is larger than SQL syntax. Modern applications need transactional updates, relational joins, graph traversal, document-style flexibility, full-text search, and vector or embedding search in the same workflows. A SQL-centered design usually responds in one of two ways: stretch tables with extensions, or split capabilities across specialized stores. Both paths push incidental knowledge into application code: which rows own which relationships, which data was copied into a search index, which JSON fields duplicate relational columns, and which vectors must be refreshed after an update. This knowledge is not essential to the business domain. As it accumulates, the application becomes harder to change and more likely to sink into the software engineering "tar pit" [1].

Datalevin's answer is to replace the table as the center of gravity with the fact. A fact can participate in a row-like record, a graph edge, document metadata, a search result, or a vector-neighbor workflow without leaving the same durable database model. Datalog then queries those facts as ordinary data rather than as strings assembled inside application code.

Datalevin combines this fact-first model with a high-performance key-value storage layer, integrated indexes, and search capabilities.

The central idea is locality. Facts, indexes, documents, search entries, and vector data stay close enough that one ACID (atomicity, consistency, isolation, and durability) transaction can keep them consistent and one query model can combine them.

This chapter explains how that design replaces SQL's table-first compromise for day-to-day application state, how Datalevin unifies multiple access patterns, and where it fits in a modern stack.

1. A Unifying Data Model

At the Datalog layer, Datalevin represents facts as EAV triples: Entity, Attribute, and Value. Each assertion is one small, independently queryable fact, called a **datom** (data atom). Modeling data in terms of datoms produces a model that is compact, sparse, and composable.

Semantically, the four terms mean:

- **Entity (E):** the thing or identity the fact is about.
- **Attribute (A):** the named property or relationship being asserted.
- **Value (V):** the value asserted for that entity and attribute.
- **Datom:** one statement that an entity has an attribute with a value.

The database meaning is more specific:

- **Entity (E):** a database identity around which facts collect. In stored datoms, this is an internal numeric entity id (eid). An entity is the set of facts sharing that id, not a row locked to one table shape.
- **Attribute (A):** a named database property, usually a namespaced keyword such as `:user/email`, `:order/customer`, or `:doc/body`. Attributes can have value types, cardinality, uniqueness, indexing behavior, documentation, and transaction-time validation.
- **Value (V):** a Datalevin value. A value may be a string, number, keyword, boolean, reference to another entity, tuple, timestamp, document value, vector, or another supported value type.

Because the relationship between entity and attribute is open, the data model can be sparse: only asserted datoms are present in the database. Missing facts are represented by the absence of a datom rather than by placeholder values. This does not prevent storing a boolean `false` as an actual value; in that case, the datom is present and its value is `false`.

When surfaced in query functions, datoms are commonly represented as `[e a v]`. For example, the fact that entity `101` has Alice's email address is the datom `[101 :user/email "alice@example.com"]`.

This tiny shape is the first new concept to internalize. A datom is smaller than a row and flatter than a document, but it can express both. A group of datoms that share the same entity id can look like a row. A graph edge can be a datom whose value is another entity id. A document-like aggregate can be modeled as ordinary facts when its fields should join, validate, reference other entities, or change independently. Chapter 14 introduces the complementary choice: storing a nested map or JSON-like payload as one document value while Datalevin indexes its internal paths. The uniform shape makes these choices composable instead of isolated. Figure 1.1 shows those three readings of the same `[e a v]` shape.

What Is an Entity?

Entities also deserve a precise definition. An entity is not a table, class, or JSON object. It is the bundle of datoms anchored by one eid: the currently asserted properties and relationships that share the same database-local handle. If entity `101` has `:user/email`, `:customer/tier`, and `:employee/manager`, Datalevin does not need three separate containers to hold those facts. They are simply assertions about the same entity. This gives the model room to evolve as the application learns more about the thing being represented.

This makes Datalevin's notion of entity close to the metaphysical idea known as bundle theory: an object can be understood as the bundle of properties that belong together, rather than as a hidden substance underneath those properties [2]. The analogy should not be pushed too far: a Datalevin eid is not the object itself.

Because eids are database-local handles, they should not be confused with stable application identifiers; two databases containing the same application data may assign different sets of eids. Application domain identifiers such as user IDs, order IDs, document IDs, UUIDs, and slugs should be stored as ordinary attribute values, often with the `:db.unique/identity` property (see Chapter 3 for details). Internal relationships can still use entity references, so the database does not have to make every relationship depend on those application-facing identifiers.

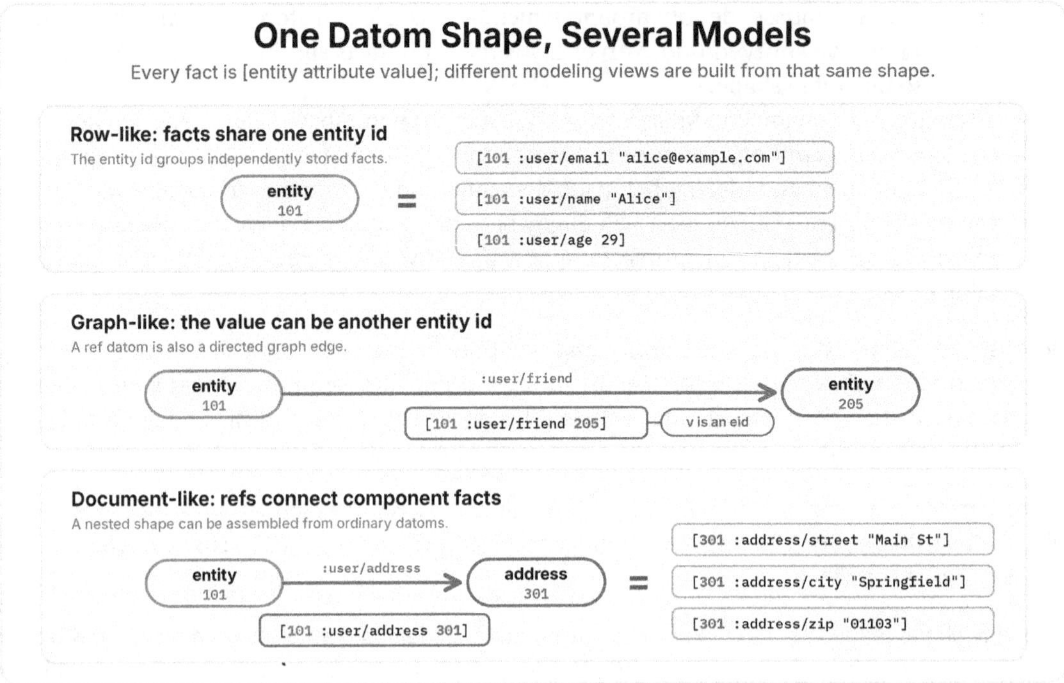

Figure 1.1: One datom shape can look row-like when facts share an entity id, graph-like when a ref datom's value is another entity id, and document-like when component facts are connected by refs

Aggregate Models vs Fact-First Model

SQL tables and document databases are typically aggregate-first: a row or document is the main unit of mutation and retrieval, and columns or nested fields are scoped under that container.

Fact-first systems start one level lower: each EAV assertion is atomic and independently queryable. Cross-aggregate joins, inference, and graph traversal can then operate over the same facts rather than over application-owned aggregates.

Instead of asking "which table owns this field?" you ask "what fact am I asserting?" Instead of asking "which document should duplicate this data?" you ask "which entity does this fact describe, and which other entities does it refer to?" You can still return an entity as a map, pull a nested shape for an API response, or index a JSON-like document for path queries. The difference is that these aggregate shapes are views over durable facts, not the only way to see the data. The same facts can also be joined relationally, traversed as a graph, searched by text, or combined with search constraints.

Compared with rigid table layouts, EAV handles sparse and evolving data naturally. The important shift is that there is no hidden wide table where every entity has a slot for every possible attribute. Datalevin stores the facts that exist. If no one has asserted `:user/timezone` for an entity, there is no `:user/timezone` datom for that entity. Queries that need the value match the fact; queries that care about absence can ask for missing facts explicitly. When the domain needs a meaningful value such as "unknown", "not applicable", or "withheld", model that value as data. Otherwise absence is simply absence, not a placeholder null.

> **Example**
> If only some users have a `:user/timezone`, only those users need that fact. If a later feature adds `:user/preferred-model`, existing entities do not need to be rewritten. The model grows by adding attributes and facts.

2. One Engine, Multiple Access Patterns

Datalevin exposes several complementary capabilities in one runtime. Each capability can be useful by itself, but the more important strength is that they compose over the same database state. As illustrated in Figure 1.2, Datalevin uses a single unified data model to support multiple database capabilities. This section gives an overview, while the remaining chapters fill in the details.

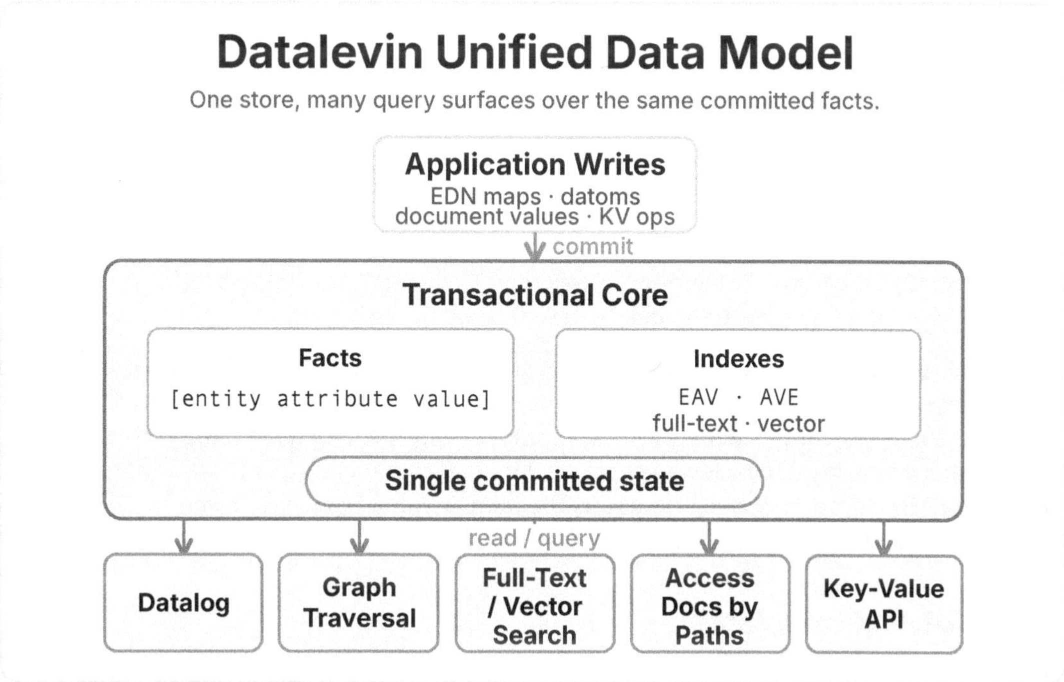

Figure 1.2: Datalevin unified data model

The same model also gives application code a clean place to observe change. When a write commits, Datalevin can report what changed and notify application listeners, so an embedded application does not have to poll or scatter change detection across search, graph, document, and cache subsystems. Chapter 6 covers that transaction-observation pattern after introducing transaction reports.

Relational and Graph Queries with Datalog

Datalog provides declarative joins, recursive rules, and reusable query composition. In practice, this lets Datalevin support both relational workloads and graph traversal workloads through one query model.

The same query language can ask for "all orders whose customer is in France" and "all dependencies reachable from this package." The first feels relational: facts are joined by shared variables. The second feels graph-oriented: entity references are followed as edges. In Datalevin both are expressed as logical patterns over datoms.

The underlying idea is classic Datalog: describe the facts that must be true, and let the engine find bindings for the variables [3]. Classic Datalog literature writes rules in Prolog syntax:

```
friend_of_friend(X, Z) :- follows(X, Y), follows(Y, Z).
```

This is a classic friend-of-friend query. The notation is abstract. The precise meanings of `:-`, `,` and `.` are not obvious from the notation alone, and the symbol followed by parentheses can look like a function call.

Datalevin keeps the logic but writes queries as EDN data structures. The same query looks like this:

```
[:find ?z-eid
 :in $ ?x-eid
 :where
 [?x-eid :person/follows ?y-eid]
 [?y-eid :person/follows ?z-eid]]
```

Read `:find` as the requested output, `:in` as the inputs supplied by the caller, and `:where` as the facts that must hold. `$` is the database input. Names that start with `?` are variables, and attributes such as `:person/follows` come from the schema. Combined, the example asks: find `?z-eid`, such that `?x-eid` follows `?y-eid`, and `?y-eid` follows `?z-eid`.

Each vector under `:where` is one fact pattern. `[?doc-eid :doc/lang "en"]` reads: bind an entity whose `:doc/lang` value is `"en"`. `[?order-eid :order/customer ?customer-eid]` reads: bind an order entity and the customer entity it points to. When the same variable appears in more than one pattern, Datalevin keeps only bindings that make all of those patterns true. In the friend-of-friend query above, `?y-eid` is the join: it is the person followed by `?x-eid` and also the person who follows `?z-eid`.

Direct Key-Value Access

The same key-value substrate that persists triples and indexes is also available directly as a durable key-value API. That substrate is DLMDB, Datalevin's extension of LMDB, the memory-mapped key-value store covered in Chapter 4. This low-level API is useful for low-latency state access, caches, and simple lookup-oriented components.

Use the key-value API when the access pattern really is just "given this key, read or update this value" or "read this key range." Examples include local application state, checkpoints, registries, counters, queues, or cached results. In those cases, forcing everything through a logical query layer would add complexity without adding much value.

Path-Indexed Documents with idoc

Datalevin supports path-indexed access to documents in EDN, JSON, and Markdown formats using the `:db.type/idoc` data type (Chapter 14). You can query nested structures with `idoc-`

`match`, including logical combinators and path predicates, while keeping the documents in the same transactional store.

This is useful because real applications rarely receive perfectly normalized facts at the boundary. Here, **normalized** means that each durable fact is stored once and related facts are connected by references, rather than copied or nested into every larger record that might need them. Applications receive JSON from APIs, EDN configuration, parsed Markdown metadata, tool outputs, and model-produced structures whose shape may change over time. An `idoc` lets you keep such nested data as a document while still indexing paths inside it.

For example, a documentation page might have structured metadata such as `{:module {:name "search" :status "stable"}}`. You may not want to promote every nested field to its own top-level schema attribute on day one. With idoc, Datalevin can still answer questions about paths inside the document and join those answers with ordinary facts about the same entity.

Idoc does not replace normal datoms. It complements them. Use datoms for core domain facts with stable meaning and relationships. Use idoc for nested, semi-structured, imported, or metadata-heavy values where path search is more important than turning every leaf into a named attribute.

Integrated Full-Text, Vector, and Embedding Search

Datalevin includes built-in full-text search, similarity search over vectors supplied by the application, and text embedding search. Because these indexes live in the same database system, you can combine search results and exact logical conditions in one Datalog query instead of synchronizing external search services.

These search modes answer different questions:

- Full-text search is lexical. It is good for matching words and phrases. Datalevin supports a variety of text analyzers and boolean search expressions. Details are in Chapter 16.
- Vector search is geometric. It is good for finding items whose vector embeddings are near a query vector under a similarity metric. See Chapter 17.
- Embedding search converts text to vectors. It is useful when the application wants meaning-based search over text fields instead of word matching.

In many systems, these capabilities live outside the primary database. The application first asks a search engine for candidate ids, then asks the database for structured facts, then performs more filtering in application code. Datalevin's integrated approach lets word search, meaning-based search, document-path search, and exact logical constraints meet inside one query.

The practical result is that an application can choose the access pattern that fits the problem without leaving the same data model. A query can ask for "documents like this question, containing this phrase, visible to this user, and marked stable" as one database question rather than as a sequence of cross-system calls.

3. Why Not Just SQL?

A possible objection to a SQL replacement built on Datalog is that modern SQL databases keep adding capabilities: JSON columns, full-text search, graph extensions, vector indexes, procedural languages, recursive CTEs, and extension ecosystems. For example, if PostgreSQL can do more and more of these jobs, why not stay with familiar SQL?

Familiarity is a real argument for SQL, but it is not the same as fit. Many teams stay with SQL because it is transactional, durable, mature, and already operationally understood. Many developers are not enthusiastic SQL users so much as resigned SQL users.

They want something more composable, more directly usable from programs, and less tied to table containers. But the available alternatives often give up too much: transactions, durability, joins, ad hoc queries, deployment simplicity, or operational maturity.

Datalevin is aimed at SQL's default application-state role. It is a database for developers who want to replace SQL in application systems, not merely wrap it, hide it behind an ORM, or add another specialized service beside it.

There are three reasons that go deeper than syntactic cosmetics, and they build on one another: SQL is awkward as an application interface, table-shaped storage makes query planning harder, and stacking SQL extensions turns every cross-cutting question into an integration problem.

SQL Is Awkward as an Application Interface

The first reason is about programming ergonomics. SQL succeeded as a standard, but standardization should not be confused with a good programming interface. SQL descends from SEQUEL, introduced as "a Structured English Query Language" [4]. That origin still shows in SQL's large English-like surface. The original ambition was to make database work feel closer to ordinary business language. That was useful for interactive querying, but it is awkward inside programs: SQL queries are usually strings assembled by a host language. Complex SQL often forces programmers to think in terms of table aliases, join syntax, and container-specific mechanics before they can express the relationship they actually mean.

The ORM and query-builder ecosystem is evidence that raw SQL is awkward at application boundaries. These tools soften many rough edges: SQL string construction, weak host-language composition, dialect differences, type mismatches, row/object mismatch, and boilerplate around joins, identity, transactions, lazy loading, and migrations.

Even tools that reject classic ORM behavior, such as query builders, LINQ-style systems, jOOQ, Ecto, and Prisma, are still trying to make SQL less like raw SQL and more like host-language data or typed expressions.

Datalog gives the programmer a smaller and more uniform surface. A query is data, not a string. It says which facts must be true, not which joins to perform step by step.

Reusing the same variable name connects facts. Rules name reusable logic. Recursive queries use the same logical form as non-recursive queries. The result is not less relational; it is a higher-level way to write relational logic. Relational algebra remains the substrate, but the user does not have to speak in join operators to ask a relational question [3].

The ergonomic difference is not limited to database specialists. In practice, experienced developers often prefer Datalog once they have used it for real application queries, because the clauses stay close to the facts they are asking for.

The contrast is even sharper for people without years of SQL habits: interns, domain experts learning to inspect data, and application developers who do not want to become query-planner whisperers before they can ask a basic question. A small set of data patterns, shared variables, and named rules is easier to teach than a large string language with dialects, aliases, join forms, grouping edge cases, and host-language impedance mismatch. This developer ergonomics goal was one of the original motivations for Datalevin.

The same simplicity matters for LLM-generated queries. Text-to-SQL is a hard problem not only because natural language is ambiguous, but also because SQL is a large, dialect-sensitive target language. A model must choose table aliases, join syntax, grouping rules, subquery shapes, vendor functions, and sometimes planner-sensitive rewrites.

Recent text-to-SQL results show the gap clearly. Spider 2.0 was designed around realistic enterprise workflows, and its authors reported much lower success rates than on older benchmarks such as Spider 1.0 and BIRD [5]. A later PExA result reported 70.2% execution accuracy on Spider 2.0 [6], which is real progress but still too much uncertainty for an automatic application interface.

Datalevin's Datalog is a more regular generation target because its surface is smaller and query structure is data. Most of the work is choosing facts, attributes, variables, predicates, and rules. The syntax is often compact compared with SQL's structured-English surface, and there are fewer dialect decisions for the model to hallucinate. Even though public Datalog training data is much smaller than SQL training data, the language itself gives both humans and LLMs a simpler target.

Datalevin's attribute-centered schema also gives an LLM better semantic handles. An attribute such as `:order/total` can carry its own type, cardinality, uniqueness, index behavior, and `:db/doc` documentation wherever that fact appears. SQL systems often support table and column comments, but they are vendor-specific metadata on table containers. In Datalevin, attribute documentation is ordinary schema metadata. It can be read, exported, and fed to a query-writing assistant along with the same attribute names used in Datalog clauses. That makes the schema more useful as prompt context and reduces the guesswork involved in choosing the right facts.

Table Storage Makes Query Planning Harder

The second reason is query planning. SQL databases are table-first systems. Data is bundled into row or column containers, and the optimizer must infer how many rows will survive predicates and joins over those containers.

Real application data is usually neither evenly distributed nor independent. A few customers may account for most orders; a few tags may appear on most documents; some vendors, regions, products, and refund patterns may strongly move together. Data distributions are called **skewed** when some values are much more common than others. They are **correlated** when one predicate being true changes the odds of another predicate being true. Skewed and correlated data make a query optimizer's job difficult because formulas based on uniformity or independence can produce wildly inaccurate estimates for them. Traditional estimators therefore depend on approximations such as histograms and, in some systems, learned models.

This difficult problem has a name: **cardinality estimation**, or estimating result sizes. It has been repeatedly identified as one of the central problems in query optimization in the database research literature [7,8]. For container-based storage, there is no cheap general answer. The counts the planner needs are not directly present in the storage layout, so the optimizer has to approximate them.

Consider a production-style question: find refunded late orders since a given time, for products from one vendor, shipped to customers in one region. In Datalevin, the logical form of that question is a set of fact patterns:

```
[:find ?order-id ?customer-email ?product-name
 :in $ ?vendor-id ?region ?since
 :where
 ;; Find the vendor and its products.
 [?vendor-eid :vendor/id ?vendor-id]
 [?product-eid :product/vendor ?vendor-eid]
 [?product-eid :product/name ?product-name]

 ;; Follow line items from products to orders.
 [?line-eid :line-item/product ?product-eid]
 [?line-eid :line-item/order ?order-eid]
 [?order-eid :order/id ?order-id]

 ;; Find the customer and restrict the shipping region.
 [?order-eid :order/customer ?customer-eid]
 [?customer-eid :customer/email ?customer-email]
 [?customer-eid :customer/address ?address-eid]
 [?address-eid :address/region ?region]

 ;; Require a late shipment for the same order.
 [?shipment-eid :shipment/order ?order-eid]
 [?shipment-eid :shipment/status :shipment.status/late]

 ;; Require a recent refund for the same order.
 [?refund-eid :refund/order ?order-eid]
 [?refund-eid :refund/created-at ?refund-time]
 [(>= ?refund-time ?since)]]
```

The actual execution plan in a database engine is not straightforward. The engine must decide whether to start from the vendor's products, the recent refunds, the late shipments, the region, or some other selective fact, and then join outward.

This kind of query is not exotic. It appears whenever an application keeps data normalized and asks compositional questions: which documents can this user see, which orders match a constraint spanning products, vendors, shipments, refunds, and customers, which clinical cases satisfy rules spanning diagnoses, providers, prescriptions, and coverage, or which papers connect authors, affiliations, topics, institutions, and collaborators. If a system never seems to have such queries, that is often because the complexity has been moved into denormalized records, application-side filtering, or cache materialization rather than disappearing.

In a SQL engine, this kind of question is usually a many-join query over table-shaped containers. The optimizer must estimate selectivity: how many rows survive each predicate and each join. The fact counts it needs are bundled inside rows, indexes, histograms, and assumptions about value independence.

When the estimates are wrong, the optimizer can start from a broad table, build huge intermediate results, and only later discover that a different predicate would have narrowed the search immediately. This makes complex SQL performance hard to predict. Large join queries become optimizer-sensitive.

For example, PostgreSQL's own documentation [9] notes that possible join orders grow exponentially, that exhaustive planning becomes impractical beyond roughly ten input tables, and that the planner switches to heuristic search for many-relation queries. It also documents techniques for constraining join order to reduce planning time. That is the practical pressure behind advice to denormalize, split a query, materialize intermediate results, or hand-guide the planner when SQL joins become too large or too sensitive to cardinality estimation [7].

A triplestore changes the shape of the problem. In Datalevin, a datom is an explicit cell: entity, attribute, value. Missing facts are absent rather than stored as nullable positions. Attributes are globally scoped. EAV (Entity-Attribute-Value) and AVE (Attribute-Value-Entity) indexes make many counts direct, and larger estimates can be sampled under the same query conditions that execution will use.

Datalevin still has to optimize queries, but it starts from a storage model that exposes the units the optimizer needs to count or sample [8]. That is why an optimizing Datalog engine on a triplestore has advantages: it can plan the complex question as one query without making the application hard-code a join order, denormalize the model, or split the question into many smaller queries.

Join Order Benchmark (JOB) is the benchmark version of this problem [10]. It is not a benchmark of simple key lookups or single-table scans, nor an academic exercise with synthetic data. JOB is based on the real IMDb dataset, where value frequencies and relationships have the same nonuniform shape seen in many production datasets. It requires joins across many related entities: 113 analytical queries, with query shapes that range from a few joins to more than a dozen. Its queries have enough joins that a poor query plan can produce orders of magnitude more intermediate rows than a good one. The background comparison is described in [11], and the current result is published in [12]. Datalevin completed the workload in 38.1 seconds, compared with the retained 128.2-second PostgreSQL baseline. SQLite took at least 821.8 seconds when each of its nine timed-out queries is counted at the 60-second cutoff. That makes Datalevin 3.37x faster than PostgreSQL and at least 21.59x faster than SQLite in this snapshot.

Stacking Extensions Becomes an Integration Tax

The third reason is integration. SQL extensions are valuable; they are one reason systems such as PostgreSQL remain useful for so many applications. The problem is not that extensions exist. The issue is composition.

Each extension often brings its own syntax, operators, indexes, cost model, and limitations. JSON paths, text search, vector similarity, recursive queries, and procedural code may all be available in one SQL database. But the application still has to compose several feature-specific surfaces and understand how the planner handles their interaction.

That flexibility is helpful when a workload needs one extension in isolation. It becomes a tax when a single application question needs search, document structure, relationships, similarity, and exact constraints to work together.

For example, suppose a documentation assistant needs to find pages that mention "vector", belong to a stable module in nested metadata, are visible to the current user, and are close to the user's question in embedding space. In a PostgreSQL-style design, this usually means combining a text-search expression, a JSON path or containment expression, a vector-similarity expression, and joins through permission tables.

Each part can be indexed, but each part also has its own operators, index choices, selectivity estimates, and tuning rules. The engineering cost is not only syntax. The application may need to maintain generated search columns or embedding columns, create several different index types, tune a query plan whose estimates cross extension boundaries, or fetch candidate ids from one subsystem and filter them in another. The runtime cost is extra memory traffic, intermediate result sets, and CPU spent reconciling candidates that were found through different access paths.

Datalevin's approach is to make these capabilities participate in one logical query model over the same facts. The same documentation question can be written as one Datalog query whose parts say, in one place: this entity matches the text, this entity has the required nested metadata, this entity is visible to the user, this entity is close in meaning to the question, and this entity has the exact attributes required by the application.

The cost shifts from application-level stitching to database-level planning over one fact-oriented model. When a bounded query has a compatible ordering, the optimizer can use the matching index incrementally rather than materializing every candidate before applying the remaining logic. Section 4 shows a smaller concrete version of this pattern.

Datalevin's argument is that a fact-first model, paired with Datalog, is a better foundation for applications whose data is relational, graph-shaped, sparse, evolving, searchable by words and meaning, and increasingly used by AI systems.

4. A Minimal Unified Query Example

Suppose you are building a documentation assistant. You want to find English documents that mention a term in free text, while also requiring a structured module status inside nested document metadata.

The examples below all perform the same four steps:

1. Declare a small schema. `:doc/body` is a full-text attribute, `:doc/idoc` is a nested-document attribute, and `:doc/lang` is an ordinary string fact.
2. Open a throwaway in-memory database.
3. Transact one document entity with a body, a language, and nested metadata.
4. Run one query that asks for document entity ids satisfying three constraints.

Read the query around `?doc-eid`. The full-text clause finds documents whose `:doc/body` mentions the search term, the idoc clause keeps documents whose nested metadata has `{:module {:status "stable"}}`, and the Datalog clause requires the exact fact `[?doc-eid :doc/lang "en"]`. The repeated `?doc-eid` variable is the join: all three clauses must describe the same document entity.

Listing 1.1. Minimal unified query across full-text, idoc, and Datalog

```
(require '[datalevin.core :as d])

;; Schema is optional, but this example declares it for type checking and
↪ indexes.
```

```
(def schema
  {:doc/id   {:db/valueType :db.type/string
              :db/unique    :db.unique/identity}
   :doc/body {:db/valueType :db.type/string
              :db/fulltext  true
              :db.fulltext/autoDomain true}
   :doc/lang {:db/valueType :db.type/string}
   :doc/idoc {:db/valueType :db.type/idoc}})

;; Create a throwaway in-memory DB connection.
(def conn (d/create-conn "/tmp/why-datalevin-search"
                         schema
                         {:kv-opts {:inmemory? true}}))

;; Transact one entity.
(d/transact! conn
  [{:db/id    -1
    :doc/id   "guide/search-overview"
    :doc/lang "en"
    :doc/body "Datalevin adds idoc indexing and vector search."
    :doc/idoc {:module {:name "search" :status "stable"}}}])

;; Run the query.
(d/q '[:find ?doc-eid
       :in $ ?term
       :where
       ;; fulltext/idoc-match return many document refs, each shaped as [e a v].
       ;; The outer vector binds each row; the inner vector destructures one
        ↪ row.
       [(fulltext $ :doc/body ?term) [[?doc-eid _ _]]]
       [(idoc-match $ :doc/idoc {:module {:status "stable"}}) [[?doc-eid _ _]]]
       [?doc-eid :doc/lang "en"]]
     (d/db conn) "vector")

(d/close conn)
```

In the query, `$` represents the database. The double brackets in `[[?doc-eid _ _]]` are a binding pattern for a function that returns multiple rows. The outer vector means "for each returned row"; the inner vector destructures one `[e a v]` row.

The key point here is composition: word search, structure-aware filtering, and exact logical predicates are executed in one query context over one database state. In a split architecture, this often requires multiple systems and intermediate joins. Here it is a single declarative query over shared facts.

5. One Model Across Deployment Modes

The design choices described above are meant to survive contact with real deployment. A query should still be data when it moves from a local prototype to a production service. Attributes should carry the same meaning in application code, shell scripts, and AI tools. The application should not have to redesign its data model merely because the operational topology changes.

Datalevin therefore treats deployment flexibility as part of the design criteria. Deployment mode should not remake the model.

You can begin with an embedded database in the same process as the application, move to client/server when sharing or centralized operations are needed, use replicas or HA[1] when availability requirements grow, and still script the same database from the command line. The facts, attributes, transactions, queries, and indexes remain the same across those modes.

Datalevin is designed to run where your application runs:

- **Embedded mode** for local, process-level access. This is the simplest mode: the application links to Datalevin and reads or writes a local database directly. Embedded access is supported from Clojure, Java, Python, and Node.js.
- **Client/server mode** for shared deployment and centralized operations. This is useful when multiple processes or machines should access the same service. Clients are available for the same supported environments.
- **Non-HA async read-only replicas** for simple read scaling on the server. Replicas can serve reads where eventual freshness is acceptable and automatic failover is not required.
- **Consensus-lease HA** for automatic leader promotion. This mode is intended for server deployments that need higher availability without changing the application data model.
- **Babashka pod mode** for scriptable workflows, automation, and lightweight command-line tools. Babashka [13] is a fast-starting Clojure scripting runtime.
- **MCP server mode** for local AI tool integration. Datalevin can expose persistent state and retrieval capabilities to MCP-compatible AI tools without requiring a separate adapter service. MCP [14] is the Model Context Protocol, a standard interface for connecting tools and data sources to AI clients.

This flexibility lets teams start small and evolve deployment architecture without changing core data and queries. The same schema, datoms, Datalog queries, idoc values, and search indexes remain identical as you move from a local prototype to a server deployment or an AI tool integration. Operational shape can change later; the conceptual model does not have to be redesigned each time. That continuity is the practical point: Datalevin is meant to feel like one database across development, production, automation, and AI integration, rather than a collection of special-case systems connected by glue code.

6. Where Datalevin Fits

Given these design choices, Datalevin fits where SQL has become the default home for application state mainly because it is familiar, durable, and operationally serious, not because SQL is the right language.

Datalevin targets that default role: the operational application database that must transact, query, search, evolve, and feed application or AI logic without scattering state across extensions, ORMs, and side services.

Datalevin is especially strong when one operational database system should support:

[1] HA means **high availability**: a deployment designed to keep the service available when a server fails. **Non-HA** replication means read-only replicas can serve reads, but they do not automatically become the writer if the primary fails. **Consensus-lease HA** is Datalevin's HA server mode: servers use a consensus protocol to agree which server holds the time-limited write lease, so another server can be promoted when the leader is no longer safe to use. Chapter 22 covers the operational details.

- application state with ACID transactions,
- relational and graph-style query patterns,
- document-style payloads with path indexing,
- search pipelines that combine full-text, vector similarity, and logic constraints.

This combination appears in more places than it may first seem. A SaaS product may need accounts, permissions, audit events, search, and recommendation features. A developer tool may need package dependency graphs, documentation search, local state, and structured metadata. An AI application may need episodic memory, semantic facts, vector recall, source documents, and strict filters that prevent irrelevant context from entering a prompt.

Datalevin is especially attractive when the boundaries between these needs are porous. If a search result must be filtered by permissions, joined to ownership metadata, checked against a lifecycle state, and then ranked with meaning-based signals, keeping those facts in one engine simplifies the application. In SQL, that kind of query often crosses ordinary joins, JSON expressions, search operators, vector extensions, permission tables, and application-side filtering. In Datalevin, the same application question can stay inside one fact-oriented query.

Another advantage is the ability to absorb change. Datalevin lets you start with a small schema, add declared structure where the application has learned enough to justify it, and keep optional facts absent rather than storing placeholder nulls. Declared schema is enforced when transactions write data. Because facts are isolated datoms and attributes are additive, new features can add new attributes without rewriting old entities or forcing every record through a table migration. Stable domain identifiers, lookup refs, and unique attributes let the application keep its own identities while Datalevin manages database-local entity ids. This makes the model well suited to systems whose shape changes as the product, data, or AI workflow matures.

Together, these choices make Datalevin a good default candidate for new applications whose data is already relational, graph-shaped, searchable, document-shaped, or AI-facing. It is also a modernization path for mature systems where SQL has become a friction point: too many handwritten joins, too much ORM ceremony, too many side indexes and auxiliary stores, or too much application code compensating for queries the database should be able to express directly.

That does not mean Datalevin should be the only database for all cases. If your workload is narrowly specialized, a single-purpose engine may still be the better choice. For example, a pure analytical warehouse, a massive append-only log pipeline, or a dedicated search cluster may be better served by systems built only for those jobs.

The target is broad but concrete: SQL's default role as the application database whose state must be modeled, queried, searched, evolved, and reasoned over in one coherent system. For developers who have been waiting for a database that keeps SQL's practical seriousness while offering a simpler, more composable query model, Datalevin is designed to be that replacement.

Summary

Datalevin is built to replace SQL database as the default center for application state when workloads cross boundaries between key-value storage, logical querying, graph relationships, document modeling, full-text search, and vector or embedding search. Its core move is to treat facts as the shared substrate and indexes as composable capabilities over that substrate. The remaining chapters develop these concepts in depth.

References

[1] Frederick P. Brooks, Jr., *The Mythical Man-Month: Essays on Software Engineering*, Anniversary Edition, Addison-Wesley, 1995.

[2] Howard Robinson and Ralph Weir, "Substance," *The Stanford Encyclopedia of Philosophy*, substantive revision May 6, 2024, especially Section 3.2, "Bundle theories versus substrata and thin particulars." URL: https://plato.stanford.edu/entries/substance/.

[3] Stefano Ceri, Georg Gottlob, and Letizia Tanca, "What You Always Wanted to Know About Datalog (And Never Dared to Ask)," *IEEE Transactions on Knowledge and Data Engineering* 1(1):146-166, 1989. URL: https://hdl.handle.net/11311/665510.

[4] Donald D. Chamberlin and Raymond F. Boyce, "SEQUEL: A Structured English Query Language," in *Proceedings of the 1974 ACM SIGFIDET Workshop on Data Description, Access and Control*, 1974, pp. 249-264. DOI: https://doi.org/10.1145/800296.811515.

[5] Fangyu Lei, Jixuan Chen, Yuxiao Ye, Ruisheng Cao, Dongchan Shin, Hongjin Su, Zhaoqing Suo, Hongcheng Gao, Wenjing Hu, Pengcheng Yin, Victor Zhong, Caiming Xiong, Ruoxi Sun, Qian Liu, Sida Wang, and Tao Yu, "Spider 2.0: Evaluating Language Models on Real-World Enterprise Text-to-SQL Workflows," arXiv:2411.07763, 2024. URL: https://arxiv.org/abs/2411.07763.

[6] Tanmay Parekh, Ella Hofmann-Coyle, Shuyi Wang, Sachith Sri Ram Kothur, Srivas Prasad, and Yunmo Chen, "PExA: Parallel Exploration Agent for Complex Text-to-SQL," arXiv:2604.22934, 2026. URL: https://arxiv.org/abs/2604.22934.

[7] Yuxing Han, Ziniu Wu, Peizhi Wu, Rong Zhu, Jingyi Yang, Liang Wei Tan, Kai Zeng, Gao Cong, Yanzhao Qin, Andreas Pfadler, Zhengping Qian, Jingren Zhou, Jiangneng Li, and Bin Cui, "Cardinality Estimation in DBMS: A Comprehensive Benchmark Evaluation," *Proceedings of the VLDB Endowment* 15(4), 2022. URL: https://arxiv.org/abs/2109.05877.

[8] Viktor Leis, Bernhard Radke, Andrey Gubichev, Alfons Kemper, and Thomas Neumann, "Cardinality Estimation Done Right: Index-Based Join Sampling," CIDR 2017. URL: https://www.cidrdb.org/cidr2017/papers/p9-leis-cidr17.pdf.

[9] PostgreSQL Global Development Group, "Controlling the Planner with Explicit JOIN Clauses," *PostgreSQL Documentation*. URL: https://www.postgresql.org/docs/current/explicit-joins.html.

[10] Viktor Leis, Andrey Gubichev, Atanas Mirchev, Peter Boncz, Alfons Kemper, and Thomas Neumann, "How Good Are Query Optimizers, Really?" *Proceedings of the VLDB Endowment* 9(3):204-215, 2015. URL: https://www.vldb.org/pvldb/vol9/p204-leis.pdf.

[11] Huahai Yang, "Competing for the JOB with a Triplestore," yyhh.org, 2024. URL: https://yyhh.org/blog/2024/09/competing-for-the-job-with-a-triplestore/.

[12] Datalevin project, "Join Order Benchmark," September 1, 2026 Datalevin run with August 23, 2026 PostgreSQL and SQLite baselines, methodology, and artifacts. URL: https://github.com/datalevin/datalevin/blob/master/benchmarks/JOB-bench/README.md#results.

[13] Babashka project, "Babashka." URL: https://babashka.org/.

[14] Model Context Protocol project, "What is the Model Context Protocol (MCP)?" URL: https://modelcontextprotocol.io/.

CHAPTER 2

Your First Datalevin Session

To get started, this chapter gives you one small interactive session with embedded Datalevin: open a database, transact facts, query them, close the connection, and reopen the same database. The remaining chapters build on this first session.

Before You Start

The printed book shows executable code examples in Clojure only. The web edition also includes Java, Python, and JavaScript examples.

Appendix A has the full installation instructions for language bindings, the standalone `dtlv` executable, Docker, Babashka, and MCP.

Add the Datalevin package for the language you want to use:

```
;; deps.edn
{:deps {org.datalevin/datalevin-embedded {:mvn/version "1.1.0"}}}
```

Datalevin uses native storage code through the JVM and requires Java 21 or newer. When starting the JVM directly, pass the required JVM options; language packages that manage the JVM for you handle these options automatically.

```
;; deps.edn
{:aliases
 {:datalevin/jvm
  {:jvm-opts ["--add-opens=java.base/java.nio=ALL-UNNAMED"
              "--add-opens=java.base/sun.nio.ch=ALL-UNNAMED"
              "--enable-native-access=ALL-UNNAMED"]}}}
```

Start a REPL with `clojure -M:datalevin/jvm`.

> **Troubleshooting**
>
> If the first connection-opening call fails, see Appendix A, Section 14 Troubleshooting, before changing the example code.

1. Open a Database

Import the public API namespace `datalevin.core`, define a directory path on disk, and define a small schema:

```
(require '[datalevin.core :as d])

(def db-path "/tmp/datalevin-guide-getting-started")
```

```
(def schema
  {:user/name {:db/valueType :db.type/string
               :db/unique    :db.unique/identity}
   :user/age  {:db/valueType :db.type/long}
   :user/city {:db/valueType :db.type/string}})
```

In all Clojure snippets, `datalevin.core` is given the alias `d` for brevity, so later calls can use `d/q` rather than `datalevin.core/q`, for example.

A Datalevin schema is a map from attribute keywords to attribute properties. Here, `:user/name` is an attribute. Its property map says that values should be strings and that the attribute is a unique identity, meaning it can identify an entity. If a later transaction uses the same user name, Datalevin can resolve it to the same entity instead of creating a duplicate entity.

Open a database connection at the `db-path` directory path defined above:

```
(def conn (d/get-conn db-path schema))
```

The path names the local database directory. The directory does not need to exist: Datalevin creates it, so the parent directory must be writable. If the database already exists, Datalevin opens it and uses the supplied schema to ensure the attributes are available.

2. Transact Facts

Transact two user entities to the database:

```
(d/transact! conn
  [{:user/name "Alice"
    :user/age  30
    :user/city "Paris"}
   {:user/name "Bob"
    :user/age  25
    :user/city "Berlin"}])
```

In the EDN notation, the transaction data is a vector of two maps, and each map describes facts about one user entity. Datalevin turns those maps into a list of datoms: entity-attribute-value facts.

Python and JavaScript build that same transaction with `tx.entity` and `tx.data`. Attribute-name strings passed to these typed builders do not need a leading colon; the builder turns them into Datalevin attribute keywords while leaving ordinary string values unchanged.

Logically, this transaction creates six datoms:

```
[10001 :user/name "Alice"]
[10001 :user/age  30]
[10001 :user/city "Paris"]

[10002 :user/name "Bob"]
[10002 :user/age  25]
[10002 :user/city "Berlin"]
```

In Datalevin, entity ids are 64-bit integers automatically assigned by the system. The exact numeric entity ids in the example above are illustrative only.

After the transaction data is converted into datoms, the datoms are encoded into binary form and inserted into the underlying key-value storage.

A successful transaction call returns a transaction report. In a REPL you may see keys such as `:db-before`, `:db-after`, `:tx-data`, and `:tempids` in the report; a transaction that introduces previously unknown attributes may also include `:new-attributes`. Some application code may want to inspect the report to see which datoms were transacted, which ids were generated, or other details worth auditing.

3. Query the Database

Read the current database state from the connection and query it:

```
(d/q '[:find ?name
       :where
       [?user-eid :user/name ?name]
       [?user-eid :user/age ?age]
       [(> ?age 28)]]
     (d/db conn))
;; => #{["Alice"]}
```

For now, read the query as: find the names of users whose age is greater than 28. Chapter 8 explains Datalog in detail. The important first point is that a query is also a piece of data: a vector of keywords and nested vectors.

The `:find` keyword names what the query should return. Here it returns `?name`. Python and JavaScript express the same query with the pure `q` builder: `q.var` creates Datalog variables, `q.datom` creates fact patterns, and `q.predicate` creates the filter clause. These builder values are ordinary, immutable host-language data; constructing them does not start the JVM.

In Clojure, the result is a set of tuples, so the one matching name prints as `#{["Alice"]}`. Java, Python, and JavaScript return the same relation as language-native list or array values. Chapter 8 covers the other result shapes.

The `:where` keyword is followed by a list of vectors, each of which is called a **where clause**. Two kinds of where clauses are shown here.

`[?user-eid :user/name ?name]` is a triple pattern, which describes facts that must be true for the query to return non-empty results. The entity position of the match pattern is a `?user-eid` variable, the attribute is given, and the value position is a `?name` variable. When a datom in the database matches this triple pattern, its entity id and value will be bound to these two variables, respectively.

`[(> ?age 28)]` is a predicate, a boolean function that must evaluate to `true` for the query to return non-empty results. > is the function name. This predicate tests whether the value bound to `?age` is greater than 28.

In Clojure, the whole query is prefixed by a quote character `'`. This prevents the query data from being evaluated as Clojure code. In Python and JavaScript, the typed `q` forms make the distinction explicit: `q.var("name")` is a Datalog variable, while the ordinary string `"?name"` remains a string value.

The function call `(d/db conn)` gives `d/q` access to the current database state, a DB object, as the read handle over the underlying storage plus cached metadata for that storage state. Get it when you are about to read, use it for that operation, and ask the connection again before making a later decision. A query runs against data, not against the connection object itself. Writes are different: `d/transact!` works through a connection because a transaction changes the database.

4. Close and Reopen

Close the connection when the process is done with it:

```
(d/close conn)
```

Because the database was persistent on disk at the specified path, the facts survive process restart. Open the same path again and query the data:

```
(def conn2 (d/get-conn db-path schema))

(d/q '[:find ?city
       :where
       [?user-eid :user/name "Bob"]
       [?user-eid :user/city ?city]]
     (d/db conn2))
;; => #{["Berlin"]}

(d/close conn2)
```

Next Steps

You have seen the core loop: open a connection, transact facts, query current database state, and close the connection. Chapter 3 explains the mental model behind those operations. Chapter 4 explains the storage layer that makes the operations fast and durable. Chapter 5 explains attributes and entities in depth. Appendix A covers details of Datalevin installation and deployment options.

CHAPTER 3

Datalevin Mental Model

Transitioning to Datalevin often requires a straightforward but fundamental shift in how you think about data. If you are coming from a SQL or document database background, your mental model likely revolves around “containers”: table rows with fixed columns or documents with nested structures. In these systems, the container is the primary unit of truth.

Datalevin asks you to look deeper, past the containers, to the atomic facts themselves. Moving from SQL to Datalevin is less like moving between different spreadsheet programs and more like moving from a spreadsheet to a knowledge graph: a network of things and relationships represented as facts. This chapter provides the narrative framework to help you navigate this shift.

The chapter moves through four stages:

- Sections 1-3 explain why Datalevin starts from facts and attributes rather than tables or documents.
- Section 4 gives the minimum transaction vocabulary needed to read the examples: assertions, retractions, references, and tempids.
- Sections 5-8 show the modeling and query habits that follow from the model.
- Section 9 ties the logical, storage, and query views together.

1. Think in Datoms, Not Containers

In traditional databases, if you want to record that Alice’s email address is `alice@example.com`, you must first decide which “box” that fact belongs in. You might create a `Users` table and add a row for Alice. The fact of her email address is then trapped within the context of that row and those other columns.

At the conceptual level, Datalevin discards these rigid boxes. Instead, it stores data as a collection of atomic facts called **datoms**. Each datom is a simple statement of truth represented as a triple: `[entity attribute value]`.

- **Entity (E):** The “who” or “what” (a unique 64-bit integer id in Datalevin).
- **Attribute (A):** The “property” (a keyword like `:user/email`).
- **Value (V):** The “data” (like `"alice@example.com"`).

From a traditional perspective, `e` and `a` are the two coordinates in a two-dimensional table view of values, and the value `v` is contained in the cell identified by the coordinates.

However, that view is limited to a single table. A better view is to imagine a giant ledger where every single cell of every spreadsheet in your entire company is cut out and pasted as an individual line. Each line tells you exactly which “row” it came from (Entity), what “column” it represents (Attribute), and what the “value” is.

This fact-first approach is powerful because it is **additive**. To add a new piece of information, like a user's preferred language, you do not need to "alter a table." You just add another datom: `[101 :user/preferred-language "en"]`.

2. Attributes Over Tables: The Fluid Entity

Because facts are stored individually, the concept of a "table" disappears. In its place is **attribute-centric modeling**: you describe the behavior of each attribute instead of defining a fixed row shape for each entity type.

In SQL, the schema belongs to the table: a `users` table might declare columns such as `id`, `email`, and `display_name`, while an `orders` table declares columns such as `id`, `customer_id`, `status`, and `total`. In Datalevin, the schema belongs to the **attribute**. A schema is a map from attribute names to behaviors: value type, cardinality, uniqueness, full-text indexing for searching text, references to other entities, and other properties. You define what a `:user/email` is, for example, a string and perhaps a value that must be unique, but you do not define what a "User" is in a rigid sense.

Three schema terms appear often:

- **Value type** describes what kind of value an attribute stores, such as a string, integer, timestamp, or reference.
- **Cardinality** describes how many values one entity can have for one attribute. `:db.cardinality/one` means at most one value; it is the default. `:db.cardinality/many` means the entity can have a set of values.
- **Uniqueness** describes whether a value can identify only one entity. `:db.unique/identity` is commonly used for natural keys such as emails, slugs, and external ids that uniquely identify an entity. It also enables **upsert**, meaning "update the existing entity if the identity already exists, otherwise insert a new one."

To update a value, for example, Alice's email address, you do not modify a row; you transact a new fact about the same attribute of the same entity. For a cardinality-one attribute, the new value replaces the old value for that entity and attribute. For a cardinality-many attribute, the new value is added to the entity's set of values. Uniqueness is a separate concern: it prevents two entities from claiming the same unique value and lets you look up an entity by a domain value such as an email address.

As illustrated in Figure 3.1, an **entity** in Datalevin is just a collection of datoms that happen to share the same entity id. A single id could have attributes associated with a `:user`, a `:customer`, and an `:employee` simultaneously. You do not "join" separate tables; you simply query for an id that possesses all those attributes.

This open-ended shape lets your data model evolve alongside your application. You never have to worry about "breaking the schema" by adding a new attribute to an existing entity, because there is no rigid "row" to break. Entity types are conventions you create by naming your attributes, such as the `person/` namespace in `:person/name`; they are not constraints enforced by the storage engine.

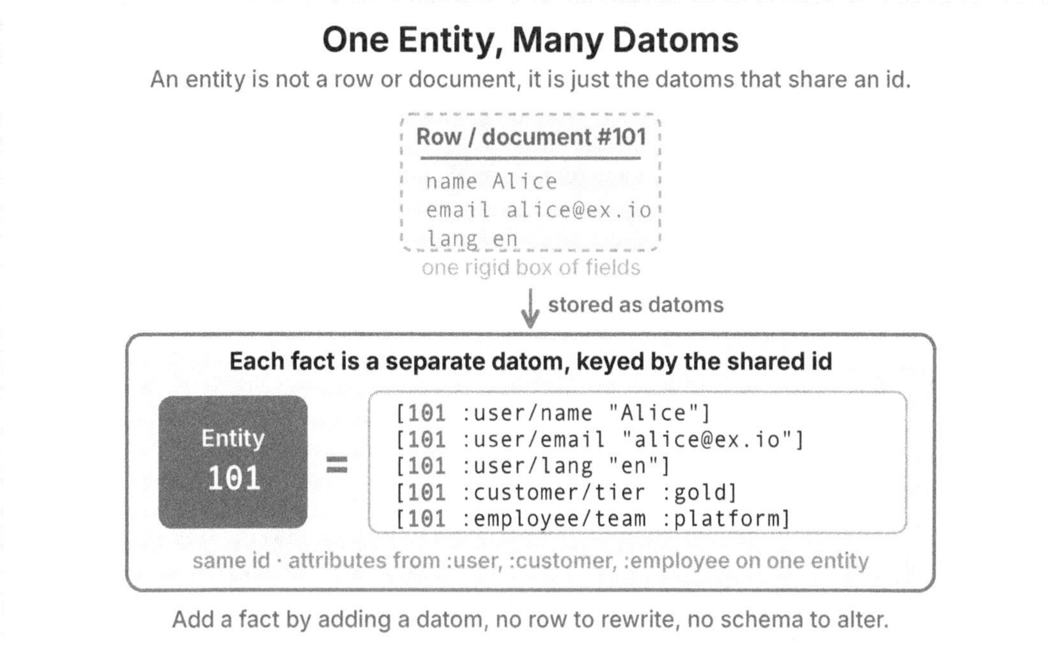

Figure 3.1: One entity as many datoms: a row or document is really the set of datoms that share an entity id

3. Data-Oriented Application Design

The previous two sections describe Datalevin's database model, but they also point to a broader application design principle: **data-oriented programming**. In this style, the essential state of the application is represented as plain, inspectable data, and behavior is expressed as functions and queries over that data rather than as hidden mutation inside object graphs [1].

Datalevin is a natural fit for this style. Schemas are maps. Transactions are data. Queries are data. Rules are data. Entity maps are data. Even a relationship is not an object pointer with private behavior; it is an explicit fact whose value is another entity id. That means the important parts of the application can be printed, logged, diffed, tested, stored, transmitted, and reasoned about with ordinary tools.

This does not mean applications have no behavior. It means behavior stays at the edges as named operations over data:

- A command handler receives plain input data and produces transaction data.
- A transaction commits facts that describe what changed.
- A query describes the facts that must be true for an answer.
- A projection or report turns query results into a shape the caller needs.
- A rule names reusable logic without hiding the underlying facts.

This is why later chapters repeatedly prefer data shapes over framework machinery. A transaction map is easier to inspect than an object with many mutating methods. A Datalog query is easier to test than a traversal hidden in a service class. A fact with provenance is easier to audit than a cached field whose origin is unclear. Part VI applies the same principle to agent memory:

episodes, facts, tasks, evidence, and working-memory slots are ordinary entities and attributes, not a special memory subsystem.

The practical payoff is not just aesthetic. Data-oriented programs are easier to debug, migrate, replay, and explain because the application state has a durable, queryable shape. Datalevin gives that style a database substrate: facts remain small, attributes remain additive, and queries compose across domains.

4. Operational Basics: Transactions and Tempids

This section switches from motivation to mechanics. It is not the full transaction API; Chapter 6 covers that. The goal here is to give you enough vocabulary to understand how facts enter the database and how new entities can refer to each other in one atomic change.

A database is not only a place to read facts. It is also a system for changing facts without leaving the application in a half-updated state. A **transaction** is a single atomic unit of change: all changes in the transaction succeed together, or none are applied. Transaction processing is the database discipline of making those changes atomic, isolated, and durable while many users or programs read and write concurrently [2].

A deeper motivation for transactions is that a database acts as a surrogate for the part of the external world your application cares about. Real-world events do not happen halfway. A customer either placed an order or did not. A payment either settled or did not. A job either moved from pending to running or it did not. The database may need several internal writes to record one such event, but the application wants the event itself to appear as one coherent change.

This is the core workload of **Online Transaction Processing (OLTP)** systems: applications record orders, messages, jobs, profile edits, tool results, and other current operational state. OLTP databases are usually optimized for many small concurrent reads and writes over current state, not for large historical scans or immutable log analysis. Datalevin is an OLTP database.

In Datalevin, a transaction is the boundary around assertions and retractions. An assertion adds a fact; a retraction removes a fact. One transaction can create entities, update cardinality-one attributes, add cardinality-many values, retract facts, and update custom key-value data in the same Datalevin store. After the transaction commits, readers see a consistent database state.

A small example makes this concrete. In transaction data, `:db/id` identifies the entity being written. If its value is a negative integer or string, Datalevin treats it as a temporary id, or **tempid**, and replaces it with a permanent entity id when the transaction commits. In the example below, MIT uses tempid `-10` because Alice's `:person/school` value points to it. Alice does not need `:db/id` because no other new entity refers to Alice.

```
;; The schema defines the behavior of attributes, not the shape of rows.
(def schema
  {:person/name    {:db/valueType :db.type/string}
   :person/email   {:db/valueType :db.type/string}
   :person/school  {:db/valueType :db.type/ref} ; A reference to the id of
    ↪ another entity
   :school/name    {:db/valueType :db.type/string}
   :school/country {:db/valueType :db.type/string}})
```

```
;; Data is written as a collection of entities here.
;; Notice how 'Alice' and 'MIT' are just sets of attributes.
(d/transact! conn
  [{:person/name "Alice" :person/email "alice@example.com" :person/school -10}
   {:db/id -10 :school/name "MIT" :school/country "USA"}])
```

Tempids are local to one transaction. In map-style transaction data, if the transaction does not need to connect new entities to one another, `:db/id` can be omitted and Datalevin will assign permanent entity ids automatically.

> **Warning**
>
> Do not put application identifiers in `:db/id`. A permanent Datalevin entity id is a system-managed `long` integer, not a UUID, email, slug, or external primary key. The value is local to one database: two databases with the same application data may assign different eids. Use a separate unique identity attribute, such as `:user/id` or `:user/email`, for application identity. Application code can then address the entity with a lookup ref instead of depending on the internal id.

Chapter 6 covers transaction data shapes, transaction reports, listeners, transaction functions, and durability settings in detail. For now, keep one mental model in view: Datalevin changes the database in coherent committed steps, not by mutating one fact at a time in isolation.

5. Store Each Fact Once, Then Link Entities

With the basic write vocabulary in place, the modeling habit worth practicing is to think in datoms: store each durable fact in one place, then connect entities with references. Suppose a customer places many orders. You could copy the customer's email address into every order, but then one email change would require many writes. In Datalevin, you usually store the customer's email once on the customer entity, and store a reference from each order to that customer.

The connection is itself a fact. If entity `1001` is a customer and entity `2001` is an order, the relationship can be represented as another datom:

```
[1001 :customer/email "alice@example.com"]
[2001 :order/number "A-100"]
[2001 :order/customer 1001]
```

The third datom says that order `2001` points to customer `1001`. Nothing is embedded inside the order, and the customer's email is not copied into the order. The facts stay separate, but the shared entity id connects the order facts to the customer facts.

Database terminology calls this **normalization** [3]: representing each durable fact once and connecting related entities by references. The opposite is **denormalization**: copying or nesting the same information into several larger records so each read can fetch a preassembled shape. Both styles have uses. If an order should always show the customer's current email address, store the email on the customer entity and link the order to that customer. If the order must preserve the email address used at purchase time, store that value on the order as its own historical fact. Datalevin's fact model makes the normalized style natural, while still letting you store intentional snapshots when the domain needs them.

The database term for following this kind of connection is a **join**. A join combines facts that share a value. In this case, a query can read the `:order/customer` value from the order, treat that value as the customer entity id, and then read the customer's email from that entity.

In Datalevin, all facts are stored in a unified set of indexes. An **index** is a sorted data access path optimized for a particular lookup order. Joining an `Order` to a `Customer` is a matter of looking up datoms that share a value.

This is why storing facts once works well in Datalevin. Updates stay granular: you can update a customer's email without touching thousands of order records. Indexes stay useful because every datom participates in the core indexes, and AVE indexing is automatic for every attribute. When each fact is stored once, the engine can use its EAV (Entity-Attribute-Value) and AVE (Attribute-Value-Entity) indexes to jump directly to the relevant facts.

Query planning also gets clearer. Datalevin's cost-based optimizer estimates how many datoms each query clause might match, then chooses an execution order. Storing each fact once gives the optimizer a cleaner picture of how to order joins for speed. This estimate is also called cardinality in query planning; do not confuse it with the schema property `:db/cardinality`, covered in Chapter 5's attribute properties section. For an example of how this affects many-join query performance, see Chapter 1's "Table Storage Makes Query Planning Harder."

6. Datalog: Querying as Logic Programming

The next three sections are query intuition, not a complete language reference (see Chapter 8). They show how the fact model affects ordinary reads, search, and reusable query logic.

If SQL is like giving the database a set of instructions on how to assemble a result table, Datalog is like giving the database a **description** of the answer you want. That declarative style comes from **logic programming** [4]: programs are expressed as logical clauses, and evaluation searches for values that make those clauses true.

You are not telling the engine to "JOIN table A with table B." You are stating: "I am looking for an entity `?user-eid` that has attribute `:user/name` with value `'Alice'`, and that same entity `?user-eid` must also have an attribute `:user/email` whose value is `?email`."

```
(d/q '[:find ?email
       :where
       [?user-eid :user/name "Alice"]
       [?user-eid :user/email ?email]]
     db)
```

The syntax has a few pieces:

- `:find` describes what the query returns.
- `:where` lists the clauses that must all be true.
- Symbols beginning with ?, such as `?user-eid` and `?email`, are variables.
- Reusing the same variable in multiple clauses creates a join by **unification**: the engine finds values that make all occurrences of that variable agree.
- `db` is a Datalevin DB object used as the query input.

The engine takes these statements and finds the set of values that makes all of them true simultaneously. Because joins are implicit (represented by the shared variable `?user-eid`), the query stays declarative as joins are added: you state the facts that must agree, and the engine chooses the index walks and join order.

Now that you have seen `db` used as a query input, it is worth being precise about what it is. Datalevin's DB object is a read handle to storage plus cached metadata about the storage state, not a durable immutable snapshot. In application code, keep and share the connection. When a Clojure read API needs a DB object, call `(d/db conn)` at the read site and use that object for the operation. Connection-level APIs in Java, Python, and JavaScript do this for ordinary reads.

The useful mental model is perception, not possession. The database is the surrogate of an external world that changes. Calling `d/db` is how Clojure code takes a read handle for the operation it is about to perform. After writes, read again from the connection instead of treating a previous read object as live application state.

Attributes are queryable too. In a datom pattern, the entity, attribute, and value positions can all be variables. For example, this query asks which attributes are present on entities that have a `:person/name`:

```
(d/q '[:find ?attr
       :where
       [?person-eid :person/name]
       [?person-eid ?attr]]
     db)
```

The first pattern finds entities with a `:person/name`; the second pattern binds `?attr` to each attribute asserted for those same entities.

7. Unified Retrieval: Plugging into the Datom Flow

One of the most powerful aspects of the Datalevin mental model is that "special" features like full-text search, vector similarity, and document indexing are not separate services running beside the database. Full-text search finds text by terms (Chapter 16), vector similarity finds nearby numeric vectors (Chapter 17), and document indexing (chapter 14) makes paths inside nested document values queryable. They are integrated directly into the Datalog flow.

When you perform a full-text search, the search function behaves just like any other Datalog clause. It returns a set of datoms (`[e a v]`) that can be immediately joined with other relational facts. The examples below destructure each returned datom as `[[?doc-eid _ _]]`: keep the entity id, and ignore the attribute and value. `_` is a placeholder symbol for any value, and Datalevin always ignores it instead of binding it to a variable.

The query assumes `:doc/body` is declared with `:db/fulltext true` and `:db.fulltext/autoDomain true`. In the `fulltext` call, `$` means "the database input to this query," the same database handle supplied to the query call.

```
;; A hybrid query combines text search with exact metadata filters.
(d/q '[:find ?title
       :where
       [(fulltext $ :doc/body "clojure") [[?doc-eid _ _]]] ; Keep eid from
 ↪   returned datoms
       [?doc-eid :doc/status "published"]                  ; Standard datom join
```

```
    [?doc-eid :doc/title ?title]]                     ; Retrieve the title
  db)
```

This unified retrieval model means you do not have to sync your database with an external search engine. With the default synchronous indexing mode, search results are transactionally consistent with your data: they reflect the same committed database state as the ordinary datoms in the query.

8. Rules: Reusable Query Logic

Rules are how you encapsulate and reuse logic in Datalevin. A **rule** is a named Datalog pattern. Query clauses can call the rule by name, and Datalevin expands that call into the rule body during query evaluation. If you find yourself frequently querying for "Active Users who have Premium subscriptions," you can define a rule called `premium-user`.

Rules are supplied to a query as an input. In `:in $ %`, `$` is the database input and `%` is Datalevin's rule-set input. The clause `(premium-user ?user-eid)` calls the rule and binds `?user-eid` to each entity that satisfies the rule body.

```
(def user-rules
  '[[(premium-user ?user-eid)
     [?user-eid :user/status :user.status/active]
     [?sub-eid :subscription/user ?user-eid]
     [?sub-eid :subscription/tier :subscription.tier/premium]
     [?sub-eid :subscription/status :subscription.status/active]]])

(d/q '[:find ?email
       :in $ %
       :where
       (premium-user ?user-eid)
       [?user-eid :user/email ?email]]
     db user-rules)
```

The rule body records what the policy means: the user is active, has a subscription entity, and that subscription is both premium and active. If the policy changes, the calling queries can stay focused on the business concept rather than repeating all of those clauses.

Rules can also be **recursive**, meaning a rule can call itself. Recursion is essential for graph traversal because it lets a query follow an unknown number of edges. A classic example is finding ancestors in a family tree:

```
(def ancestry-rules
  '[[(ancestor ?x ?y)         ; Rule name and arguments
     [?x :person/parent ?y]] ; Case 1: y is the direct parent of x

    [(ancestor ?x ?y)         ; Case 2: Recursion
     [?x :person/parent ?z] ; z is the parent of x
     (ancestor ?z ?y)]])     ; y is an ancestor of z
```

By defining this rule, the database has a named `ancestor` relation. The query below calls that rule as `(ancestor 42 ?ancestor)`:

```
(d/q '[:find ?ancestor
       :in $ %
       :where
       (ancestor 42 ?ancestor)]
     db ancestry-rules)
```

You can now ask "Who are all the ancestors of entity 42?" and the engine will navigate the graph for you. Chapter 9 discusses rules and their use in detail.

9. Three Views of the Same Database

After seeing transaction data, queries, and rules, it helps to name the layers Datalevin keeps separate. Most application code lives in the logical view: datoms, schema, transactions, and queries. Beneath that, Datalevin stores those facts in sorted indexes. At runtime, the Datalog engine connects the two. These are not three different databases. They are three ways of looking at the same set of facts, as shown in Figure 3.2.

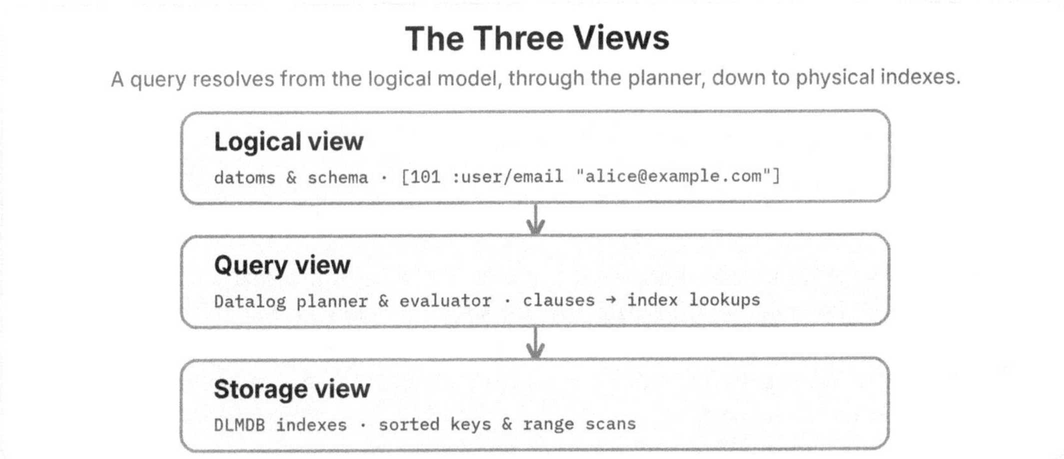

Figure 3.2: The three views of a Datalevin database: a query resolves from the logical view (datoms and schema), through the query view (the Datalog planner and evaluator), down to the storage view (DLMDB indexes of sorted keys read with range scans)

Logical View: Datoms and Schema

This is the view you use when designing your application. You think in entities, attributes, values, references, and schema. At this level, a fact such as "Alice's email is alice@example.com" is simply a datom: `[101 :user/email "alice@example.com"]`.

The logical view is intentionally small. It gives you a uniform way to describe people, documents, orders, permissions, embeddings, and relationships without inventing a new container shape for every case.

Storage View: Indexes and Sorted Keys

The storage view is where logical facts become sorted indexes. Datalevin persists them in DLMDB, its extension of LMDB, so common access patterns can be served by ordered key lookups and range scans.

You do not normally model at this layer. It matters when you want to understand performance or use Datalevin's direct key-value API for simple, low-level state.

Query View: Datalog, Planning, and Evaluation

The query view is the **Datalog engine.** It evaluates Datalog queries by translating logical clauses into index lookups and joins over the storage layer. A cost-based optimizer decides the order for retrieving and combining facts.

This view is what lets you stay focused on the shape of the answer instead of manual traversal. You describe the relationships that must hold, and the engine handles joins, filtering, recursion, and lookup order.

Summary

The Datalevin mental model is about moving from "data-in-boxes" to "data-as-facts."

- **Datoms** are indivisible entity-attribute-value facts, the basic units for modeling information in Datalevin.
- **Schema attributes** define the behavior of the attribute position in facts: value type, cardinality, uniqueness, references, and indexing behavior.
- **Data-oriented application design** keeps important state plain, inspectable, and queryable.
- **Transactions** move the database from one coherent state to the next.
- **Storing each fact once and linking entities** keeps the model clean and the engine fast.
- **Datalog** lets you query by describing the logic of your answer.
- **Unified retrieval** lets search, vector, document, and exact filters compose as Datalog clauses over the same facts.
- **Rules** let you build complex, reusable, and recursive logic.
- **The logical, storage, and query views** are different ways to understand the same database.

Using this model gives you flexibility that is harder to get from fixed table-shaped or document-shaped records.

References

[1] Yehonathan Sharvit, *Data-Oriented Programming*, Manning, 2022.

[2] Jim Gray and Andreas Reuter, *Transaction Processing: Concepts and Techniques*, Morgan Kaufmann, 1993.

[3] E. F. Codd, "Further Normalization of the Data Base Relational Model," IBM Research Report RJ909, August 31, 1971. Republished in Randall J. Rustin, ed., *Data Base Systems: Courant Computer Science Symposia Series 6*, Prentice-Hall, 1972.

[4] Robert A. Kowalski, "Predicate Logic as Programming Language," IFIP Congress, 1974, pp. 569-574. URL: https://www.doc.ic.ac.uk/~rak/papers/IFIP74.pdf.

CHAPTER 4

Storage Fundamentals

Datalevin's query model is fact-centric, but its performance and reliability are grounded in its storage architecture. Unlike many databases that build custom buffer managers and page caches, Datalevin leverages the operating system's existing strengths by building on **LMDB** (Lightning Memory-Mapped Database). Implementation-wise, this storage foundation is a native layer exposed upward through the JVM; Datalevin's Clojure database layer builds schema, Datalog, and indexing semantics on top while leveraging LMDB's transaction machinery.

This chapter explores why Datalevin chose this foundation, how it extends LMDB into **DLMDB**, and how the physical layout of data supports efficient Datalog queries, secondary indexes, and high-throughput writes.

1. The Foundation: Why LMDB?

LMDB is a small embedded key-value store with a deliberately narrow design. Its philosophy is to do as little as possible in user space, offloading memory management to the operating system (OS) kernel. LMDB does not know Datalevin attributes, datoms, or schema. It stores and orders raw bytes. Datalevin encodes typed values into those bytes and decodes them back into typed values.

1.1 Why Memory-Mapping Fits Datalevin

LMDB uses **memory-mapping (`mmap`)**. It treats the database file as if it were a large array in the process's virtual address space. This does not mean the whole database file is loaded into memory. When Datalevin reads data, the OS kernel handles fetching the required pages from disk into the file-system **page cache**. This is the operating system's cache of file-backed pages in main memory, not the CPU's L1/L2/L3 hardware caches. A page is a fixed-size block of virtual memory that the OS maps, caches, and evicts as a unit.[1] The memory map reserves a range of virtual address space; physical memory is used for pages that are actually touched, and clean pages can be evicted by the kernel when memory is needed elsewhere. LMDB deliberately made this architectural choice, because the OS already manages file-backed pages across all processes on the machine, and it can evict, prefetch, and share those pages with knowledge of the whole system's workload.

Traditional databases often manage their own "buffer pool," a chunk of memory where they keep data pages. That design makes sense when the database is the main program on the machine and wants tight control over every page of memory. But many modern databases do not own the whole computer. They run in cloud VMs and containers beside other services, on desktops

[1] Common page sizes are 4KB on many Linux and Windows systems and 16KB on macOS with Apple Silicon. The exact size is platform dependent; the conceptual point is that memory-mapped files are brought into memory in page units.

beside user applications, or embedded inside application servers. In those environments, the OS has a broader view of memory pressure than any one database process can have.

A database-owned buffer pool sees only the database's allocation. It may keep a page in its own cache while the OS also keeps the same page in the file-system cache, causing "double buffering." It also competes with the rest of the machine through a fixed memory budget chosen ahead of time. With mmap, Datalevin lets the OS page cache be the buffer cache. If another process needs memory, clean file-backed pages can be reclaimed by the kernel and read again later from the database file.

This does not mean mmap is always the right design for every database. It shifts some control from the database engine to the OS kernel, so systems that need custom I/O scheduling, specialized eviction policies, or tight control over page-fault latency may choose a private buffer manager. Datalevin's use case is different: it favors an embedded, low-administration engine that cooperates well with the rest of the host. This design has several benefits:

- **Zero-Copy Reads**: On the read hot path, Datalevin avoids copying bytes from the mapped database file into a separate JVM heap byte array. Because LMDB memory-maps the file, the native layer hands the address of the LMDB value memory to Java as a `DirectByteBuffer`. Datalevin's Clojure runtime then reads from that buffer and decodes the bytes into typed Datalevin values. This bypasses allocation of an intermediate byte array in Datalevin's JVM heap and minimizes garbage-collection overhead.
- **OS-Wide Cache Management**: The page cache is shared with the rest of the machine. This is a strong fit for cloud deployments, desktop applications, and embedded applications, where Datalevin is one component among many. The kernel can make better global decisions about clean page eviction than a database process that only sees its own heap and buffer pool.
- **Less Configuration**: A private buffer pool needs a size. Set it too small, and the database misses cache opportunities. Set it too large, and it starves the application or the rest of the host. mmap lets Datalevin rely on the OS to grow and shrink the effective cache according to real memory pressure.
- **Immediate Recovery for Direct Commits**: In LMDB's direct commit mode, crash recovery for the LMDB file does not require a long log-replay process. Section 1.2 explains the copy-on-write root-switching mechanism this relies on; Datalevin's optional WAL mode is a separate layer discussed later in this chapter.

In the default file-backed mode, this memory map is **read-only**. That detail is important because many objections to mmap focus on writable mappings: accidental memory corruption, dirty-page writeback surprises, or weaker control over when modified pages reach storage. Default write transactions use LMDB's copy-on-write commit path rather than updating pages through that mapping. `:writemap` is a separate opt-in with additional safety and performance caveats; combining it with `:mapasync` also weakens durability.

1.2 Read-Heavy Concurrency

LMDB provides **MVCC (Multi-Version Concurrency Control)** [1]. Readers never block writers, and writers never block readers. An illustration of LMDB's MVCC is in Figure 4.1. The easiest way to understand this, especially for Clojure programmers, is to think of LMDB's B+Tree as a persistent immutable data structure [2].

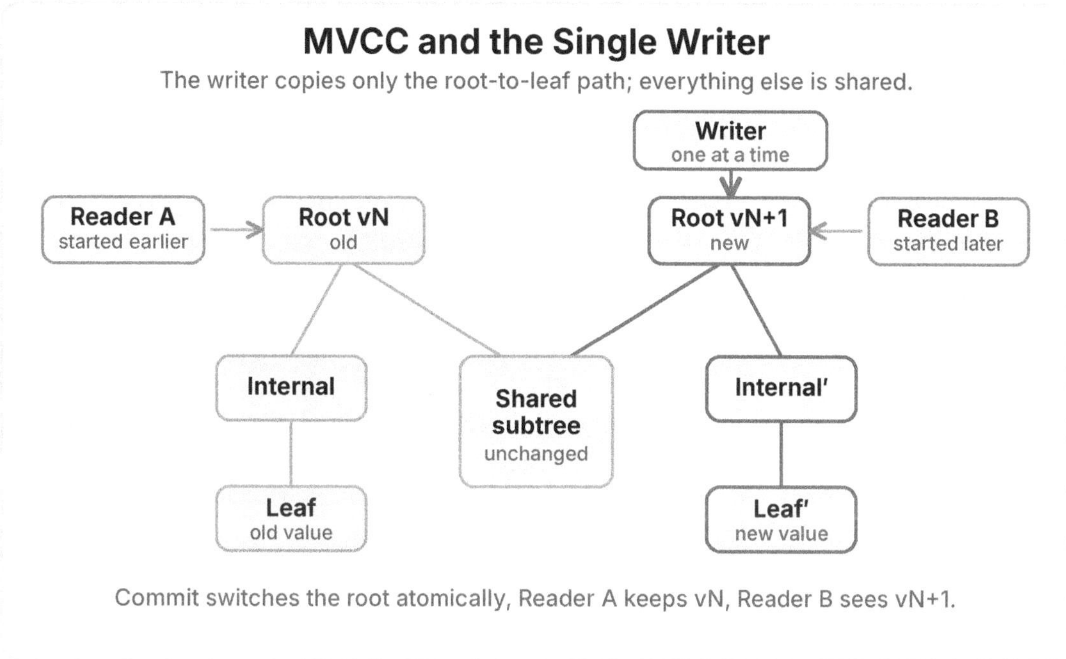

Figure 4.1: MVCC and the single writer: old and new B+Tree roots share unchanged pages while readers stay pinned to their snapshot

When a read transaction starts, it sees one stable root of the tree. That root is a snapshot: the reader can keep using it while other transactions commit newer versions. A writer does not mutate the pages that existing readers may still be using. Instead, it copies and modifies only the path from the root to the changed leaf pages. The unchanged parts of the tree are shared between the old and new versions.

At commit time, LMDB publishes the new version by switching the database root to the newly written tree. Readers that started earlier keep following the old root; new readers see the new root. This is structural sharing at storage-engine scale: small updates copy a small part of the tree, large unchanged regions are shared, and readers do not need locks on the writer's working pages.

The root switch is the publication point, not the whole physical write. New pages may have been written before the switch, but they are not the database state until the metadata that points to the new root is committed. If a crash happens after some new pages reach disk but before that metadata update, recovery uses the previous valid root. The transaction's changes are lost from the application's point of view, exactly as if the transaction had not committed. The physical bytes for the unpublished pages may still be present in the file, but no committed root points to them, so they are unreachable pages and can be overwritten or reused by later writes. Chapter 19 separates this atomic visibility rule from the disk-sync rules that determine when a commit is durable.

This design gives Datalevin fast and safe read concurrency. Reads are zero-copy and lockless, while writes remain atomic because a committed transaction becomes visible only when the root switch succeeds. In read-heavy workloads, throughput is often limited by CPU, memory bandwidth, or storage latency rather than by reader/writer lock contention.

2. B+Trees vs. LSM-Trees

Database storage engines generally fall into two families: sorted tree designs and log-structured designs. LMDB stores data in a B+Tree. Systems such as PostgreSQL and Oracle also rely heavily on B-tree-family indexes, while systems such as Cassandra, RocksDB, and LevelDB use log-structured merge-trees (LSM-trees) as their main storage pattern.

A **B+Tree** stores sorted keys in fixed-size pages [3]. A lookup starts at the root, walks through internal pages, and ends at a leaf page that contains the target key or, if the key is missing, the place where it would be. Because leaf entries are ordered by key, range scans can move through adjacent entries in sorted order. An **LSM-Tree** takes a different path [4]: it buffers writes and flushes sorted runs to disk, then merges those runs in the background. This can make writes very fast, but reads may need to consult several levels, and background compaction can add latency and CPU variation. The table below summarizes the performance characteristics of these two approaches.

Feature	B+Tree (Datalevin/LMDB)	LSM-Tree (RocksDB)
Read Performance	Excellent (O(log N) point/range)	Variable (may check multiple levels)
Write Performance	Good (constrained by random I/O)	Excellent (sequential appends)
Complexity	Simple (no background compaction)	High (compaction, bloom filters)
Large Values Handling	Good (little write amplification)	Poor (may rewrite data repeatedly during compaction)
Space Efficiency	Lower (due to fragmentation)	Higher (compressed on disk)
Read Amplification / Tuning	Low and predictable for point and range reads	Can require bloom filters, block cache, and compaction tuning

Datalevin chose B+Trees because the Datalog query engine relies heavily on **range scans** and **index joins**. A B+Tree keeps data in key order and provides predictable read latency. By using LMDB, Datalevin avoids the "write stalls" and CPU spikes associated with LSM background compaction, making it a better fit for predictable, real-time applications. In addition, the multi-paradigm nature of Datalevin demands secondary indexes (Section 5), which often require storing **large values**, a pattern less suitable for LSM storage because such values can be rewritten repeatedly during compaction.

3. From LMDB to DLMDB

Standard LMDB is a general-purpose tool, but Datalevin uses a specialized fork called **DLMDB** to support advanced Datalog features. DLMDB improves LMDB in two ways that matter directly to query execution.

3.1 Order Statistics

A standard B+Tree cannot count every range without walking that range. DLMDB adds **counted B+Trees**, where each node in the tree stores the count of items in its subtree. Datalevin uses this metadata for indexes where fast counts, samples, and rank-based access matter. The counted B+Tree gives Datalevin these advantages:

- **Cheap Counts**: Prefix and range counts can be answered from tree metadata instead of by materializing matching datoms, which would be an O(n) operation. Common attribute/value-bound counts are cheap. For example, `(d/count-datoms db nil :person/age 30)` can be answered in O(1) while `(d/count-datoms db nil :person/age nil)` can be answered in O(log n).
- **Fast Pagination**: You can skip the first 1,000,000 items in a sorted range and start reading the 1,000,001st item instantly using "rank-based" lookups.
- **Quick Sampling**: Given an ordered sequence of positions to sample, a counted DB allows a cursor to move up and down the tree to reach each position directly without going through all the keys in between. In those cases, sampling becomes O(log n) instead of O(n).

These advantages translate directly into **efficient query planning**: the Datalog engine uses these counts and samples to decide the most efficient join order, such as whether to filter by age first or by city first.

3.2 Prefix Compression

DLMDB also uses page-level prefix compression. Datalog indexes often contain neighboring keys that share an encoded prefix. In EAV, nearby keys may begin with nearby entity ids. In AVE, many neighboring keys may begin with the same encoded attribute id before the encoded value differs. Prefix compression avoids storing those shared leading bytes over and over.

The practical effects are smaller pages and better locality. More index entries fit in the same amount of storage, and more useful keys fit in memory pages and processor caches, which can make index walks and range scans cheaper. The benefit is workload-dependent. Long structured KV keys can see larger savings. Datalog indexes benefit too, but the shared prefix is often a compact encoded id, so the effect is usually more modest than the best KV cases. Chapter 19 returns to the tuning implications, and Chapter 15 explains how DUPSORT nesting avoids repeating larger EAV and AVE prefixes in the first place.

4. From Datoms to EAV and AVE

At the logical level, a Datalevin fact is a datom: `[entity attribute value]`. At the storage level, the same datoms are encoded into sorted key-value entries. Datalevin stores the primary Datalog indexes in two orders:

- **EAV** (Entity-Attribute-Value): optimized for finding the facts about one entity.
- **AVE** (Attribute-Value-Entity): optimized for finding entities that have a particular attribute/value pair, and for range scans over attribute values.

As shown in Figure 4.2, Datalevin puts the same set of datoms into both indexes simultaneously. The important point for this chapter is the access pattern: EAV keeps facts for one entity close

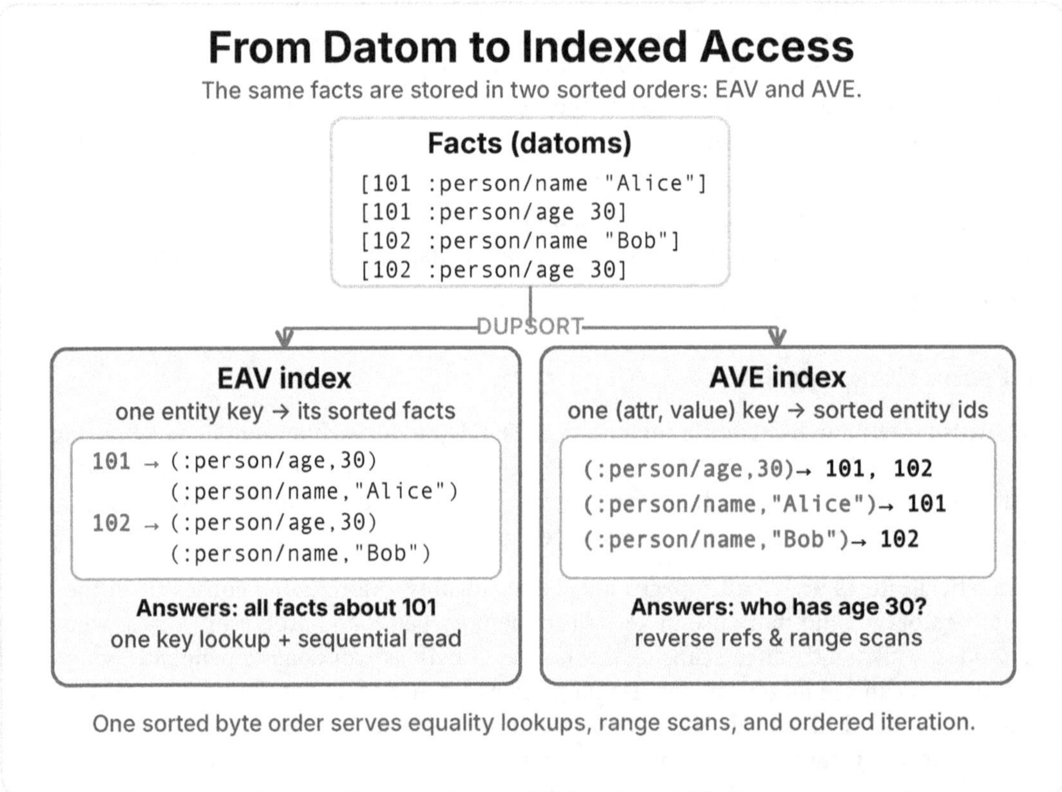

Figure 4.2: From a datom to indexed access in the EAV and AVE orders

together, while AVE keeps facts for one attribute/value lookup or range close together. Chapter 15 explains the physical DUPSORT nesting behind these indexes, how the key/value split saves space by avoiding repeated prefixes, and why Datalevin's core Datalog indexes are EAV and AVE.

Datalevin encodes triples into raw bytes, using a header byte to distinguish data types. Datalevin's binary encoding ensures that the binary sort order matches the logical data sort order. For example, encoded numbers sort as numbers, encoded strings sort as strings, and encoded instants sort by time. This property lets the storage engine use the same byte ordering for equality lookup, range lookup, and ordered iteration. This consistency between binary and logical ordering is the key design feature that leads to Datalevin's simplicity.

5. Secondary Indexes and Blob Storage

Beyond the primary triple indexes, Datalevin supports specialized secondary indexes such as **full-text search (FTS)**, **vector search**, and **indexed document (idoc) indexes**. An idoc is Datalevin's indexed document value: a nested map-like document whose paths can be queried efficiently. These features are not external plugins; they are integrated directly into the same LMDB environment.

5.1 Leveraging the KV Substrate

Secondary indexes are implemented as additional named sub-databases (DBIs) within the same storage file. A **DBI** is a named logical key-value space inside one LMDB environment. It lets Datalevin keep the primary datom indexes, full-text postings, vector metadata, document path indexes, and internal bookkeeping in one transactional storage substrate.

By default, Datalevin updates these indexes synchronously in the same transaction as the source datoms, preserving read-your-writes behavior for query functions such as `fulltext`, `vec-neighbors`, `embedding-neighbors`, and `idoc-match`. The secondary indexes are available for query immediately after commit by default.

For high-throughput data ingestion, though, full-text, vector, and embedding indexes can also opt into `:indexing-mode :async`. In async mode, the source datoms and a durable secondary-index job are committed atomically, then an in-process worker applies the index update after the transaction returns. Queries over that index become eventually consistent until the worker catches up. This is useful when indexing is expensive, especially for embedding providers that may call a local model or remote API.

5.2 Blob Storage Capabilities

While a Datalog triple is often a small piece of data, Datalevin also needs to handle large values. There are two common cases.

- **Large secondary-index structures**: Full-text, vector, embedding, and idoc indexes may need to store postings lists, vector metadata, serialized document structures, or other large internal values. Datalevin can store these as blobs in specialized DBIs instead of scattering them through the main triple indexes. For example, a search index might store a large compressed postings bitmap as a single KV value, then retrieve and process it with the same zero-copy efficiency as a simple triple.

- **Large datom values**: A user datom can also have a large value, such as a document, binary payload, or long text value. Datalevin does not put the whole payload inline in EAV and AVE. Instead, it stores the large value in a dedicated `datalevin/giants` DBI, keyed by an auto-incrementing integer **giant id**. The EAV and AVE indexes store that giant id, not the full value. When the datom is read, Datalevin follows the giant id to retrieve the original value from `datalevin/giants`. This is automatic internal storage behavior during transaction processing for values that exceed Datalevin's inline encoding threshold. Application code chooses the attribute value and, when needed, its schema type; it does not opt into `datalevin/giants` with a separate schema property or API call.

This indirection keeps the primary indexes compact and ordered even when some values are large. It also lets Datalevin leverage LMDB's ability to store values up to 2 GB in size while preserving efficient equality lookup, range scans, and secondary-index maintenance.

6. The Key-Value API: Direct Access

While Datalog is the primary interface, Datalevin exposes the underlying KV store as a first-class API. This is a deliberate design choice: the Datalog engine is built *on top* of this KV layer, rather than as a separate opaque engine.

Datalog and custom KV data can live in the same LMDB file. The Datalog engine uses named DBIs for indexes such as EAV and AVE, and application code can use other named DBIs in the same store for direct key-value data. Those DBIs share the same LMDB environment, durability settings, backup/copy behavior, and transaction boundary.

> **Best Practice**
>
> Datalevin's internal DBIs are the named key-value spaces the Datalog engine maintains for its own indexes and metadata, such as `datalevin/eav`, `datalevin/ave`, `datalevin/giants`, and DBIs created for secondary indexes. Treat these DBIs as owned by the engine; use separate, application-named DBIs for your own KV structures.

Exposing the KV layer lets application code bypass the triple model when it is not the best fit, for example when building custom indexes, high-frequency counters, or large binary blob stores. This multi-paradigm approach means you are not forced to fit every data shape into a Datalog triple if a simple key-value pair is more efficient.

The details of the KV API and its practical usage are covered in Chapter 10.

7. Persistence and Durability: WAL Mode

By default, local embedded Datalevin stores use LMDB's direct commit path. This is safe and fast for many workloads, and, as described above, LMDB's copy-on-write root switching gives the LMDB file immediate recovery after a crash.

For write-intensive workloads, Datalevin can add **Write-Ahead Log (WAL) mode** above LMDB. A WAL is an append-only transaction log. Datalevin first records the transaction in that log, then applies the change to LMDB in a faster, non-durable mode. Depending on the WAL durability profile, Datalevin may sync WAL records immediately or batch them before syncing. Once a WAL record is durable, it is the recovery source, while LMDB remains the indexed B+Tree state used for reads. For a new local store, when WAL mode is enabled and no profile is specified, the default profile is `:strict`, so each commit waits for the WAL record to reach the operating system's `fsync` durability boundary. The `:relaxed` profile is an explicit opt-in that batches syncs and trades a short crash-loss window for higher throughput.

Application code reads the same committed state in WAL mode. If `transact!` returns success, the datoms have been applied to LMDB's indexes and the next query on the connection can see them, just as in non-WAL mode. What WAL mode changes is when the log and LMDB file are synced to durable storage; it does not require a different read-after-write pattern. In the explicitly selected `:relaxed` profile, a crash before the next WAL sync, or before a snapshot/checkpoint that forces one, can lose the most recent acknowledged transactions, but while the process is running those transactions are part of the current LMDB-backed database state.

WAL mode also supports multi-threaded write concurrency and lays the foundation for data replication and high availability (HA) server behavior. In this chapter, the important point is the storage shape of WAL. The operational details are covered in Part V.

7.1 How WAL Mode Works

At a high level, a WAL transaction follows this sequence:

1. **Application Request**: The application requests a transaction.
2. **WAL Append**: The change is encoded and appended to a sequential log segment file.
3. **Position Tracking**: Every transaction receives a strictly increasing log position. Later chapters call this position a log sequence number (LSN).
4. **Fast LMDB Update**: The change is applied to the LMDB B+Tree, so ordinary reads can see the committed data. In WAL mode this LMDB update may be non-durable until a checkpoint or sync; durability comes from the WAL record rather than from forcing the LMDB file itself to storage.
5. **Durability Scheduling**: Depending on the chosen durability profile, Datalevin either syncs the WAL record immediately or batches it for a later sync.
6. **Acknowledgment**: Under `:strict` and `:extra`, acknowledgment waits for the WAL record to reach durable storage. Under `:relaxed`, acknowledgment can occur before the next group sync, leaving a small gap between acknowledged commit and durable sync.

On restart after a crash, Datalevin scans the WAL segments and replays any committed transactions that were not yet fully persisted in the LMDB file.

WAL mode is enabled according to the Datalevin surface or deployment mode you are using. For local stores, WAL is an explicit opt-in:

- **Datalog databases**: Local embedded databases disable WAL by default; enable it with `{:wal? true}` in `create-conn` or `get-conn` options.
- **Key-value stores**: Disabled by default; enable with `{:wal? true}` in `open-kv` options.

For server deployments, WAL is part of the operational design:

- **Async read replicas**: A non-HA replica requires WAL on the primary so it can bootstrap from a copy and then tail durable records.
- **Consensus-lease HA**: HA forces WAL on.

7.2 Replication and Operations

WAL mode has operational APIs such as `create-snapshot!`, `gc-txlog-segments!`, `txlog-watermarks`, and `open-tx-log`. Those operations matter for long-running services, replication, low-level change capture, and production tuning. They are covered in detail in Chapters 19, 20, and 22.

Summary

Datalevin's storage layer focuses on pragmatic engineering:

- It uses **LMDB** for stable embedded storage and zero-copy reads.
- It uses **B+Trees** to ensure predictable read performance for complex Datalog queries.
- It adds **DLMDB** extensions like order statistics to power the query planner.
- It stores datoms in **EAV** and **AVE** orders, covering entity-local reads and attribute/value lookups with compact sorted indexes.
- It integrates full-text, vector, embedding, and idoc indexes into the same LMDB environment, with synchronous updates by default and async indexing as an explicit ingestion option.
- It keeps large values out of the primary datom indexes by storing them through giant-value indirection.
- It exposes the underlying **KV API** for application-owned DBIs that need direct key-value access in the same store.
- It provides a **WAL mode** to scale write performance without sacrificing the B+Tree model.

By understanding these fundamentals, you can better tune your schema and queries for maximum performance.

References

[1] Howard Chu, "LMDB," *The Databaseology Lectures*, Carnegie Mellon University, Fall 2015, YouTube video. URL: https://www.youtube.com/watch?v=tEa5sAh-kVk.

[2] Chris Okasaki, *Purely Functional Data Structures*, Cambridge University Press, 1998. DOI: https://doi.org/10.1017/CBO9780511530104.

[3] Rudolf Bayer and Edward M. McCreight, "Organization and Maintenance of Large Ordered Indexes," *Acta Informatica* 1:173-189, 1972. DOI: https://doi.org/10.1007/BF00288683.

[4] Patrick E. O'Neil, Edward Cheng, Dieter Gawlick, and Elizabeth J. O'Neil, "The Log-Structured Merge-Tree (LSM-Tree)," *Acta Informatica* 33:351-385, 1996. DOI: https://doi.org/10.1007/s002360050048.

CHAPTER 5

Attributes, Namespaces, and Entities

Chapter 4 covered how Datalevin stores triples (datoms) in DLMDB. This chapter moves up the stack to the logical model. Datalevin's data model is built on three pillars: **Attributes**, **Namespaces**, and **Entities**.

Datalevin's schema model is flexible by default and provides powerful controls when you need performance and integrity. Declared attribute properties control how values are stored, constrained, indexed, and queried, while namespaces keep the vocabulary readable and entities give facts a stable place to collect.

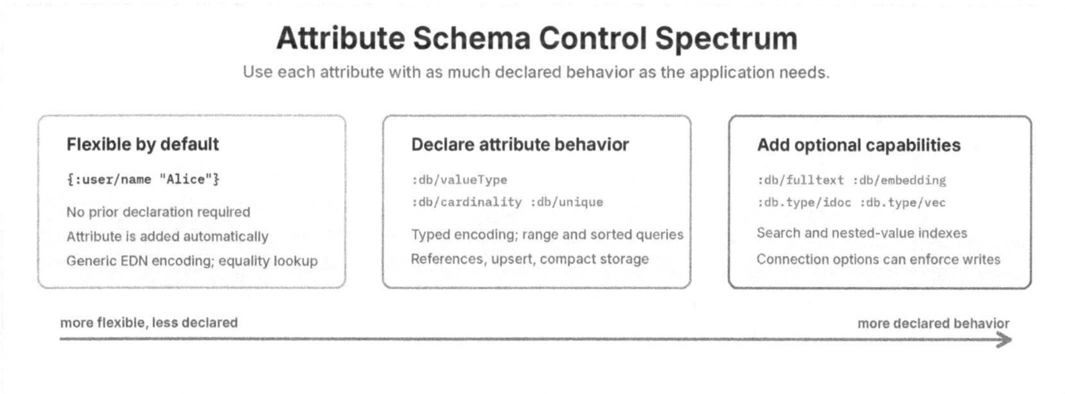

Figure 5.1: Attribute schema control spectrum: undeclared attributes are flexible and added automatically; declared attribute behavior such as :db/valueType, :db/cardinality, and :db/unique adds typed encoding, range queries, references, upsert, and compact storage; optional capabilities such as :db/fulltext, :db/embedding, and :db.type/idoc add search and nested-value indexes; connection-wide options such as :validate-data? and :closed-schema? add write-time checks

Figure 5.1 shows Datalevin's schema controls as a spectrum. An attribute can start with no declaration, but you can add declared behavior when the application needs typed storage, range access, identity, references, specialized indexes, or stricter write-time checks. The sections below explain each control.

1. Attributes: Flexible Schema

An attribute (the "A" in EAV) defines a property that can be associated with an entity. Attributes are the main subjects of a Datalevin schema.

1.1 Automatic Attribute Creation

By default, you do not need to "create a table" or "define a schema" to start using Datalevin. When you transact a new attribute keyword that the database has not seen before, Datalevin automatically adds it to the schema for you.

```
;; This works even if :user/name was never defined
(d/transact! conn [{:user/name "Alice"}])
```

> **Info**
>
> In Clojure source, attribute names such as `:user/name` are EDN keywords. Python and JavaScript schema maps and legacy native forms spell them as colon-prefixed strings. The typed `q` and `tx` builders also accept `"user/name"` and convert it in attribute positions. They do not reinterpret values: use `q.kw("user.status/active")` when the stored value itself must be a keyword.

This convenience has a flip side: a misspelled attribute does not cause an error. Transacting `:user/emial` quietly creates a new attribute alongside `:user/email`, and queries against the correct name will not see the misspelled facts. During development this is a minor annoyance; in production you can open the connection with the option `{:closed-schema? true}`, which rejects any transaction that mentions an attribute not already defined in the schema. Chapter 11 covers this and other write-time validation options.

1.2 The Default Type: EDN Encoding

If an attribute is added automatically, Datalevin treats its value as a generic **EDN-encoded value**, which has some trade-offs.

- **Pros**: You can store any EDN data structure (maps, vectors, sets) directly.
- **Cons**: Because the values are stored as opaque binary blobs, the database cannot perform efficient **range queries**, such as "find all orders with total greater than 100", because it does not know how to sort the binary data.

Exact-value matching on untyped attributes is byte-encoded matching, not a semantic equality layer for every EDN shape. Simple values that encode the same way can still be matched exactly, so a query such as "find the order whose total is exactly 100" can use the index. Do not rely on untyped maps, sets, or nested values for canonical equality lookup: Datalevin does not sort or normalize all EDN structures on every transaction before serializing them. What you lose is meaningful *ordering*, range scans, sorted results, and comparisons. You also cannot enable specialized indexes that require typed values. Embedding search requires a declared string type. Full-text indexing is more permissive: Datalevin calls `str` on the transacted value before indexing it, though string attributes are still the usual choice for human text.

1.3 Why Explicit Types Are Preferred

To enable range queries, validation, compact storage, and specialized indexes, you are strongly encouraged to specify a data type. Explicit types allow Datalevin to use specialized codecs for sorting values in the B+Tree instead of treating values as opaque EDN-encoded blobs.

This is a pay-as-you-go tradeoff. During early exploration, undeclared attributes let you move quickly. Once an attribute becomes part of an important access path or application contract, declaring its type gives Datalevin the information it needs to compare, range-scan, validate, and index that value predictably.

Common Datalog value types for `:db/valueType`:

Type	Use
`:db.type/string`	Strings. Use for names, emails, external ids, and text fields.
`:db.type/keyword`	Keywords. Good for enums such as `:order.status/paid`.
`:db.type/boolean`	`true` or `false`.
`:db.type/long`	64-bit signed integers.
`:db.type/bigint`	Large integer values.
`:db.type/double`	64-bit floating-point numbers.
`:db.type/instant`	Instants, as Unix epoch time.
`:db.type/uuid`	UUID values.
`:db.type/bytes`	Byte arrays.
`:db.type/ref`	Entity references. Required for reverse attribute navigation and component relationships.

Raw keyword values are useful for small symbolic fields. When the enum value should itself be an entity with identity, documentation, sort order, lifecycle, or additional facts, model it as a `:db/ident` entity and point to it with a `:db.type/ref` attribute. Chapter 11 covers that pattern.

Beyond these scalar types, Datalevin offers specialized value types: `:db.type/tuple` for tuple values, `:db.type/vec` for similarity-indexed vectors (Chapter 17), and `:db.type/idoc` for path-indexed nested documents (Chapter 14). Composite indexes over several attributes use `:db/tupleAttrs` (Chapter 11). For a complete list of acceptable value types, see Appendix C.

2. Attribute Properties

While attributes are created automatically, you can provide a schema map when opening a connection with `d/get-conn` or `d/create-conn`, or later with `d/update-schema`, to define behavior for specific attributes. The table below lists some example attribute properties.

Property	Description
`:db/valueType`	The data type, such as `:db.type/long`.
`:db/cardinality`	`:db.cardinality/one` (allow single value, default) or `:db.cardinality/many` (allow multiple values).
`:db/unique`	`:db.unique/identity` (unique, upsert when duplicate) or `:db.unique/value` (unique, reject duplicate).
`:db/fulltext`	Set to `true` to enable full-text search. Values are converted to strings before indexing.
`:db/embedding`	Set to `true` on string attributes to maintain an embedding similarity index.
`:db/idocFormat`	Format for `:db.type/idoc` attributes: `:edn`, `:json`, or `:markdown`.
`:db/doc`	Human-readable documentation string for the attribute.

Appendix C includes a complete reference for the schema properties that Datalevin interprets.

> **Info**
>
> Datalevin stores attributes in indexes by an internal 32-bit attribute id, exposed in schema inspection as `:db/aid`. A database can therefore have at most about 2.1 billion distinct attribute ids, including built-in attributes. That ceiling is far above ordinary application schemas, but it is a reminder that attributes are schema vocabulary, not an unbounded place to put user-generated keys. If a key space can grow without control, model the key as data, an enum/value, an idoc path, or a referenced entity instead of minting a new Datalevin attribute for every key.

Here is a schema that uses several of these properties. The `:db/doc` entries describe the intended meaning of each attribute:

```
(def schema
  {:user/email       {:db/valueType :db.type/string
                      :db/unique    :db.unique/identity
                      :db/doc       "Login and notification email; unique
                      ↪  identity for a user."}
   :user/login-count {:db/valueType :db.type/long
                      :db/doc       "Number of successful logins recorded for
                      ↪  the user."}
   :user/tags        {:db/valueType   :db.type/string
                      :db/cardinality :db.cardinality/many
                      :db/doc         "Labels assigned to the user for
                      ↪  segmentation or search."}
   :user/bio         {:db/valueType :db.type/string
                      :db/fulltext  true
                      :db/embedding true
                      :db/doc       "Profile biography used for full-text and
                      ↪  embedding search."}})
```

2.1 Cardinality: One Value or a Set of Values

Cardinality controls how many values one entity may hold for one attribute.

- `:db.cardinality/one` is the default: an entity holds at most one value, and transacting a new value **replaces** the old one. There is no separate "update" operation: asserting `{:db/id 1 :user/email "new@example.com"}` retracts the old email and asserts the new one in a single step.
- `:db.cardinality/many`: transacting a new value **accumulates** instead. The values behave as a set: duplicates are ignored, and there is no ordering.

The example below uses the schema from Section 2: `:user/tags` is declared as a string-valued `:db.cardinality/many` attribute.

```
;; Assumes :user/tags is declared string-valued and :db.cardinality/many.
(d/transact! conn [{:db/id 1 :user/tags "clojure"}])
(d/transact! conn [{:db/id 1 :user/tags ["databases" "clojure"]}])
(d/transact! conn [{:db/id 1 :user/tags ["databases" "datalevin"]}])
;; Entity 1 now has tags "clojure", "databases", and "datalevin": a set, no
↪  duplicates.
```

This collection syntax is specific to map-style entity transaction data. For a cardinality-many attribute, a scalar value adds one value, and a non-map collection adds each element as a separate value. If `:user/tags` were undeclared, it would be an untyped cardinality-one attribute. The vector would then be stored as one EDN value and would replace the previous value; it would not become two string tags. In raw datom operation vectors, the value position is always one value. To add several values with raw datom operations, write one operation per value, such as `[:db/add 1 :user/tags "databases"]` and `[:db/add 1 :user/tags "clojure"]`.

If you need an ordered or duplicate-preserving collection, model it differently: store a small typed vector as a `:db.type/tuple` when the positions and element types are fixed, store an arbitrary vector in an EDN-valued attribute, or model the list items as entities that carry a position attribute (a variant of the join-entity pattern in Chapter 11). Tuple values have their own schema rules and size limits; see Appendix C, Section 8.

2.2 Uniqueness: Identity vs. Value

Both `:db/unique` settings enforce that no two entities share a value for the attribute, and both can be used in lookup refs. They differ in how a duplicate write is handled:

- `:db.unique/value` is a pure constraint: transacting a duplicate is an **error**. Use it when a value must be unique, but a duplicate write means "reject this data", not "find that entity and update it." One-time invitation codes and password-reset tokens are good examples: the token can be used in a lookup ref, but if newly generated token data collides with an existing token, the application should fail the transaction or generate a different token, not merge fields into the existing token entity.
- `:db.unique/identity` treats the value as the entity's identity: transacting a map with an existing value **upserts** by updating the existing entity instead of creating a new one. Use it for natural keys such as email addresses, slugs, or external ids that application code uses to identify an entity.

Chapter 6 shows lookup refs and upserts in transactions; Chapter 11 goes deeper into identity modeling.

3. Namespaces: Semantic Grouping

In Datalevin, attribute names are keywords, and by convention, they almost always include a **namespace**.

In EDN, `:user/email` is a qualified keyword. The part before the slash, `user`, is the keyword namespace. The part after the slash, `email`, is the keyword name. The same syntax is used for attribute names and for keyword values, so the examples below include both.

Attribute names are not throwaway local names. In SQL, a column name is often mediated through aliases, views, ORM fields, or API serializers. In Datalevin, the attribute name usually appears directly in transactions, queries, pull patterns, logs, migrations, and sometimes API payloads. Treat an attribute such as `:user/email`, `:doc/body`, or `:invoice/total` more like a public data key than a local variable: it should clearly represent the domain because it becomes part of the schema vocabulary.

```
:user/email          ;; namespace: user, name: email
:line-item/quantity  ;; namespace: line-item, name: quantity
:order.status/paid   ;; namespace: order.status, name: paid
:invoice/line.total  ;; namespace: invoice, name: line.total
```

For attributes, the namespace is part of the attribute's identity. `:user/name`, `:product/name`, and `:company/name` are three different attributes. They may all store strings called "name" in English, but Datalevin stores and indexes them as distinct facts.

A dot inside the namespace or name is just another keyword character. Datalevin does not treat `:order.status/paid` as nested under `:order/status`, nor `:invoice/line.total` as nested under `:invoice/line`; the slash is the delimiter that separates namespace from name. Dotted namespaces are a common convention for enum value domains, such as order statuses, but Datalevin assigns no special meaning to the dot.

3.1 Namespaces Are Not Tables

Namespaces are a naming convention, not a storage container or type constraint. An entity can have attributes from more than one namespace:

```
{:user/email "ada@example.com"
 :account/id "acct-1"
 :account/plan :account.plan/pro}
```

That is legal because entities are collections of facts, not rows in a table. The namespace tells readers what the attribute means; it does not prevent the attribute from appearing on a particular entity.

> **Best Practice**
>
> If your application needs to enforce that only certain entities have certain attributes, enforce that where transactions are built: command handlers, transaction functions, or higher-level validation. For transaction-time checks that need the would-be database state, use `:db/ensure`.

For checks that must see the entity after all transaction data has been processed, use `:db/ensure` in the transaction. It lets a predicate inspect the would-be database state before commit, which is the right place for entity-shape invariants that namespaces alone do not enforce. Chapter 6 covers the transaction form in detail, and Chapter 11 covers schema-design uses.

3.2 Why Namespaces Are Preferred

Namespaces give Datalevin's flexible schema enough structure to remain maintainable:

- **Collision avoidance**: `:user/id`, `:order/id`, and `:product/id` can coexist without fighting over a generic `:id`.
- **Query readability**: `[?order-eid :order/customer ?user-eid]` says more than `[?order-eid :customer ?user-eid]`.
- **Schema evolution**: A well-named attribute can be moved, specialized, or renamed deliberately. Chapter 11 shows the `update-schema` workflow for schema changes and attribute renames.

3.3 Attribute Namespaces and Value Namespaces

The namespace on an attribute describes the fact. The namespace on a keyword value describes the value domain. These often differ:

```
{:order/status :order.status/paid}
```

Here, `:order/status` is an attribute on an order. The value `:order.status/paid` is an enum-like keyword from the order-status value domain. This pattern keeps the attribute and its possible values distinct. It is a naming convention, not a database requirement.

> **Info**
>
> System namespaces such as `:db`, `:db.type`, `:db.cardinality`, `:db.unique`, and `:datalevin` are used by Datalevin itself. Treat them as reserved for schema properties, built-in values, and Datalevin-specific metadata.

3.4 Datalevin Attributes and RDF Predicates

Readers who know RDF will notice the similarity: RDF uses subject-predicate-object triples, while Datalevin uses entity-attribute-value datoms. That resemblance is useful, but the design goals are different [1].

The terminology difference is deliberate. An **entity** is a database identity whose facts may accumulate over time; it is not necessarily a globally named subject. An **attribute** is a schema-bearing property: it can have a value type, cardinality, uniqueness, indexing behavior, documentation, and validation. Datalevin avoids calling this position a predicate because, elsewhere in database and query discussions, a predicate usually means a test that returns true or false. A **value** is also broader than an RDF object: it may be a scalar, a reference to another entity, a tuple, a document value, a vector, or another supported Datalevin value type.

In RDF, predicates are often IRIs, Internationalized Resource Identifiers, chosen from shared vocabularies or ontologies. An IRI is the Unicode-capable generalization of a URI; RDF uses the more precise term. RDF predicates are meant to carry portable meaning across datasets, organizations, and reasoning systems. RDF Schema and OWL can then attach richer ontology semantics to those predicates: class hierarchies, property relationships, domain/range constraints, and inference rules [2,3].

Datalevin attributes are intentionally more local and operational. An attribute such as `:order/placed-at` or `:invoice/issued-at` is not meant to be a universal ontology term for time. It is a precise fact name inside your database. Its schema properties tell Datalevin how to store, compare, index, and validate values; they do not assert a global ontology.

That difference makes Datalevin's model more open in day-to-day application design. You can create many specific attributes when they make the data clearer:

```
:order/placed-at
:invoice/issued-at
:shipment/delivered-at
:support-ticket/closed-at
```

Those attributes may all be `:db.type/instant`, but they mean different things to the application. You do not need to collapse them into a generic `:event/time` attribute just to make the vocabulary look smaller. In Datalevin, a larger set of clear, specific attributes is often better than a small set of overloaded general attributes.

This does not prevent ontology-like modeling when you need it. You can still use enum entities, references, rules, and Datalog queries to represent domain knowledge. The point is that Datalevin does not require every attribute to be a general-purpose ontology predicate before it can be useful.

3.5 Rules of Thumb

The following are conventions and suggestions, not hard requirements. Use them unless your domain has a strong reason to differ:

1. Use qualified keywords for application attributes: `:user/email`, not `:email`.
2. Use singular domain nouns for namespaces: `:user/email`, not `:users/email`. The attribute name can still be plural when it represents a cardinality-many value, such as `:user/emails`.
3. Choose the namespace that reflects where the relationship fact belongs in your model: `:order/customer`, `:comment/post`, `:role-assignment/user`.
4. Use separate value namespaces for enums: `:order.status/paid`, `:order.status/cancelled`.
5. Avoid using namespaces as security or type boundaries. They clarify meaning; they do not enforce authorization or entity classes.

Ownership is a modeling choice, not a storage container. The attribute name tells readers where the fact belongs and usually matches the shape you want to write, pull, and validate.

For a one-to-one relationship, `:user/profile` says the user owns the profile link; use that shape when the profile is normally reached from the user. For a one-to-many relationship, `:comment/post` puts the parent pointer on each comment. A post's comments can still be found by querying for comments whose value is the post entity id, but each comment can be added, removed, or filtered independently. For a many-to-many relationship, `:user/roles` can be fine for a small owned set. When the relationship has its own facts, make the relationship an entity, with attributes such as `:role-assignment/user`, `:role-assignment/role`, and `:role-assignment/assigned-at`.

The namespace does not enforce the relationship. Cardinality, reference types, uniqueness, and transaction-time validation do that. Chapter 11 covers those schema choices in detail.

4. Entities and Ids

An **entity** (the "E" in EAV) is a collection of datoms that share the same entity id. There is no separate entity record in storage: the id is the only thing tying the datoms together. Assert a datom with a new id and an entity comes into existence; retract the last datom carrying an id and that entity is gone. As Chapter 3 discussed, an entity has no declared type, only whatever attributes have been asserted about it.

Datalevin uses 64-bit integers for entity ids. An entity id is the stable internal handle for the entity inside the database. The system attribute `:db/id` is the transaction-data notation for addressing an entity, but it is not application identity. Chapter 6 covers the transaction mechanics.

Because entity ids are assigned by the database, they carry no stable meaning outside it. Avoid exposing them in URLs, API payloads, or other systems: data that is exported and re-imported, merged from another database, or rebuilt from source may end up with different ids. Two databases can contain the same application data while using different eids for the corresponding entities.

For a stable, externally visible identity, give the entity a natural key declared as `:db.unique/identity`, such as an email, a slug, or a UUID. The example schema in Section 2 declares `:user/email` this way, which lets the rest of the application address that user as `[:user/email "alice@example.com"]` without ever knowing the internal id. Chapter 11 also introduces `:db/ident`, a built-in identity attribute for system-wide named entities such as enums.

5. Schema Workflow in Practice

In practice, the schema workflow often follows this path:

1. **Prototyping**: Start without a schema and transact maps.
2. **Relations**: Use `:db.type/ref` to connect entities, forming the graph that Datalog traverses efficiently.
3. **Optimization**: Once you know your access patterns, add `:db/valueType` to enable range queries on specific attributes.
4. **Integrity**: Add `:db.unique/identity` for fields like emails or slugs so you can use them as lookup refs, such as `[:user/email "alice@example.com"]`.
5. **Lockdown**: In production, consider opening the connection with `{:closed-schema? true}` to reject unknown attributes, and `{:validate-data? true}` to check values against their declared types.

Because schemas are EDN data, production projects often keep them in a file such as `resources/app-schema.edn` and commit that file to version control. That makes schema changes reviewable like code and gives migrations an explicit artifact to point at.

Chapter 11 covers this workflow in depth, including `update-schema`, typed-value migration, closed schemas, and write-time validation.

Summary

Datalevin's approach to schema is **pay-as-you-go**. You get the speed of a schemaless store during development, but the power of a typed, indexed database as your application matures. Namespaces keep your attributes organized. System-managed ids keep references stable across transactions, while unique identity attributes, not raw ids, provide the stable names the outside world should use.

References

[1] Richard Cyganiak, David Wood, and Markus Lanthaler, "RDF 1.1 Concepts and Abstract Syntax," W3C Recommendation, February 25, 2014. URL: https://www.w3.org/TR/rdf11-concepts/.

[2] Dan Brickley and R. V. Guha, "RDF Schema 1.1," W3C Recommendation, February 25, 2014. URL: https://www.w3.org/TR/rdf-schema/.

[3] W3C OWL Working Group, "OWL 2 Web Ontology Language Document Overview (Second Edition)," W3C Recommendation, December 11, 2012. URL: https://www.w3.org/TR/owl2-overview/.

PART II

Core APIs: Datalog First, KV When Needed

CHAPTER 6

Transactions and Atomic Updates

Every read or write to a Datalevin database happens within a **transaction.** This chapter focuses on write transactions, so "transaction" here means a write transaction.

Transactions are the cornerstone of database reliability, ensuring that your data moves from one consistent state to another, even in the face of concurrent operations or system crashes. Datalevin provides ACID (Atomicity, Consistency, Isolation, Durability) guarantees [1], and this chapter explores how. It closes by relating the transaction modes to current write-benchmark results.

Figure 6.1 gives an overview of a Datalog transaction's lifecycle. The sections below explain the details.

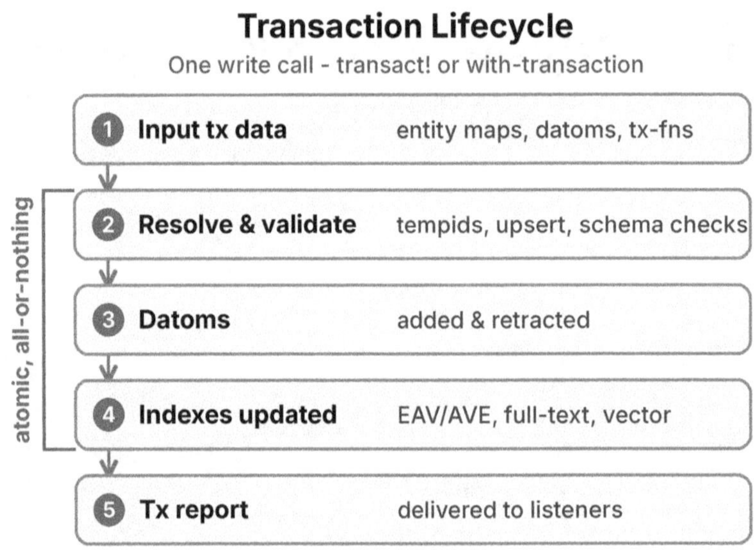

Figure 6.1: The transaction lifecycle: input data, resolution and validation, datom changes, index updates, and the resulting report

1. The Transaction Model

Datalevin supports several transaction modes with different properties.

1.1 Default Mode: Synchronous LMDB Transactions

In its default (non-WAL) mode, a Datalevin transaction is a direct mapping to an underlying **LMDB transaction**, which has strict ACID guarantees.

- **Atomicity**: All changes within a single `transact!` call are applied as a single, atomic unit. They either all succeed or all fail.
- **Durability**: By default, LMDB writes changed pages with `pwrite` and calls `fsync` before confirming the transaction. This guarantees that once a transaction is committed, it is durable and will survive a system crash. The `fsync` call is often expensive, and its cost depends on the hardware and workload.

> **Store-scoped atomicity**
>
> A transaction is scoped to one Datalevin store: one LMDB environment whose primary data lives in one memory-mapped data file. In the default non-WAL mode, a Datalog transaction is ultimately a KV transaction over that store.
> It can atomically update the Datalog indexes and any other DBIs that live in the same store, but it cannot atomically write to two separately opened Datalevin stores or two different LMDB environments. If you need one atomic unit across multiple logical datasets, put those datasets in the same store; separate files require application-level coordination.

1.2 Durability Settings

For use cases where maximum write speed is more important than crash-proof durability, such as bulk loading or caching, you can relax commit-time `fsync` by setting LMDB flags during connection creation. For example, using `:nosync` can significantly improve write throughput at the cost of durability: how frequently dirty pages are synced is left to OS discretion, and an untimely system crash can result in a corrupted database. There are also other non-durable flags; see Chapter 10 for details.

Apart from controlling durability during connection creation, you can explicitly force a flush with `d/sync`, and you can change LMDB environment flags at runtime with `d/set-env-flags`. Both functions operate on the KV store handle; for a Datalog connection, use the backing KV handle returned by `d/datalog-kv`.

```
(def kv (d/datalog-kv conn))

;; Force a synchronous flush to disk after a burst of faster writes.
(d/sync kv)

;; Change sync behavior for subsequent transactions.
(d/set-env-flags kv #{:nosync} true)  ; relax commit-time sync
(d/set-env-flags kv #{:nosync} false)
```

Change runtime flags only in controlled ingestion or maintenance workflows. They affect subsequent transactions on that store, so they should not be used as incidental per-request toggles.

1.3 Asynchronous Transactions

For independent high-volume writes, asynchronous transactions let Datalevin batch work internally and improve throughput. Asynchronous transactions change when the caller waits, not what a transaction means. The async transaction API submits ordinary Datalevin transaction data and returns a future or promise immediately. The transaction still commits as one atomic unit, produces the usual transaction report, and either succeeds or fails as a whole.

> **Async is not fire-and-forget**
>
> If later code depends on the write, wait for the returned future or promise before reading or issuing dependent writes. Treating async transactions as fire-and-forget can hide validation errors, constraint violations, and write failures until too late.

1.4 Nested Transaction Calls

Nested transactions in Datalevin are not independent subtransactions. When code is already inside a write transaction, another synchronous transaction call uses that active transaction. There is still only one transaction, not a stack of separately committed transactions. If the inner code aborts or throws, the whole active transaction aborts; Datalevin does not provide nested savepoints.

2. Transacting Data

The primary function for writing Datalog data is `d/transact!`. It takes a connection and a collection of transaction values, called the **transaction data**. A transaction value is one item that Datalevin can prepare into datoms. Transaction values can take several forms:

- Entity maps, such as `{:user/name "Alice"}`.
- Raw datom operation vectors, such as `[:db/add eid attr value]`.
- Transaction function calls, such as `[:db.fn/call f arg1 arg2]`.
- Transactable Entity objects, a Clojure API convenience covered in Chapter 7.

Maps are the clearest shape for ordinary creates and updates, so this chapter starts with maps before moving on to the lower-level forms.

In Python and JavaScript, the typed `tx` namespace builds these same transaction values without hand-writing operation keywords. It converts attribute-name arguments to keywords, preserves ordinary values, and produces immutable data that the connection lowers when the transaction is submitted.

The order of transaction data matters during transaction preparation. Datalevin processes the collection in order, so a transaction function sees the database at the point where that form is expanded, and later forms such as `:db/cas` can depend on facts produced by earlier forms in the same transaction. Tempid resolution is transaction-wide, so tempids can still be used before the map that introduces them appears. The commit is atomic: all prepared changes succeed together, or none are applied.

A successful `transact!` returns a transaction report; Section 2.11 examines that report after the write forms. If the transaction fails validation, violates a constraint, aborts, or throws from transaction code, it does not commit and no transaction report is returned.

2.1 Entity Maps

The most common way to express an entity is using a map. Each map represents an entity with its attributes and values:

```
(d/transact! conn
  [{:user/name "Alice" :user/email "alice@example.com"}
   {:user/name "Bob" :user/email "bob@example.com"}])
```

In map-style transaction data, omitting `:db/id` means "create a new entity and let Datalevin assign its entity id." This convenience applies to entity maps, not to raw datom operation vectors, where the entity position is always explicit. The assigned entity id is a Datalevin internal `long`. Once assigned, it is the permanent handle for that entity inside the database: updating or adding attributes never changes the id, so references between entities remain stable across transactions.

When new entities need to refer to one another in the same transaction, use a tempid instead of omitting `:db/id`. A negative integer or string in `:db/id` is a tempid. Datalevin still replaces it with a new permanent entity id. Positive ids are not tempids; the next section describes how they are handled.

2.2 Updating Existing Entities

To update an existing entity, include its entity id in the map:

```
(d/transact! conn
  [{:db/id 101, :user/active? false}])
```

This updates entity id 101 to set `:user/active?` to false. Positive `:db/id` values are concrete entity ids, not tempids. If the current maximum entity id is `1000` and you transact `{:db/id 2000 :foo "bar"}`, Datalevin asserts facts for entity `2000`, advances the internal maximum entity id to `2000`, and the next automatically assigned entity id will be `2001`. It does not create entities `1001` through `1999`.

Use explicit positive ids only for controlled imports, restores, or other cases where you deliberately manage id allocation. In ordinary application transactions, omit `:db/id` for a new entity. Use a negative number or string tempid only when new entities need to refer to one another inside the same transaction, as shown in Section 2.4.

2.3 Automatic Entity Timestamps

Many applications want to know when an entity was first created and when it was last modified. Datalevin can maintain this information automatically when the connection is opened with `:auto-entity-time? true`:

```
(require '[datalevin.core :as d])

(def conn
  (d/get-conn
    "/tmp/entity-time-demo"
    {:user/email {:db/valueType :db.type/string
                  :db/unique    :db.unique/identity}}
    {:auto-entity-time? true}))
```

```
(d/transact! conn
  [{:user/email "alice@example.com"
    :user/name  "Alice"}])

(d/transact! conn
  [{:user/email "alice@example.com"
    :user/name  "Alice Smith"}])
```

With this option enabled, a newly created entity receives both `:db/created-at` and `:db/updated-at`. Later transactions that modify the entity update `:db/updated-at` while preserving `:db/created-at`. The values are stored as Unix epoch milliseconds using `:db.type/long`.

These attributes describe Datalevin's system-level mutation time. They are not a replacement for domain event times. If your application needs to know when an order was placed, when an invoice was paid, or when an episode happened, model that explicitly with attributes such as `:order/placed-at`, `:invoice/paid-at`, or `:episode/timestamp`. Automatic entity timestamps answer "when did this entity change in the database?", not "when did the real-world event occur?"

Automatic timestamps also store only the current created/updated values. For a full audit trail, model audit events explicitly or use transaction-log-based operations, if WAL mode is enabled, such as the txlog tools covered in Chapter 19.

2.4 Tempids

When creating new entities that will be referenced by other entities in the same transaction, use **tempids**: temporary entity ids that act as placeholders. Negative numbers and strings can both serve as tempids; the examples below use negative numbers. They also assume `:user/friend` is declared as a `:db.type/ref` attribute. In an untyped attribute, a negative number in value position is just data.

In map-style transaction data, put the tempid in `:db/id`:

```
(d/transact! conn
  [{:db/id -1, :user/name "Alice" :user/friend -2}
   {:db/id -2, :user/name "Bob"}])
```

In this example, `-1` references `-2` as a friend through the ref-valued `:user/friend` attribute. Datalevin resolves these tempids during the transaction, replacing them with real entity ids.

String tempids work the same way and can make larger transactions easier to read. This example assumes `:person/school` is declared as a `:db.type/ref` attribute:

```
(d/transact! conn
  [{:db/id "alice"
    :person/name "Alice"
    :person/email "alice@example.com"
    :person/school "mit"}
   {:db/id "mit"
    :school/name "MIT"
    :school/country "USA"}])
```

Here, "alice" and "mit" are transaction-local placeholders. They do not become entity ids or stored string values; Datalevin resolves them to permanent integer entity ids during the transaction.

The same transaction can be written with raw datom operation vectors. In that form, there is no omitted :db/id; the entity position is explicit in every operation:

```
(d/transact! conn
  [[:db/add -1 :user/name "Alice"]
   [:db/add -1 :user/friend -2]
   [:db/add -2 :user/name "Bob"]])
```

A tempid is meaningful only within a single transaction; using -1 again in a later transaction denotes a different new entity. String tempids are also only placeholders. They are resolved to Datalevin-managed integer entity ids; they do not become string-valued entity ids. To learn which permanent ids were assigned, inspect the :tempids map in the transaction report, covered in Section 2.11.

Tempids are especially useful when the data has cycles. Two new entities can refer to each other before either one has a permanent entity id:

```
(d/transact! conn
  [{:db/id -1
    :user/name "Alice"
    :user/friend -2}
   {:db/id -2
    :user/name "Bob"
    :user/friend -1}])
```

Both references are resolved in the same transaction, so the database never needs a half-built intermediate state where only one side of the cycle exists.

2.5 Lookup Refs

Instead of knowing the entity id, you can use a **lookup ref** to identify an existing entity by a unique attribute. A lookup ref is a vector [attribute value]. It can be used anywhere transaction data expects an entity id, including the :db/id field of an entity map:

```
(d/transact! conn
  [{:db/id [:user/email "alice@example.com"]
    :user/active? false}])
```

This finds the entity with :user/email equal to "alice@example.com" and updates it. If no entity exists with that email, the transaction will fail.

Lookup refs are particularly useful when you only have a unique identifier (like an email) but not the entity id.

2.6 Unique Attributes and Upsert

When an attribute is marked as :db.unique/identity, Datalevin automatically performs an **upsert**: if an entity with that value exists, it updates that entity; otherwise, it creates a new one.

Declare the unique identity attribute in the schema before relying on upsert. You can include that schema when opening the connection, or add it to an open connection with `update-schema`:

```
;; First, make :user/email a unique identity attribute.
(d/update-schema conn
  {:user/email {:db/valueType :db.type/string
                :db/unique    :db.unique/identity}})

;; Now upsert based on email
(d/transact! conn
  [{:user/email "alice@example.com" :user/name "Alice v2"}])
```

If `"alice@example.com"` already exists, this updates the existing entity. If not, it creates a new one. This eliminates the need to check for existence before transacting.

2.7 Raw Datom Operation Vectors

For fine-grained control, you can express changes as raw datom operation vectors `[op entity attribute value]`:

```
(d/transact! conn
  [[:db/add -1 :user/name "Alice"]
   [:db/add -1 :user/email "alice@example.com"]
   [:db/retract 101 :user/active? true]])
```

The operations are:

- `:db/add` - adds or updates an attribute value.
- `:db/retract` - removes an attribute value.

2.8 Retracting Attributes and Entities

Use `:db/retract` when you want to remove one attribute value. Use `:db.fn/retractAttribute` when you want to remove an entire attribute from an entity, regardless of its current value. Use `:db/retractEntity` when you want to remove an entire entity.

For cardinality-one attributes, retracting the known current value and retracting the whole attribute usually have the same final effect. The difference matters most for cardinality-many attributes: `:db/retract` removes one selected value from the set, while `:db.fn/retractAttribute` removes all values for that attribute on that entity.

```
;; Remove one known value.
(d/transact! conn
  [[:db/retract [:user/email "alice@example.com"]
                :user/status
                "inactive"]])

;; Remove the whole :user/status attribute from Alice.
(d/transact! conn
  [[:db.fn/retractAttribute [:user/email "alice@example.com"]
                            :user/status]])

;; Remove Alice as an entity.
```

```
(d/transact! conn
  [[:db/retractEntity [:user/email "alice@example.com"]]])
```

`retractEntity` accepts an entity id or a lookup ref. It retracts facts where the entity is the subject, and it also retracts declared `:db.type/ref` facts that point to the entity. If the entity owns child entities through attributes declared with `:db/isComponent true`, those component children are retracted recursively. This is usually the right operation for deleting an entity that may be referenced elsewhere.

For example, with this schema:

```
{:order/id    {:db/valueType :db.type/string
               :db/unique    :db.unique/identity}
 :order/items {:db/valueType   :db.type/ref
               :db/cardinality :db.cardinality/many
               :db/isComponent true}
 :line/sku    {:db/valueType :db.type/string}}
```

Deleting the order also deletes its owned line-item entities:

```
(d/transact! conn
  [[:db/retractEntity [:order/id "o-1001"]]])
```

Component relationships are lifecycle ownership. Mark a ref attribute with `:db/isComponent true` only when the referenced entity has no useful lifecycle apart from the owner. The flag matters when `:db/retractEntity` is applied to the owning entity: owned children, such as line items inside an order, are retracted recursively. It is not the mechanism for merely removing a link. To remove one relationship value, use `:db/retract`; to remove all values for a relationship attribute, use `:db.fn/retractAttribute`.

For example, `:order/items` is a component relationship because a line item belongs to one order and is not useful after that order is deleted. By contrast, `:order/customer`, `:issue/assignee`, and `:team/member` should usually not be components: deleting an order should not delete the customer, changing an assignee should not delete the user, and removing a member from a team should not delete the member entity. If either side of a non-component relationship is later retracted with `:db/retractEntity`, Datalevin removes the reference datoms that point at the retracted entity; it does not recursively delete the other independent entity.

2.9 Mixed Forms

You can mix all these forms in a single transaction:

This example assumes `:user/friend` is a ref attribute; because it receives a collection of lookup refs, a real schema would normally declare it as `:db.cardinality/many`.

```
(d/transact! conn
  [{:user/email "new@example.com" :user/name "New User"}  ; upsert
   [:db/add [:user/email "alice@example.com"] :user/status "active"] ; lookup
↪  ref
   {:db/id -1, :user/friend [[:user/email "bob@example.com"]]} ; tempid + lookup
↪  ref
   [:db/add -1 :user/notes "Added via datom vector"]]) ; raw datom
```

This lets you choose the most convenient form for each operation in your transaction.

2.10 Typed Values, Validation, and Coercion

Datalevin works best when important attributes have explicit value types in the schema, such as `:db.type/long`, `:db.type/string`, `:db.type/boolean`, `:db.type/uuid`, or `:db.type/instant`. A declared `:db/valueType` tells Datalevin how to encode values, compare them, build indexes, and canonicalize transaction input.

Strict input validation is controlled separately by `:validate-data?`:

`:db/valueType`	`:validate-data?`	Write behavior
Not declared	N/A	Stored as serialized EDN; flexible, but weak for typed range access and predictable cross-language input.
Declared	`false` (default)	Encoded with the declared type; simple values are coerced or canonicalized before storage.
Declared	`true`	Checked against the declared type before storage; loose conversions accepted in default mode are rejected.

The automatic coercions are deliberately simple. Declared string attributes use `str`; numeric attributes use the corresponding numeric cast (`long`, `float`, `double`, `bigint`, or `bigdec`); UUID attributes accept UUID values or parseable UUID strings in the default mode; instant attributes accept instant values and, in the default mode, integer epoch milliseconds; keyword and symbol attributes normalize with `keyword` and `symbol`; byte attributes store byte arrays; boolean attributes use Clojure truthiness; and tuple attributes are stored as vectors. Reference attributes resolve tempids, lookup refs, and entity ids before storing the referenced id as a long.

When you transact data without a value type in schema, Datalevin serializes it as an EDN binary blob. In Clojure/JVM code, where numeric literals and casts can produce different boxed numeric types, this can sometimes behave unexpectedly:

```
;; Transacting with an explicit int
(d/transact! conn [{:db/id -1, :my-entity/val (int 42)}])

;; Querying with a plain integer literal (defaults to long)
(d/q
  '[:find ?e :in $
    :where [?e :my-entity/val 42]]
  (d/db conn))
;;=> #{}  ;; Unexpected! No results because the untyped encodings differ.
```

The query returns no results because the untyped value was serialized as a 32-bit integer, but the query literal is encoded as a 64-bit long. Clojure numeric equality is not the issue here; the untyped encodings differ at the index boundary.

However, if you specify a type for `:my-entity/val`, such as `:db.type/long`, Datalevin does type coercion for you, so your `(int 42)` will be stored as a `long` instead. If you open the database with `{:validate-data? true}`, the same declared schema becomes stricter: malformed input fails before it is written instead of being corrected where possible.

Use coercion for trusted internal inputs and imports where convenience matters. Use `:validate-data? true` at system boundaries where malformed user, JSON, or remote-client data should fail early. Chapter 11 covers schema validation, coercion, and schema evolution in more detail.

2.11 Transaction Reports

Every successful `transact!` returns a **transaction report.** If a transaction fails or is aborted, the call throws instead; for async transaction APIs, the future or promise completes with that failure. In that case there is no committed transaction and no transaction report. A successful report contains the database handles before and after the transaction, the datoms that were actually added or retracted, the tempid resolution map, and any transaction metadata supplied by the caller. When a transaction introduces attributes that were not previously present in the database schema, the report also includes `:new-attributes`. The main report keys are:

Report key	Value
`:db-before`	The Datalog Db handle used as the transaction input.
`:db-after`	The Datalog Db handle produced by the transaction.
`:tx-data`	The full datom objects actually added or retracted by the transaction; Chapter 15 explains their `:e`, `:a`, `:v`, `:tx`, and `:added` fields.
`:tempids`	Map from transaction tempids to assigned entity ids.
`:tx-meta`	The metadata value supplied as the optional third argument to `transact!`, or `nil`.
`:new-attributes`	Optional vector of attribute idents first introduced by this transaction. This appears only when the transaction adds facts for attributes that were not already in the schema.

The `:db-before` and `:db-after` fields are mainly for code that needs the immediate transaction boundary for the datoms affected by this transaction. `:db-after` lets the caller or an embedded listener interpret the state produced by the transaction, for example after resolving tempids. `:db-before` lets tools compare the affected input state with the affected output state. These fields are not a general history API, and they are not meant to answer arbitrary questions such as "what would this unrelated query have returned before the transaction?" They are transaction-report context, not an application snapshot model. For normal application reads, read through the connection and take the current Db handle at the point of use.

A common use is a simulated report: ask Datalevin what a transaction would do and leave the live connection unchanged. The simulated report has the same boundary fields as a committed transaction report. Across Clojure, Java, Python, and JavaScript, `:db-before` is the input Db handle and `:db-after` is a readable post-transaction Db handle for inspection with the normal read helpers in that language. The `:db-after` handle represents the would-be state; it has not been committed.

```
(let [preview (d/tx-data->simulated-report
                (d/db conn)
                [{:db/id -1
                  :user/email "preview@example.com"
                  :user/name "Preview User"}])
      eid     (get (:tempids preview) -1)]
  {:preview-user (d/pull (:db-after preview)
                   [:user/email :user/name]
```

```
                  eid)
   :live-user    (d/q '[:find ?e .
                        :where [?e :user/email "preview@example.com"]]
                      (d/db conn))})
;=> {:preview-user {:db/id 101
;                   :user/email "preview@example.com"
;                   :user/name "Preview User"}
;    :live-user nil}
```

For a committed transaction, the raw `:tx-data` shows datoms added or retracted; the two Db handles let you ask the same domain question over the affected data before and after the transaction. Assume account entity `101` currently has status `"open"`. The Clojure and Java examples compare the report Db handles directly; the Python and JavaScript examples show the portable committed-state read from the connection:

```
(let [report (d/transact! conn
               [{:db/id 101
                 :account/status "closed"}])]
  {:before (d/pull (:db-before report)
                   [:account/status]
                   101)
   :after  (d/pull (:db-after report)
                   [:account/status]
                   101)
   :datoms (:tx-data report)})
;=> {:before {:db/id 101, :account/status "open"}
;    :after  {:db/id 101, :account/status "closed"}
;    :datoms [#datalevin/Datom [101 :account/status "open" ... false]
;             #datalevin/Datom [101 :account/status "closed" ... true]]}
```

You can also attach application metadata to a transaction. This metadata is not transaction data and does not create datoms; Datalevin carries it through to the transaction report as `:tx-meta`. It is useful for request ids, audit context, or listener callbacks that need to know why the write happened.

```
(def report
  (d/transact! conn
    [{:db/id -1
      :user/email "alice@example.com"
      :user/name "Alice"}]
    {:source :signup-form}))

(select-keys report [:tempids :tx-meta])
;=> {:tempids {-1 101}
;    :tx-meta {:source :signup-form}}
```

The example above selects two keys of the map to show, because the full report is large.

3. Observing Committed Transactions with `listen!`

Many applications need to react after a write commits: invalidate a cache, notify a UI session, enqueue a background job, update an in-process projection, or append an audit record in another system. The listener API registers an in-process callback on a Datalevin connection. After a transaction commits, Datalevin calls the callback with the transaction report:

```
(def listener-key
  (d/listen!
    conn
    :audit-log
    (fn [{:keys [tx-data tx-meta tempids]}]
      (println "committed" (count tx-data) "datoms")
      (println "metadata" tx-meta)
      (println "tempids" tempids))))

(d/transact!
  conn
  [{:db/id -1
    :user/email "ada@example.com"
    :user/name "Ada"}]
  {:request-id "req-123"})

(d/unlisten! conn listener-key)
```

The listener registration call returns the listener key. When you pass a key explicitly, as above, that same key is returned and can be passed to the matching unlisten operation. When you register without a key, Datalevin generates and returns one for you. The callback receives the same transaction report returned by `transact!`.

The listener key can be any unique value meaningful to the application. Registering another listener with the same key replaces the previous callback, so listener registration is idempotent:

```
(d/listen! conn :cache (fn [report] (invalidate-cache! report)))
(d/listen! conn :cache (fn [report] (enqueue-cache-refresh! report)))
```

Durable side effects use an outbox

Listeners observe committed transactions; they do not make the listener's side effects part of the original transaction. If a callback writes to a log file, publishes to a queue, or calls another service, that work can fail separately from the Datalevin commit. Keep listener callbacks small and predictable. For durable domain events, transact the event as data in the same Datalevin transaction, then use the listener only to wake a publisher or projector. The event is first stored in the database, so publishing can be retried from durable state if the callback, message queue, or downstream service fails.

A minimal outbox shape looks like this:

```
(d/transact!
  conn
  [{:user/email "ada@example.com"
    :user/name "Ada"}
   {:event/type :event.type/user-created
    :event/user [:user/email "ada@example.com"]
    :event/request-id "req-123"}])
```

```
(d/listen!
  conn
  :event-publisher
  (fn [_report]
    (publish-pending-events! conn)))
```

4. Atomic Read-Modify-Write with `with-transaction`

Often, you need to read a value, modify it, and write it back as a single atomic operation. That pattern has a classic race condition unless it is protected. Clojure, Java, and Python expose explicit Datalog transaction callbacks for this purpose:

```
(d/with-transaction [cn conn]
  (let [current-balance (d/q '[:find ?bal . :where [101 :account/balance ?bal]]
↪   (d/db cn))
        new-balance (- current-balance 100)]
    (d/transact! cn [{:db/id 101, :account/balance new-balance}])))
```

The transaction callback keeps the read and write inside one isolated read/write transaction. Within one store, Datalevin serializes committed writes, so no other writer can commit in the middle of that read-modify-write sequence.

`q`, `pull`, and the other read APIs do not have separate isolation rules. They read through the connection or Db handle you pass:

Read target	What the read sees
Transaction-bound connection or Db inside `with-transaction`	The active transaction, including writes already performed in the same body. If the body aborts, those writes never become visible after the callback.
Ordinary connection or Db outside the active transaction	The committed state available to that handle; it does not see uncommitted writes from the active transaction.
`:db-before` or `:db-after` from a transaction report or simulated report	Exactly the boundary state described by that report. For a simulated report, `:db-after` is a would-be state only.

Outside `with-transaction`, a query runs against the Db supplied to it. That read view is stable for the query and isolated from concurrent writes, but it is not an application-level read/write transaction unless the related reads and writes are placed inside `with-transaction`.

The explicit Datalog abort helper, `abort-transact`, is meaningful only while an explicit Datalog transaction is active.

The JavaScript binding exposes the abort bridge for API parity, but does not expose an explicit Datalog transaction callback; use `:db/cas` or transaction functions for conditional Datalog writes instead. More generally, when a host API lacks an explicit callback, encode the conditional write in one transaction or run the command inside the Datalevin-hosting process when the read-modify-write logic must be serialized.

Explicit transaction callbacks can also take a timeout for the callback body:

```
(d/with-transaction [cn conn {:timeout-ms 5000}]
  (d/transact! cn [{:db/id 101 :account/checked true}]))
```

The default explicit transaction timeout is unset. It can be read or changed through the explicit transaction timeout helpers. A per-call timeout overrides the default; a nil/null value disables the default for that call. If the timeout expires, Datalevin interrupts the transaction body, aborts the transaction, and reports `:type :transaction/timeout` with `:timeout-ms`. Code that ignores interruption may continue running until it returns, so transaction callbacks should remain small and bounded.

> **Not a retry loop**
> `with-transaction` is a serialization tool, not a general retry loop. If the body throws, the transaction aborts. Logical failures such as CAS mismatch, lookup-ref miss, validation failure, or unique conflict are returned to the caller. Chapter 22 pulls those cases together into a production retry and error-handling policy.

4.1 Mixing Datalog and KV Writes

Most applications can stay entirely in Datalog transaction data. Mixing Datalog and KV writes is useful when the Datalog facts are the source of truth, but you also maintain a custom KV structure for a workflow that benefits from direct key-value access: an audit or outbox DBI, an ingestion checkpoint, a compact deduplication table, or a small side index for application code. In those cases, the Datalog facts and the KV entry often describe one business event and should commit or abort together.

A local Datalog database and its custom KV DBIs live in the same Datalevin store. That means a Datalog update and a KV update can commit as one atomic unit, but only if both writes run inside the same transaction.

Suppose marking an order paid must also write an audit entry to a custom `"audit-log"` DBI. The shape to avoid is: commit the Datalog update through the Datalog connection, then write the audit entry through a separately opened `open-kv` handle on the same directory. Those are two independent transactions. A crash or exception between them can leave the order marked paid without the audit entry.

```
;; Avoid this shape: two separate transactions.
(d/transact! conn
  [{:order/id     "o-1001"
    :order/status :order.status/paid}])

(let [separate-kv (d/open-kv db-dir)] ; the same directory used by conn
  (d/transact-kv separate-kv "audit-log"
    [[:put "o-1001" {:event/type :order/paid
                     :order/id   "o-1001"}]]
    :string :data))
```

The correct shape is to use the KV handle that belongs to the Datalog connection, open the application DBI during setup, and then perform both writes while one transaction is active. The examples below use `with-transaction` because it gives application code a transaction-bound connection. Inside the block, get the transaction-bound KV handle from that connection and use it for the KV write. The borrowed KV handle is owned by the Datalog connection; close the Datalog connection, not the borrowed handle.

Listing 6.1. Mixing Datalog and key-value writes in one transaction

```
(let [kv (d/datalog-kv conn)]
  ;; DBI opening is idempotent, so this is safe during application startup.
  (d/open-dbi kv "audit-log")

  (d/with-transaction [cn conn]
    (let [cn-kv (d/datalog-kv cn)]
      (d/transact! cn
        [{:order/id     "o-1001"
          :order/status :order.status/paid}])

      (d/transact-kv cn-kv "audit-log"
        [[:put "o-1001"
          {:event/type :order/paid
           :order/id   "o-1001"}]]
        :string :data))))
```

If the block returns normally, both the Datalog datoms and the KV entry are committed. If the block throws or the active Datalog transaction is explicitly aborted, both parts are aborted. This only works for DBIs that live in the same local Datalevin store; it is not a cross-file transaction mechanism.

5. Transaction Functions

Transaction functions are transaction data forms that run while a transaction is being prepared and expand to more transaction data. They run against the current DB object and are committed atomically with the rest of the transaction. Transaction functions are expressed as transaction data vectors, not as a bare symbolic list form such as `(my-tx-fn arg)`.

If a transaction function calls `q`, `pull`, or another read API, that read uses the DB object passed to the function. This is the database at the point where the function is expanded during transaction preparation, not the final post-transaction database.

Transaction-time vector forms include:

- `[:db.fn/call f arg ...]` calls an inline function, an installed function, or a user-defined function (UDF) descriptor.
- `[:some/installed-fn arg ...]` calls an installed transaction function whose entity has `:db/ident :some/installed-fn`.
- `[:db/cas e a old new]` or `[:db.fn/cas e a old new]` performs compare-and-swap.
- `[:db.fn/retractAttribute e a]`, `[:db.fn/retractEntity e]`, and `[:db/retractEntity e]` are built-in transaction functions for retraction.
- `[:db.fn/patchIdoc ...]` patches an idoc value; idoc document modeling is covered in Chapter 14.
- `[:db/ensure pred arg ...]` validates a post-condition that runs after all transaction data has been expanded and applied, but before commit.

5.1 Compare-and-Swap

Use `:db/cas` when the write should succeed only if the current value is what you expect. The entity position may be an entity id or a lookup ref.

```
(d/transact! conn
  [[:db/cas [:user/email "alice@example.com"]
    :account/balance
    100
    75]])
```

If Alice's balance is not currently `100`, the whole transaction fails. This is useful for conditional updates that must not silently overwrite newer data. A CAS failure is a logical conflict, not an automatic retry signal; Chapter 22 shows how to re-read and retry when that is the correct application behavior.

5.2 Transaction UDFs

There are two ways to install a named application transaction function. Choose based on where the implementation lives:

Mechanism	Stored attribute	Use when
Descriptor-backed UDF	`:db/udf`	The function must be callable from Java, Python, JavaScript, client/server deployments, or another runtime that resolves descriptors through a registry.
Interpreted Clojure function	`:db/fn`	The function is Clojure code for embedded Clojure-style environments and can be represented with `datalevin.interpret` helpers.

Both mechanisms install an entity with `:db/ident`, and both can be invoked with `[:some/ident arg ...]` or through `:db.fn/call`. The difference is not the call shape; it is how the implementation is stored and resolved.

For arbitrary user-defined functions (UDFs), Datalevin allows descriptor-backed transaction functions. The same descriptor and registry model works in client/server deployments as long as the runtime that executes the transaction can resolve the UDF implementation.

Store the descriptor in `:db/udf`, open the database with a runtime UDF registry or resolver, and call it by descriptor or by installed `:db/ident`. Transaction UDF descriptors use `:udf/kind :tx-fn`. Post-condition predicates used by `:db/ensure` follow the same registry model with `:udf/kind :predicate`; Section 5.5 shows that form. The `:udf/lang` value is part of the descriptor identity and resolver namespace. Use the same descriptor when registering and calling the implementation. The examples below use `:clojure` for the Clojure function, `:java` for Java, `:python` for Python, and `:javascript` for JavaScript.

Listing 6.2. Descriptor-backed transaction function

```
(require '[datalevin.udf :as udf])

(def descriptor
  {:udf/lang :clojure
   :udf/kind :tx-fn
   :udf/id   :user/bootstrap})

(def registry
  (doto (udf/create-registry)
    (udf/register! descriptor
      (fn [_db email name]
        [{:db/id       -1
          :user/email  email
          :user/name   name}]))))

(def conn
  (d/create-conn
    "/tmp/tx-udf-demo"
    {:user/email {:db/valueType :db.type/string
                  :db/unique    :db.unique/identity}}
    {:runtime-opts {:udf-registry registry}}))

;; Call the descriptor directly.
(d/transact! conn
  [[:db.fn/call descriptor "ada@example.com" "Ada"]])

;; Or install the descriptor and call it by :db/ident.
(d/transact! conn
  [{:db/ident :user/bootstrap
    :db/udf   descriptor}])

(d/transact! conn
  [[:user/bootstrap "bob@example.com" "Bob"]])
```

Only the descriptor is stored in the database. The implementation comes from the runtime registry or resolver, so server processes must be configured with the same UDF implementation before they can execute the transaction function.

5.3 Inline Transaction Functions

In embedded Clojure, `:db.fn/call` can call a regular Clojure function. The function receives the current DB object as its first argument and must return transaction data.

```
(defn rename-user [db email new-name]
  (if-let [eid (d/q '[:find ?e .
                      :in $ ?email
                      :where [?e :user/email ?email]]
                    db email)]
    [[:db/add eid :user/name new-name]]
    (throw (ex-info "No user with email" {:email email}))))
```

```
(d/transact! conn
  [[:db.fn/call rename-user "bob@example.com" "Bob V2"]])
```

The returned transaction data is prepared and committed as part of the same write transaction.

5.4 Installed Transaction Functions

For stored transaction functions written in Clojure, install an entity with `:db/ident` and `:db/fn`. The function value should be created with `datalevin.interpret` helpers such as `inter-fn` or `definterfn`, which keeps it usable in native image, server, and Babashka contexts.

Use this mechanism when the transaction logic is naturally Clojure and the environment can evaluate Datalevin interpreted functions. If the same installed function must be resolved by different server processes or runtimes, use the `:db/udf` descriptor mechanism from Section 5.2 instead.

```
(require '[datalevin.interpret :as i])

(def rename-user*
  (i/inter-fn [db email new-name]
    (if-some [ent (d/entity db [:user/email email])]
      [[:db/add (:db/id ent) :user/name new-name]]
      (throw (ex-info "No user with email" {:email email})))))

(d/transact! conn
  [{:db/ident :user/rename
    :db/fn     rename-user*}])

(d/transact! conn
  [[:user/rename "bob@example.com" "Bob V2"]])
```

You can also call an installed function through `:db.fn/call`:

```
(d/transact! conn
  [[:db.fn/call :user/rename "bob@example.com" "Bob V3"]])
```

5.5 Post-Conditions with `:db/ensure`

Use `:db/ensure` when the invariant must be checked after the entire transaction has been expanded and applied, but before commit. Most transaction functions see the database at the point where they are expanded. `:db/ensure` is different: it runs after transaction functions, UDF-backed calls, CAS operations, and ordinary datom additions have produced the would-be database.

The first argument is a predicate: a direct function, a var, a qualified symbol, or a descriptor-backed UDF predicate. A UDF predicate uses `:udf/kind :predicate`, is registered in the runtime UDF registry, and can be passed directly in the `:db/ensure` form. It receives the would-be database as its first argument, followed by the resolved ensure arguments. Datalevin resolves tempid arguments through the transaction report before calling the predicate.

The same transaction shape works in each embedded binding. The implementation below is written in the host language of that binding, registered under the same stable UDF id, and returns truthy only when the account is open and not locked. Each example deliberately locks the account after creating it as open, so the ensure predicate aborts before commit.

In Python and JavaScript, an immutable `UdfDescriptor` can be used directly for both registry operations and typed transaction data.

```
(require '[datalevin.core :as d]
         '[datalevin.udf :as udf])

(def account-ok?
  {:udf/lang :clojure
   :udf/kind :predicate
   :udf/id   :account/open-and-unlocked?})

(def registry
  (doto (udf/create-registry)
    (udf/register! account-ok?
      (fn [db eid]
        (let [account (d/entity db eid)]
          (and (= :open (:account/status account))
               (not (:account/locked? account))))))))

;; Open the connection with {:runtime-opts {:udf-registry registry}}.
(d/transact! conn
  [{:db/id "acct"
    :account/status :open}
   [:db/add "acct" :account/locked? true]
   [:db/ensure account-ok? "acct"]])
```

`:db/ensure` forms are not included in `:tx-data`. If the predicate returns false or nil, or if it throws, the transaction aborts.

6. Choosing and Tuning Transaction APIs

The transaction APIs differ mostly in where the transaction boundary is placed and when the caller waits. For live commits, they share the same core rule: within one Datalevin store, committed writes have one serial order.

Use this table as a recap:

API or pattern	Best use	Transaction boundary	Reads inside
`transact!`	Ordinary application writes.	One collection of transaction data commits or aborts as one unit; the call returns after commit.	Application code does not run between tx-data forms, but transaction functions in the data run inside the same transaction.

API or pattern	Best use	Transaction boundary	Reads inside
`transact-async`	High-volume independent writes where the caller can wait later.	Same transaction semantics as `transact!`, but the caller receives a future or promise immediately.	Dependent code must wait for the future or promise before reading or submitting dependent writes.
`with-transaction`	Read-modify-write logic and multi-step writes that must be one atomic operation.	One explicit read/write transaction around the callback, with an optional timeout. Nested synchronous transaction calls inside the callback reuse this transaction instead of opening another one.	Queries through the transaction-bound connection have read-your-writes behavior. No other writer can interleave.
Transaction functions such as `:db/cas` or `:db.fn/call`	Conditional or server-side write logic expressed as transaction data.	The function runs inside the containing transaction and returns more transaction data.	The function receives the current Db for that transaction. Its result is committed atomically with the rest of the transaction.
`tx-data->simulated-report` and variants	Tests, dry runs, validation previews, and explanations of what a transaction would do.	No live commit; Datalevin prepares the transaction against the supplied Db and returns a report shape.	Reads the supplied Db only. It does not serialize with a live connection.

7. Write Performance in Context

Datalevin's write benchmark [2] measures end-to-end throughput while inserting one million four-attribute person records into a fresh database. It includes data generation, transaction processing, encoding, and result decoding. The SQLite baseline uses the same records, transaction boundaries, and lookup indexes, although the physical layouts differ.

The following snapshot was measured on a 2023 Macbook Pro with 36GB memory. Values are completed records per second. `Default` and `Strict WAL` use synchronous durability settings.

Durability	Batch	Datalog sync	SQLite	Datalog async
Default	1	2,380/s	1,814/s	37,357/s
Default	10	5,157/s	6,233/s	114,960/s
Default	100	12,546/s	14,171/s	282,470/s
Default	1,000	49,436/s	30,722/s	284,146/s
Strict WAL	1	8,186/s	8,822/s	102,133/s
Strict WAL	10	26,519/s	17,609/s	195,976/s
Strict WAL	100	44,747/s	22,840/s	245,485/s

Strict WAL	1,000	114,739/s	32,105/s	238,974/s

Key takeaways:

- Larger transactions amortize commit cost; both Datalevin modes improved sharply as the batch grew from 1 to 1,000.
- Strict WAL is the default WAL profile. Here it outperformed SQLite WAL with `synchronous=FULL` at batch sizes 10 through 1,000.
- `transact-async` gains throughput by combining independent queued writes. Wait for its future or promise, or a queue barrier, before dependent work.

With four strict-WAL blocking writers, Datalevin measured 19,003 records/s at batch 1 and 151,622 at batch 1,000, versus SQLite's 7,557 and 50,735. In a 50/50 read/upsert test, Datalevin and SQLite measured 5,616 and 5,041 pairs/s under strict WAL, and 1,552 and 1,515 under default direct durability.

Results vary with hardware, data, indexes, contention, and durability settings, so use the benchmark README [2] for the full protocol and re-run it with a workload shaped like your application.

Summary

Datalevin's transaction model scales from simple, safe synchronous writes to high-performance asynchronous batching. With `listen!` for committed-write observation, `with-transaction` for atomic updates, and `transact-async` for high throughput, applications can stay both correct and fast.

Transaction functions are the other main tool for keeping write logic close to the transaction boundary. Use `:db/cas` for compare-and-swap updates, `:db/ensure` for post-conditions over the would-be database, and descriptor-backed UDFs or installed functions when application-specific logic must run inside the transaction.

References

[1] Theo Haerder and Andreas Reuter, "Principles of Transaction-Oriented Database Recovery," *ACM Computing Surveys* 15(4):287-317, 1983. DOI: https://doi.org/10.1145/289.291.

[2] Datalevin project, "Write Benchmark," benchmark implementation, methodology, and results. URL: https://github.com/datalevin/datalevin/tree/master/benchmarks/write-bench.

CHAPTER 7

Reading Data

Once data is in your database, you need a way to retrieve it in the shape your application actually uses. Datalevin gives you several read APIs, each with a different job:

- **Lookup refs** let you address an entity by a domain identifier such as an email, SKU, username, or order number.
- **`d/entity`** returns a lazy, map-like entity object for interactive or conditional navigation.
- **`d/pull`** fetches one entity as a plain map in a declared shape, including nested references and recursive trees.
- **`d/q`** finds sets of values and entities using Datalog constraints.

This chapter covers the first three. Chapter 8 focuses on Datalog query syntax, but queries appear briefly here when they help explain lookup refs.

1. Reading by Identity with Lookup Refs

Datalevin stores entities internally by numeric entity id. Application code, however, usually knows a natural key: a user's email, an account number, a URL slug, or an external system id. A **lookup ref** is the bridge between those two worlds.

> **Use lookup refs at application boundaries**
>
> Prefer lookup refs and unique domain attributes at application boundaries because entity ids are database-local handles; the same logical entity may have different numeric ids in two databases.

A lookup ref is a two-element vector:

```
[:user/email "alice@example.com"]
```

The first element must be an attribute marked `:db/unique`; Chapter 11 explains how to choose between uniqueness modes when designing identity attributes. The second element is the value of that unique attribute. Datalevin resolves the vector to the matching entity id when an API expects an entity.

Python's `tx.lookup_ref` and JavaScript's `tx.lookupRef` are typed wrappers for that same two-element form. Despite the namespace, the result is also valid for read APIs; it is not tied to a transaction.

```
(def schema
  {:user/email     {:db/valueType :db.type/string
                    :db/unique    :db.unique/identity}
   :user/name      {:db/valueType :db.type/string}
   :user/friends   {:db/valueType   :db.type/ref
```

```
                    :db/cardinality :db.cardinality/many}
   :order/id       {:db/valueType :db.type/string
                    :db/unique    :db.unique/identity}
   :order/customer {:db/valueType :db.type/ref}})

(d/transact! conn
  [{:user/email "alice@example.com"
    :user/name  "Alice"}
   {:user/email "bob@example.com"
    :user/name  "Bob"
    :user/friends [[:user/email "alice@example.com"]]}
   {:order/id "o-1001"
    :order/customer [:user/email "alice@example.com"]}])

(def db (d/db conn))
```

Lookup refs work in the same places where you would otherwise pass an entity id:

```
;; Resolve to the numeric entity id when you explicitly need it.
(d/entid db [:user/email "alice@example.com"])
;; => 1

;; Pull by lookup ref.
(d/pull db [:user/name :user/email] [:user/email "alice@example.com"])
;; => {:user/name "Alice", :user/email "alice@example.com"}

;; Navigate by lookup ref.
(def alice (d/entity db [:user/email "alice@example.com"]))
(:user/name alice)
;; => "Alice"
```

Lookup refs are also valid inputs to Datalog queries. Here, the `:in` clause declares the query's inputs: the database $, followed by the `?customer` parameter. Chapter 8 covers query inputs in detail. The query engine resolves lookup refs when they occur in positions that require an entity id:

```
(d/q '[:find ?order-id
       :in $ ?customer
       :where
       [?order :order/customer ?customer]
       [?order :order/id ?order-id]]
     db
     [:user/email "alice@example.com"])
;; => #{["o-1001"]}
```

If a lookup ref does not resolve, `d/entid`, `d/entity`, and `d/pull` return the host language's nil value: `nil`, `null`, or `None`.

2. Entity API: Lazy Map-Like Navigation

`d/entity` returns a map-like entity object tied to the Datalog Db handle used to create it. Entity navigation starts from a known entity and reads attributes as the program asks for them. In

Clojure, Python, and JavaScript, the connection-level entity APIs navigate lazily. Java examples use `entityMap` when they want ordinary `Map` values at the connection level. Pass either a lookup ref or a numeric entity id:

```
(def alice (d/entity db [:user/email "alice@example.com"]))

(:user/name (d/entity db 1))
;; => "Alice"
```

An entity behaves like a map for attribute lookup:

```
(:db/id alice)
;; => 1

(:user/name alice)
;; => "Alice"

(alice :user/email)
;; => "alice@example.com"

(contains? alice :user/name)
;; => true
```

For lazy entity APIs, reading `(:user/name alice)` or `alice.get(":user/name")` fetches and caches that attribute. Reading another attribute may touch storage again. Java's `entityMap` example below has already materialized the map, but it serves the same conditional-navigation use case:

```
(when-let [alice (d/entity db [:user/email "alice@example.com"])]
  (if (:user/active? alice)
    (:user/name alice)
    "inactive"))
```

> **Performance: lazy entity reads**
>
> Entity navigation does not go through a query-planning path: a forward attribute read is an indexed pattern lookup for that entity and attribute, and a reverse reference read is an indexed lookup for entities that point to the current entity.
> The performance tradeoff is the laziness itself. A single lookup is predictable, but walking many entities or many references can become many small reads whose count depends on application control flow. For bulk reads and API responses, prefer `d/q` with `pull`, `d/pull`, or `d/pull-many` so the read shape is explicit.

Use `d/touch`, `.touch()`, or the Java `entityMap` convenience when you intentionally want to realize all attributes of an entity for debugging, logging, or exploratory REPL work:

```
(pr-str (d/entity db [:user/email "alice@example.com"]))
;; => "{:db/id 1}"

(d/touch (d/entity db [:user/email "alice@example.com"]))
;; prints like:
;; {:db/id 1,
```

```
;;  :user/name "Alice",
;;  :user/email "alice@example.com"}
```

In Clojure, `d/touch` realizes the attributes, but it does not turn the entity into a plain persistent map. The value is still an entity object. That distinction matters with functions such as `merge` that construct maps by conjoining entries:

```
(def alice-ent (d/touch (d/entity db [:user/email "alice@example.com"])))

;; Do this when ordinary map operations should own the value:
(merge (into {} alice-ent)
       {:api/source "profile"})
```

2.1 Entity References

When an entity attribute is declared as `:db.type/ref`, lazy entity navigation returns entity objects for the referenced entities.

```
(def bob (d/entity db [:user/email "bob@example.com"]))

(:user/friends bob)
;; => a set containing Alice's entity

(map :user/name (:user/friends bob))
;; => ("Alice")
```

For lazy entity APIs, a cardinality-one reference returns one entity and a cardinality-many reference returns a set of entities. Those referenced entities are lazy too, so this kind of navigation can be very convenient in a REPL or in small conditional branches. Java examples here use pull for the same result shape because connection-level Java reads commonly return materialized maps.

2.2 Reverse Navigation on Entities

Entity navigation also supports reverse attributes:

```
(def alice (d/entity db [:user/email "alice@example.com"]))

(map :order/id (:order/_customer alice))
;; => ("o-1001")
```

Reverse attributes use the same spelling everywhere they appear:

- `:order/customer` becomes `:order/_customer`
- `:friend` becomes `:_friend`

Reverse attributes return the entities that point to the current entity through that reference. For non-component references the result is a set. Component references can return a single owner because a component has at most one owner in a well-formed model.

Datalevin does not enforce that ownership invariant when writing refs. If several parents point to the same component child, reverse component navigation returns one owner and the model should be treated as malformed. Use an ordinary reference attribute for shared targets.

2.3 DB Handles and Entity Objects

Datalevin's `db` is a read handle used by read APIs, not durable application state. Treat it as an operation-local handle. In Clojure, call `(d/db conn)` at the read site. Connection-level methods in the other bindings read through the connection for you.

Use DB-taking forms when you are inspecting a transaction-report boundary, such as `:db-before` or `:db-after` from a simulated transaction report: `d/pull` in Clojure, `Datalevin.pull(db, ...)` or `DatabaseValue.pull(...)` in Java, and `db.pull(...)` on Python or JavaScript `Database` objects.

> **Do not cache entity objects as live state**
>
> An entity object is tied to the read handle used to create it. Do not save a `db` handle or an entity object and expect it to follow later transactions. A materialized entity map is just a copy and will not update either. If current data matters after a write, read again from the connection.

The following example shows the boundary:

```
(def alice-v1 (d/entity (d/db conn) [:user/email "alice@example.com"]))

(d/transact! conn
  [[:db/add [:user/email "alice@example.com"] :user/name "Alice A."]])

(:user/name alice-v1)
;; => "Alice"

(:user/name (d/entity (d/db conn) [:user/email "alice@example.com"]))
;; => "Alice A."
```

2.4 Using `d/entity` in Transaction Functions

Transaction functions often use entity navigation for read-side checks before returning transaction data:

```
(defn rename-user [db email new-name]
  (if-some [ent (d/entity db [:user/email email])]
    [[:db/add (:db/id ent) :user/name new-name]]
    (throw (ex-info "No user with email" {:email email}))))
```

The transaction function receives a Db handle at the point where the function is expanded. Use that handle when the function must inspect current data before returning transaction data. Java uses `Datalevin.entid` here because only the id is needed. The Python and JavaScript registries expose that handle as their `Database` wrapper, so both callbacks can use the ordinary entity API and return typed transaction data.

Chapter 6 covers transaction functions in more detail.

2.5 Transactable Entities

In Clojure, Datalevin Entity objects are not plain maps, but they support a small staging interface for writes. `assoc`, `update`, `d/add`, `d/retract`, and `dissoc` return a new Entity value that carries pending changes. The original entity and the Db handle it came from are not modified. The staged changes become transaction data only when the staged entity is passed to `d/transact!` . This staging interface is Clojure-only; in other bindings, express writes as ordinary transaction maps or datom vectors.

```
(def alice (d/entity (d/db conn) [:user/email "alice@example.com"]))

(def staged-alice
  (-> alice
      (assoc :user/name "Alice A.")
      (d/add :user/friends [:user/email "bob@example.com"])))

(:user/name staged-alice)
;; => "Alice A."

(:user/name alice)
;; => "Alice"
```

At this point, only `staged-alice` sees the pending name change. The database does not change until the staged entity is transacted:

```
(d/transact! conn [staged-alice])

(d/pull (d/db conn)
        [:user/name {:user/friends [:user/name]}]
        [:user/email "alice@example.com"])
;; => {:user/name "Alice A.",
;;     :user/friends [{:user/name "Bob"}]}
```

Use `d/add` when adding a value to a cardinality-many attribute. Use `d/retract` with a value when removing one value, or `dissoc`/`d/retract` without a value when retracting an entire attribute:

```
(let [db     (d/db conn)
      alice2 (d/entity db [:user/email "alice@example.com"])
      bob    (d/entity db [:user/email "bob@example.com"])]
  (d/transact! conn
    [(d/retract alice2 :user/friends bob)]))
```

This feature is useful at the REPL and in small update flows because the code can look like ordinary map transformation while still producing normal transaction data. For larger workflows or validation-heavy updates, explicit transaction maps and datom vectors are usually easier to audit and test.

3. Pull: Fetching a Declared Shape

`d/pull` is the easiest way to fetch a plain data structure for one entity. You provide a Db handle, a **pull pattern**, and an entity id or lookup ref:

```
(d/pull db pattern eid-or-lookup-ref)
```

The pattern is a vector describing which attributes to retrieve.

```
(d/pull db [:user/name :user/email]
        [:user/email "alice@example.com"])
;; => {:user/name "Alice", :user/email "alice@example.com"}
```

If you already have the numeric entity id, pass it directly to `d/pull`. You do not need a Datalog query just to retrieve one known entity:

```
(d/pull db [:user/name :user/email] 1)
;; => {:user/name "Alice", :user/email "alice@example.com"}
```

A numeric id is treated as an id, not as an existence check. `d/entid` returns numeric ids unchanged; use a lookup ref, a query, or a specific attribute read when you need to verify that an entity currently exists.

Use `[*]` to fetch all forward attributes of the entity:

```
(d/pull db ['*] [:user/email "alice@example.com"])
;; => {:db/id 1,
;;     :user/name "Alice",
;;     :user/email "alice@example.com"}
```

Wildcard pull is convenient for exploration, but production APIs should usually name the attributes they return. Explicit patterns keep responses stable as the schema grows.

3.1 Nested Pull

Pull patterns can follow reference attributes. If `:order/customer` has `:db/valueType :db.type/ref`, a nested map in the pattern tells Datalevin how to shape the referenced entity:

```
(d/pull db
        [:order/id {:order/customer [:user/name :user/email]}]
        [:order/id "o-1001"])
```

Result:

```
{:order/id "o-1001",
 :order/customer {:user/name "Alice",
                  :user/email "alice@example.com"}}
```

Without the nested pattern, an ordinary reference is returned as a small entity reference:

```
(d/pull db [:order/id :order/customer] [:order/id "o-1001"])
;; => {:order/id "o-1001", :order/customer {:db/id 1}}
```

That behavior is intentional. Pull only expands what the pattern asks it to expand.

Component references are the important exception: a bare component reference can expand recursively. Section 3.5 shows how to keep component responses explicit.

3.2 Reverse Pull

A reference attribute points forward from one entity to another. For example, an order has `:order/customer` pointing to a user. Sometimes you want to start from the user and ask, "which orders point here?"

Pull supports this with **reverse attributes**. For a namespaced attribute like `:order/customer`, the reverse attribute is `:order/_customer`. For an unqualified attribute like `:friend`, the reverse attribute is `:_friend`.

```
(d/pull db
        [:user/name {:order/_customer [:order/id]}]
        [:user/email "alice@example.com"])
;; => {:user/name "Alice",
;;     :order/_customer [{:order/id "o-1001"}]}
```

As with forward references, a bare reverse attribute returns only small entity reference maps:

```
(d/pull db
        [:user/name :order/_customer]
        [:user/email "alice@example.com"])
;; => {:user/name "Alice",
;;     :order/_customer [{:db/id 3}]}
```

Plain forward pull returns the stored value even when an attribute is untyped. The `:db.type/ref` declaration matters when the pattern asks Datalevin to traverse: nested pull such as `{:order/customer [...]}` and reverse pull such as `:order/_customer` require `:order/customer` to be declared as a ref. Otherwise Datalevin rejects the pull pattern instead of treating an arbitrary value as an entity id.

3.3 Pulling Many Entities

When you already have several entity ids or lookup refs, `d/pull-many` applies one pull pattern to all of them and returns the maps in the same order. In embedded code this is mostly a clearer shape with a small benefit: Datalevin prepares the pattern once for the batch. In client/server code it is more important because it avoids one round trip per entity.

```
(d/pull-many db
             [:user/name]
             [[:user/email "alice@example.com"]
              [:user/email "bob@example.com"]])
;; => [{:user/name "Alice"} {:user/name "Bob"}]
```

3.4 Pull Pattern Options

Pull supports attribute options for common response-shaping needs.

3.4.1 Limiting Cardinality-Many Attributes

Pull returns at most 1000 values for a cardinality-many attribute by default. Use `:limit` to choose a different limit, or `nil` for no limit. The option is written as an attribute expression:

```
(d/pull db [[:user/friends :limit 5]]
        [:user/email "bob@example.com"])
```

The same attribute expression can be used as the key in a nested map spec:

```
(d/pull db
        [:user/name {[:user/friends :limit 5] [:user/name]}]
        [:user/email "bob@example.com"])
```

3.4.2 Default Values

Use `:default` when the response should include a value even if the entity does not currently have that attribute:

```
(d/pull db
        [:user/name [:user/active? :default true]]
        [:user/email "alice@example.com"])
;; => {:user/name "Alice", :user/active? true}
```

3.4.3 Attribute Renaming

Use `:as` when the external response should use a different key:

```
(d/pull db
        [[:user/name :as :name]
         [:user/email :as :email]]
        [:user/email "alice@example.com"])
;; => {:name "Alice", :email "alice@example.com"}
```

Renaming is a presentation choice. It does not change the stored attributes or the schema.

3.4.4 Transforming Pull Values

Use `:xform` when a value should be transformed before it is placed in the pull result. The transform may be a function value or a resolvable symbol, such as `vector` or `clojure.string/upper-case`:

```
(d/pull db
        [[:user/name :as :name :xform 'clojure.string/upper-case]]
        [:user/email "alice@example.com"])
;; => {:name "ALICE"}
```

For pull patterns that may be stored or sent across process boundaries, prefer resolvable symbols and make sure the runtime can load them.

If an attribute is missing and no default is supplied, the transform receives `nil`:

```
(d/pull db
        [[:user/nickname :xform vector]]
        [:user/email "alice@example.com"])
;; => {:user/nickname [nil]}
```

If `:default` is present and the attribute is missing, the default wins and is returned as-is:

```
(d/pull db
        [[:user/nickname :default "unknown" :xform vector]]
        [:user/email "alice@example.com"])
;; => {:user/nickname "unknown"}
```

For reference, reverse-reference, and nested pull attributes, `:xform` receives the pulled value for that attribute. This is useful for small response-shaping steps. For transformations with business rules, error handling, or portability requirements, prefer ordinary application code after the pull.

3.5 Recursive Pull and Component Trees

Nested pull is not limited to one fixed level. When a reference points to the same kind of entity repeatedly, such as a category tree, organization chart, filesystem, or bill of materials, use a recursive pull map. The map value can be `...` for unbounded recursion, or a positive number for a bounded depth.

The examples below assume `:category/slug` is unique and `:category/children` is a reference attribute.

Use a bounded depth for user-facing APIs unless the graph is naturally small:

```
(d/pull db
        [:category/name {:category/children 2}]
        [:category/slug "databases"])
```

Use `...` when the recursive structure itself is the response, and the data model guarantees a sensible boundary:

```
(d/pull db
        [:category/name {:category/children '...}]
        [:category/slug "databases"])
```

Recursive pull also works in reverse. If `:category/children` points from a parent to a child, `:category/_children` walks from a child back toward its parents:

```
(d/pull db
        [:category/name {:category/_children 3}]
        [:category/slug "lmdb"])
```

Datalevin tracks entities already seen during a recursive pull. If a cycle is encountered, the repeated entity is represented by `{:db/id ...}` instead of being expanded forever. That makes recursion safe, but it does not make every recursive response small. Choose explicit depth limits for large graphs.

Component attributes deserve special care. In a well-formed component model, a component child is owned by one parent, so a bare component reference can expand recursively. That is useful for logical documents, but it can also return more than an API endpoint is meant to expose. Prefer an explicit pattern when the external shape matters:

```
(d/pull db
        [:order/id
         {:order/items [:line/sku :line/qty]}]
        [:order/id "o-1001"])
```

Appendix E has the pull pattern syntax reference table.

4. Choosing the Right Read API

Use these rules of thumb:

Need	Use
You know a natural key and need one entity	Lookup ref with `d/pull` or `d/entity`
You know an entity id and need its data	`d/pull` by id, or `d/entity` by id for lazy navigation
You need a stable response map	`d/pull` with an explicit pattern
You need many response maps from known ids or lookup refs	`d/pull-many`
You need a bounded tree or hierarchy	`d/pull` with a recursive pattern such as `{:category/children 2}`
You are navigating conditionally	Entity API for lazy navigation, or `d/pull` when you need a realized map
You are debugging an entity at the REPL	`d/entity` plus `d/touch`
You want to stage a small update from an entity in Clojure	Transactable Entity with `assoc`, `d/add`, `d/retract`, or `dissoc`
You need to discover matching entities	`d/q`

4.1 Read Performance

These APIs differ mostly in how much of the result shape is declared up front and how much per-call work they do.

`d/pull` is a good fit when you already know one entity and want one explicit map shape. Calling `d/pull` in a loop is correct, but it repeats the API call for each entity. In embedded use that overhead is often small, especially when the same pull pattern is reused and can be prepared from cache. In client/server use, a loop can be much more expensive because each call may be a separate request.

`d/pull-many` is the batch form for known entity ids or lookup refs. It prepares one pull pattern for the batch, returns results in the same order as the input ids, and uses one client/server request. Internally it still pulls each entity; it is not a different query optimizer. Prefer it over a loop when many known entities need the same shape.

`d/entity` is lazy navigation from one known entity. A single attribute read is an indexed lookup and the result is cached on that entity object. This can be efficient when control flow may read only one or two attributes. It is a poor fit for bulk response construction, because walking many entities or references can become many small indexed reads whose total count is less visible in the code.

`d/q` is the right starting point when the database must discover which entities match constraints. Among these read APIs, it is the one that uses the query optimizer: the engine can reorder clauses and use indexes to narrow the candidate set before returning values.

> **Pull is a shape tool, not always a shortcut**
>
> Do not treat `d/pull` as a guaranteed performance shortcut over `d/q`. Datalevin's query engine is optimized, and many pull-shaped reads can be expressed as Datalog queries that are just as efficient, or more efficient.
> Internally, `d/pull` works over datoms, which have instantiation cost: attributes are realized into keywords and all values are materialized in memory, whereas `d/q` works on relations and does not realize data that is not bound to variables.

The main distinction is the shape of the problem:

- `d/entity` is for lazy navigation from one known entity.
- Transactable Entity is for staging small updates before `d/transact!`.
- `d/pull` is for returning a declared nested map for known entities.
- `d/q` is for finding facts and entities that satisfy constraints.

Summary

Datalevin gives you several ways to read the same logical facts. Lookup refs let you use domain identifiers instead of internal ids. Entity APIs provide navigation for code that wants to move through references one step at a time, and transactable entities let that same Entity API stage small updates. Pull returns explicit plain maps for application boundaries. The APIs overlap by design, but they are not redundant: choose the one that matches whether you are identifying, shaping, discovering, navigating, or updating data.

CHAPTER 8

Datalog Fundamentals

Datalevin's primary query language is Datalog: a declarative way to ask which facts must be true for an answer to exist.

In Datalevin, a Datalog query is EDN data. It is built from vectors, keywords, symbols, lists, and maps, using the classic Datalog ideas of variables, shared constraints, predicates, rules, and recursive reasoning.

Python and JavaScript expose the same grammar through the pure `q` namespace. The examples use `q.var` for variables, `q.datom` for fact patterns, and other typed helpers for clauses and result shapes. Calling `as_data()` in Python or `asData()` in JavaScript reveals the corresponding Datalog form without starting the JVM. EDN strings remain supported when an application already has a static query in that representation.

This chapter starts with basic filters and joins, then builds toward predicates, rules, recursion, query planning, and performance.

1. Basic Attribute Filters

Datalog is declarative. The `:where` clauses define a *set of constraints*: conditions the answer must satisfy, not a step-by-step procedure for the database to follow.

1.1 A Single Filter on an Entity

A Datalog query usually starts with `:find`, which says what to return, and `:where`, which says what facts must match. The most common `:where` clause is a triple pattern:

```
(d/q '[:find ?user-eid
       :where [?user-eid :user/name "Alice"]]
     db)
```

A triple pattern is positional: entity, attribute, then value. The pattern above asks for every entity `?user-eid` that has the attribute `:user/name` with value `"Alice"`.

1.2 Two Filters on the Same Entity

To add another condition on the same entity, reuse the same variable:

```
(d/q '[:find ?user-eid
       :where [?user-eid :user/name "Alice"]
              [?user-eid :user/city "London"]]
     db)
```

This query asks for entities that have a name of "Alice" and a city of "London". Each occurrence of ?user-eid must refer to the same entity within a result row.[1] That shared variable is the logical connection between the two clauses.

The clause order is not a procedure. The query does not mean "first find people named Alice, then filter by city." It means both constraints must hold. Switching the order of these clauses does not change the result; the optimizer is free to choose the same physical plan either way.

1.3 Leaving Out Trailing Positions

Datalevin allows trailing positions to be omitted when you do not need them. The pattern [?user-eid :user/name] means "some datom exists for entity ?user-eid and attribute :user/name, regardless of value."

```
(d/q '[:find ?user-eid
       :where [?user-eid :user/name]]
     db)
```

This is different from binding the value to a variable. Use [?user-eid :user/name ?name] when the value matters later in the query or should be returned from :find.

1.4 Ignoring Positions with _

When the position you want to ignore is not trailing, use _ as a non-binding placeholder. For example, this query asks for all user names. Each matching entity must have some entity id, but the query does not need to name it:

```
(d/q '[:find ?name
       :where [_ :user/name ?name]]
     db)
```

Each _ is independent. In [?user-eid _ _], the first _ means "some attribute" and the second _ means "some value"; they are not the same variable. If two positions need to be the same, or if a value must be reused in another clause, give it a normal variable name:

```
(d/q '[:find ?edge-eid
       :where [?edge-eid :edge/from ?node-eid]
              [?edge-eid :edge/to ?node-eid]]
     db)
```

Use _ when a position must be present but should not bind a variable. Use omission when later positions of a triple pattern are simply not part of the question.

2. Joins Across Entities

A join is just another use of shared variables. When two data patterns mention the same variable, they must agree on its value. If those patterns describe different entities, the shared variable relates those entities.

[1] In logic programming, making repeated variables agree is called **unification**. In database language, a shared variable across clauses acts as a join condition, and keeping only the requested :find values is a **projection**.

Before adding more query features, it is useful to walk through one ordinary Datalog query mechanically.

Assume `:order/customer` is a ref attribute and the database contains these facts[2], shown as `[e a v]` datoms. The names `alice`, `bob`, `cara`, and `order-1` through `order-4` are explanatory labels; Datalevin stores numeric entity ids internally.

```
[[alice   :user/name      "Alice"]
 [alice   :user/city      "London"]
 [bob     :user/name      "Bob"]
 [bob     :user/city      "London"]
 [cara    :user/name      "Cara"]
 [cara    :user/city      "Paris"]
 [order-1 :order/customer alice]
 [order-1 :order/total    125]
 [order-2 :order/customer alice]
 [order-2 :order/total    60]
 [order-3 :order/customer bob]
 [order-3 :order/total    200]
 [order-4 :order/customer cara]
 [order-4 :order/total    300]]
```

Now ask for London users with orders over 100. For now, read `[(> ?total 100)]` as a filter clause; Section 3.2 introduces these more formally as predicate clauses.

```
(d/q '[:find ?name ?total
       :where
       [?user-eid :user/city "London"]
       [?user-eid :user/name ?name]
       [?order-eid :order/customer ?user-eid]
       [?order-eid :order/total ?total]
       [(> ?total 100)]]
     db)
;; => #{["Alice" 125] ["Bob" 200]}
```

Conceptually, read the clauses as constraints over a set of variable bindings. The optimizer may evaluate the physical query in a different order, but the logical meaning is the same.

Figure 8.1 illustrates a possible variable binding flow that resolves this query. The steps below trace that relation one clause at a time.

The first clause finds entities whose city is London:

```
after [?user-eid :user/city "London"]
{?user-eid alice}
{?user-eid bob}
```

The second clause reuses `?user-eid` and adds `?name`. Each occurrence of `?user-eid` must refer to the same entity within a row.

```
after [?user-eid :user/name ?name]
{?user-eid alice, ?name "Alice"}
{?user-eid bob,   ?name "Bob"}
```

[2]Datalog literature often calls stored input facts the **EDB** (extensional database). Relations derived by rules are the **IDB** (intensional database). Chapter 9 uses those terms when it discusses rules and recursion.

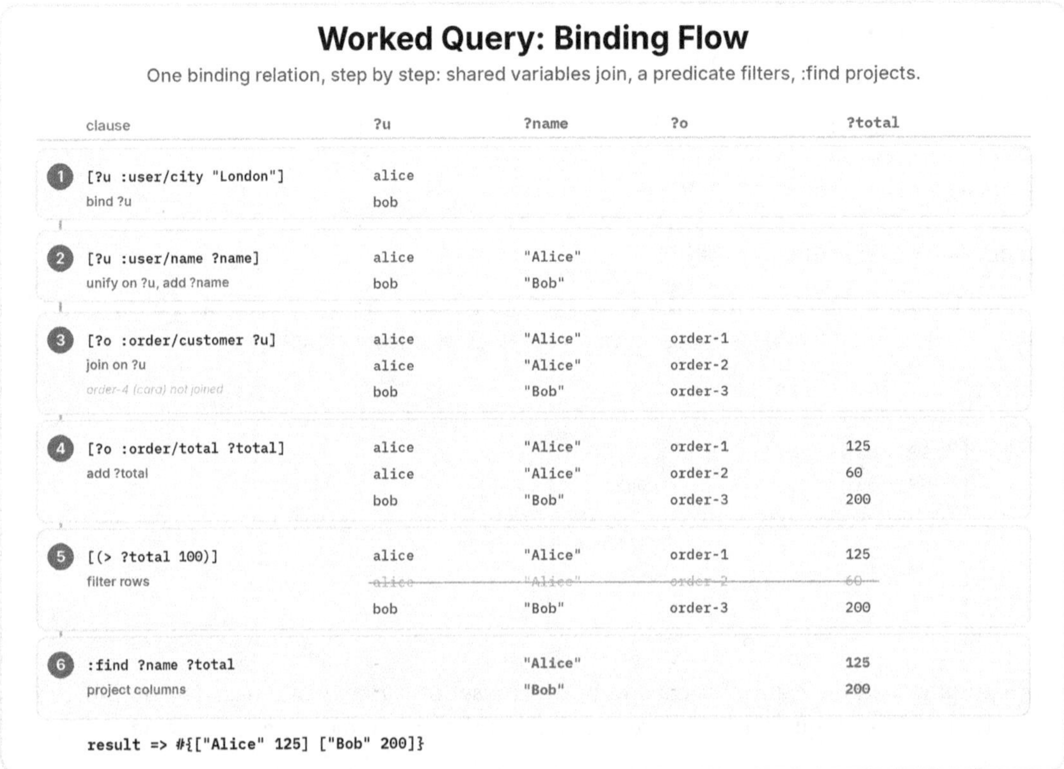

Figure 8.1: Worked query binding flow: the binding rows gain a ?user-eid column, reuse ?user-eid to add ?name, keep orders whose customer is the same ?user-eid (dropping order-4 whose customer is not a London user), add ?total, filter out the row whose total is 60, and keep ?name and ?total in the result

The third clause finds orders whose customer is the same ?user-eid. The ?user-eid in [:order/customer ?user-eid] must agree with the ?user-eid already bound by the user clauses.

```
after [?order-eid :order/customer ?user-eid]
{?user-eid alice, ?name "Alice", ?order-eid order-1}
{?user-eid alice, ?name "Alice", ?order-eid order-2}
{?user-eid bob,   ?name "Bob",   ?order-eid order-3}
```

order-4 is not present because its customer is cara, and cara was not in the London-user rows.

The fourth clause adds totals for those orders:

```
after [?order-eid :order/total ?total]
{?user-eid alice, ?name "Alice", ?order-eid order-1, ?total 125}
{?user-eid alice, ?name "Alice", ?order-eid order-2, ?total 60}
{?user-eid bob,   ?name "Bob",   ?order-eid order-3, ?total 200}
```

The final filter keeps only rows whose total is greater than 100:

```
after [(> ?total 100)]
{?user-eid alice, ?name "Alice", ?order-eid order-1, ?total 125}
{?user-eid bob,   ?name "Bob",   ?order-eid order-3, ?total 200}
```

Finally, :find ?name ?total keeps only the requested values from each surviving row. Intermediate variables such as ?user-eid and ?order-eid helped express the query, but they are not returned:

```
["Alice" 125]
["Bob"   200]
```

This is the core Datalog mental model: data patterns create candidate bindings, shared variables keep those bindings consistent, filter clauses remove rows, and :find shapes the surviving values into the result.

The join also affects result cardinality. Alice has two orders in the source facts, so the intermediate relation has two Alice rows after joining users to orders. The later predicate removes the smaller order, but without that predicate both Alice orders would contribute rows.

3. Variables and Functions in :where

Triple patterns are the foundation, but :where clauses can also use variables in the attribute position, call predicates to filter rows, and call functions that bind new values.

3.1 Variable Attributes

The attribute position can be a variable, so you can ask questions about the shape of your data:

```
(d/q '[:find ?attr
       :where
       [?person-eid :person/name]
       [?person-eid ?attr]]
     db)
```

The first clause finds people with a `:person/name`. The second clause reuses the same entity variable and binds `?attr` to every attribute asserted for those entities. With `:find ?attr`, the result is a one-column relation. If you also need the values, write `[?person-eid ?attr ?value]` and include `?value` in `:find`.

3.2 Predicate Clauses

A predicate clause filters results based on a function that returns `true` or `false`.

```
;; Find users older than 30
(d/q '[:find ?name ?age
       :where [?user-eid :user/name ?name]
              [?user-eid :user/age ?age]
              [(> ?age 30)]]
     db)
```

This returns only users who are older than 30. Notice that the single predicate function call is wrapped in a vector form. Common built-in predicates include comparisons such as =, <, >, <=, and >=, and Datalevin-specific helpers such as `like` and `in`. Appendix D lists the built-in query and aggregate functions.

3.3 Binding Clauses

You can also use functions to compute and bind new variables.

```
;; Compute an order total from subtotal and tax, and bind the result to ?total
(d/q '[:find ?order-id ?total
       :where [?order-eid :order/id ?order-id]
              [?order-eid :order/subtotal ?subtotal]
              [?order-eid :order/tax ?tax]
              [(+ ?subtotal ?tax) ?total]]
     db)
```

`?total` is a new variable that takes the result of `(+ ?subtotal ?tax)` as its value. This binding clause vector has two elements. Read from left to right: the value flows from the function call into the bound variable.

Some query functions return tuples or relations rather than a scalar value. Use a binding pattern on the right side to describe how the returned values should bind to variables. _ works in these binding patterns too:

```
;; Assumes :doc/body has :db/fulltext true and :db.fulltext/autoDomain true.
(d/q '[:find ?doc-eid
       :where [(fulltext $ :doc/body "clojure") [[?doc-eid _ _]]]]
     db)
```

The `fulltext` function is covered in Chapter 16. It appears here because its return binding is compact: bind the entity id to `?doc-eid`, and ignore the attribute and value positions.

3.4 Built-In Logic Helpers

Several built-in functions are especially useful when a query needs to reason about optional facts, literal values, and tuple-shaped values.

Use get-else, get-some, get-some-else, and missing? for attributes that may or may not be present. The predicate name is missing?, with a question mark.

```
(d/q '[:find ?name ?display ?contact-attr ?contact
       :where [?user-eid :user/name ?name]
              [(get-else $ ?user-eid :user/display-name "Anonymous") ?display]
              [(get-some $ ?user-eid :profile/email :user/email :user/phone)
               [?contact-attr ?contact]]
              [(missing? $ ?user-eid :user/deleted-at)]]
     db)
```

To keep a user in the result when none of the contact attributes is present, replace the get-some binding clause with its total variant:

```
[(get-some-else $ ?user-eid "No contact"
                :profile/email :user/email :user/phone)
 [?contact-attr ?contact]]
```

When no listed attribute exists, this binds ?contact-attr to nil and ?contact to "No contact" instead of removing the user's row.

get-else binds the attribute value when present; otherwise it binds the non-nil default. get-some tries the listed cardinality-one attributes in order and returns [attribute value] for the first one present. If none is present, that binding clause produces no row. get-some-else takes a fallback before the attribute list and returns [nil fallback] instead, so the entity remains in the result; the fallback may be nil. missing? filters for entities that do not have a value for the attribute.

Use ground, tuple, and untuple when a query needs literal data or an explicit tuple-shaped value. These are query-time value helpers; they are different from schema features such as :db/tupleAttrs and :db.type/tuple.

```
(d/q '[:find ?sku ?region
       :where [(ground [["book" "us"] ["pen" "eu"]])
               [[?sku ?region] ...]]
              [(tuple ?sku ?region) ?key]
              [(untuple ?key) [?sku2 ?region2]]
              [(= ?sku ?sku2)]
              [(= ?region ?region2)]])
```

ground returns its argument unchanged. The argument may be a literal, as above, or an otherwise-bound variable; an unbound argument is an insufficient binding. Here the collection binding on the right expands the literal pairs into rows. tuple constructs one vector from its arguments and binds that value to ?key. untuple also returns its argument unchanged; the tuple binding [?sku2 ?region2] on the right is what destructures the value. In other words, the functions produce values, while the binding shapes control how those values enter the query.

3.5 User-Defined Query Functions

For portable custom logic, register a descriptor-backed UDF in the runtime registry and call it through Datalevin's built-in udf function. The same query shape works in embedded bindings and server deployments:

```
(require '[datalevin.udf :as udf])

(def adult-pred
  {:udf/lang :clojure
   :udf/kind :predicate
   :udf/id   :user/adult?})

(def registry
  (doto (udf/create-registry)
    (udf/register! adult-pred #(>= % 18))))

;; Open the connection with {:runtime-opts {:udf-registry registry}}.
(d/q '[:find ?name
       :in $ ?adult-pred
       :where [?user-eid :user/name ?name]
              [?user-eid :user/age ?age]
              [(udf ?adult-pred ?age)]]
     db adult-pred)
```

For a custom binding function, use the same `udf` form with a descriptor whose `:udf/kind` is `:query-fn`, then bind its return value with a clause such as `[(udf ?normalize ?raw) ?normalized]`. The typed equivalents are `q.bind_udf(...)` in Python and `q.bindUdf(...)` in JavaScript.

In embedded Clojure, a Clojure function can appear directly in a predicate or binding clause, but it must be fully qualified with its namespace:

```
(ns my-app.queries
  (:require [datalevin.core :as d]))

(defn adult? [age]
  (> age 18))

(d/q '[:find ?name
       :where [?user-eid :user/name ?name]
              [?user-eid :user/age ?age]
              [(my-app.queries/adult? ?age)]]
     db)
```

The `adult?` function is not built in, so the query includes `my-app.queries/adult?`. Without that namespace, Datalevin cannot resolve the function. This shortcut is useful inside a trusted embedded Clojure process.

Direct host-function resolution is an embedded-mode feature. Remote server queries run in server-safe resolver mode: they may call Datalevin's built-in query functions and aggregates, and they may call descriptor-backed UDFs through the built-in `udf` function, but they do not resolve arbitrary host-language vars, method calls, or function-valued query inputs as call targets. This prevents a client query from becoming arbitrary code execution inside the server process. For custom query logic in server deployments, register a UDF on the server and call it with `udf`.

3.6 Predicate Performance

Predicate clauses are constraints, but not all constraints give the optimizer the same information. When a simple built-in comparison applies to a value from a triple pattern, Datalevin can turn it into an index range:

```
(d/q '[:find ?name
       :where [?user-eid :user/name ?name]
              [?user-eid :user/age ?age]
              [(> ?age 18)]]
     db)
```

Here `[(> ?age 18)]` can narrow the scan over `:user/age` values. A predicate that involves one data variable can also be pushed down to the scan as a value filter, even when it is a custom predicate. The difference is that a custom predicate is opaque to the range planner:

```
(d/q '[:find ?name
       :in $ ?adult-pred
       :where [?user-eid :user/name ?name]
              [?user-eid :user/age ?age]
              [(udf ?adult-pred ?age)]]
     db adult-pred)
```

In this query, the descriptor input is already bound by `:in`; the data variable being filtered is `?age`. Datalevin can attach this predicate to the `:user/age` scan, so it is evaluated as candidate age values are read. It still cannot look inside the registered predicate implementation and convert its body into an index range. Predicates that involve more than one data variable, such as `[(udf ?eligible-pred ?age ?status)]`, cannot be attached to a single attribute scan and are handled later, after the needed variables have been bound.

> **Performance: predicates and indexes**
>
> Use built-in comparisons, equality, `like`, and `in` for simple value tests when possible. Keep custom predicates for logic that cannot be expressed directly, and combine them with selective data patterns so they run over a small candidate set.

4. Alternatives and Negation

By default, all `:where` clauses are joined with an implicit `and`. Datalog also provides forms for alternatives and exclusion:

- `or`: one of several alternative branches may match.
- `or-join`: like `or`, but explicitly lists the variables that connect the alternatives to the surrounding query.
- `not`: excludes rows that match a negative pattern.
- `not-join`: like `not`, but explicitly lists the variables that connect the negative pattern to the surrounding query.

4.1 or

Use `or` when each branch is an alternative way to bind the same logical variables. For example, this query finds users whose status is either `:user.status/pending` or `:user.status/flagged`:

```
(d/q '[:find ?user-eid
       :where (or [?user-eid :user/status :user.status/pending]
                  [?user-eid :user/status :user.status/flagged])]
     db)
```

When an alternative needs more than one clause, wrap those clauses in `(and ...)`:

```
(d/q '[:find ?user-eid
       :where
       (or (and [?user-eid :user/status :user.status/pending]
                [?user-eid :user/priority :high])
           (and [?user-eid :user/status :user.status/flagged]
                [?user-eid :user/review-required? true]))]
     db)
```

Plain `or` requires the branches to use the same set of free variables. That is what lets the rest of the query continue with a consistent binding shape.

4.2 or-join

Use `or-join` when an alternative branch needs variables that should stay local to that branch. The vector after `or-join` names the variables that connect the alternatives to the surrounding query.

```
(d/q '[:find ?user-eid
       :where [?user-eid :user/name]
              (or-join [?user-eid]
                [?user-eid :user/status :user.status/flagged]
                (and [?user-eid :user/manager ?manager-eid]
                     [?manager-eid :user/status :user.status/flagged]))]
     db)
```

This query finds users who are flagged directly or whose manager is flagged. The variable `?manager-eid` is needed only inside the second branch. With plain `or`, the first branch would bind `?user-eid` while the second branch would bind both `?user-eid` and `?manager-eid`, so the branch shapes would not match. `or-join [?user-eid]` states that only `?user-eid` is the branch result that must unify with the rest of the query.

4.3 not

Use `not` to remove rows that match a negative pattern. A `not` clause is a test against bindings already produced by surrounding positive clauses; it does not introduce new rows by itself. At least one variable in the negated pattern must already be bound by the surrounding query.

```
(d/q '[:find ?user-eid
       :where [?user-eid :user/name]
              (not [?user-eid :user/is-admin? true])]
     db)
```

A single `not` form with multiple clauses negates their conjunction: the row is excluded only when all clauses inside the `not` match together.

4.4 `not-join`

Use `not-join` when the negative pattern needs local variables or when only some variables should connect the negative pattern to the surrounding query. The vector after `not-join` names those connecting variables. Variables not listed there are local to the negative pattern. The listed variables must already be bound before the `not-join` runs. Section 6.3 shows a complete example.

5. Query Inputs

So far, most examples have omitted `:in`. When a query has no explicit `:in`, Datalevin behaves as if it had `:in $`, where `$` names the default data source.

Real application queries usually take values from function arguments, request parameters, or UI state. Use `:in` when the query needs inputs beyond the default data source. The simplest explicit declaration with a scalar input is `:in $ ?city`:

```
(d/q '[:find ?name
       :in $ ?city
       :where [?user-eid :user/city ?city]
              [?user-eid :user/name ?name]]
     db
     "London")
```

The `$` symbol names the default data source: the database, or other EAV data source, being queried. Variables after `$` are input bindings. They are bound from the remaining arguments in the same order as they appear in `:in`. In the Clojure call above, `db` fills `$` and `"London"` fills `?city`. In the connection APIs, the connection supplies `$`, and the extra arguments fill the remaining inputs.

Inputs are just bindings. Once `?city` or `?min-total` is bound, the rest of the query treats it like any other variable:

```
(d/q '[:find ?name ?total
       :in $ ?city ?min-total
       :where [?user-eid :user/city ?city]
              [?user-eid :user/name ?name]
              [?order-eid :order/customer ?user-eid]
              [?order-eid :order/total ?total]
              [(> ?total ?min-total)]]
     db
     "London"
     100)
```

Data sources are inputs too. The default source is named `$`, and additional sources can be given their own source symbols, such as `$users`, `$orders`, or `$data`. Section 10 returns to that general case after the rest of the basic query forms have been introduced.

6. When You Think You Need a Subquery

Readers coming from SQL often reach for a subquery when they want to filter outer rows by related facts. Datalevin can run a nested `q`, but ordinary queries are usually clearer and faster when you express that logic directly in Datalog. Use `not-join` for "no matching related fact" checks, and `or-join` for branching existence checks with branch-local variables. Nested `q` is useful when the problem genuinely has two query stages, however. A common example is computing an aggregate result in an inner query and joining that result to facts in the outer query.

6.1 The Performance Trap: Correlated Nested q

It is technically possible to run a nested `q` subquery inside a predicate, but avoid correlating it with every candidate when ordinary Datalog clauses express the same operation. The example below asks for London users with no orders by running an inner query from a predicate clause. This is the shape to avoid: the inner query runs once for each candidate tuple from the outer query instead of being planned as part of one Datalog query.

```
;; Anti-pattern: runs the inner query for every candidate user.
(d/q '[:find ?user-eid
       :where [?user-eid :user/city "London"]
              [(empty? (q (quote [:find ?order-eid
                                  :in $ ?user-eid
                                  :where [?order-eid :order/customer
                                   ↪ ?user-eid]])
                           $ ?user-eid))]] ; This subquery runs for each London
 ↪ user
     db)
```

That repeated work is usually both harder to read and much more expensive than expressing the logic directly in Datalog. The problem is the per-candidate correlation, not the mere presence of `q` inside another query.

6.2 Factoring an Aggregate Result

A nested `q` is appropriate when the inner query produces a small result that the outer query consumes as a relation. For example, finding the minimum age is an aggregate operation, while finding every user having that age is a fact lookup. Keeping those stages separate avoids accidentally grouping by user in the aggregate query:

```
;; Find every user tied for the minimum age.
(d/q '[:find ?user-eid
       :where [(q (quote [:find (min ?age)
                          :where [_ :user/age ?age]])
                   $)
               [[?minimum-age]]]
              [?user-eid :user/age ?minimum-age]]
     db)
```

The inner query is not correlated with `?user-eid`; it computes a one-column relation containing the global minimum, and `[[?minimum-age]]` binds that relation into the outer query. The outer fact clause then returns all ties. In contrast, putting both `?user-eid` and `(min ?age)` in the same `:find` would group by `?user-eid` and would not compute the global minimum.

This pattern also applies when an inner query deliberately factors out a small aggregated or otherwise shaped result relation for a later join. Variables used only inside the nested query are locally scoped; the result binding is the explicit boundary between the two stages. Keep the inner result small, and check whether a direct join, rule, aggregate, `not-join`, or `or-join` expresses the operation more simply before introducing that boundary.

Python and JavaScript use `q.quote(inner)` to preserve the EDN list required by the nested query form. The inner query remains an ordinary typed `q.query` value, so variables, clauses, and result bindings can still be assembled with host-language control flow.

6.3 Excluding Matches with `not-join`

For this kind of "find things with no matching related facts" question, express the exclusion directly with `not-join`. The optimizer can plan it together with the surrounding data patterns.

```
;; Find users in London with no orders
(d/q '[:find ?user-eid
       :where [?user-eid :user/city "London"]
              (not-join [?user-eid]
                [?order-eid :order/customer ?user-eid])]
     db)
```

Here, `not-join` declares that `?user-eid` is the variable shared with the surrounding query. Datalevin can plan the negative pattern together with the positive clauses, so it can exclude users with matching orders without invoking a separate query for each candidate user. This is much more efficient than the nested `q` approach.

Figure 8.2 shows the difference visually: the nested `q` version runs one inner query per candidate user, while the `not-join` version keeps the exclusion in one planned Datalog query.

6.4 Branching Checks with `or-join`

Use `or-join` when you want to keep an outer row if any of several related patterns exists. The branch-local variables stay inside the branch, and the declared join variables connect the branch result back to the surrounding query.

```
;; Find London users who either have a large order or an open support ticket.
(d/q '[:find ?user-eid
       :where [?user-eid :user/city "London"]
              (or-join [?user-eid]
                (and [?order-eid :order/customer ?user-eid]
                     [?order-eid :order/total ?total]
                     [(> ?total 1000)])
                (and [?ticket-eid :ticket/customer ?user-eid]
                     [?ticket-eid :ticket/status :open]))]
     db)
```

Here `?order-eid`, `?total`, and `?ticket-eid` are local to their branches. Only `?user-eid` is joined back to the outer query. A nested `q` version would ask a separate question for each candidate user; `or-join` keeps the alternatives in one query plan.

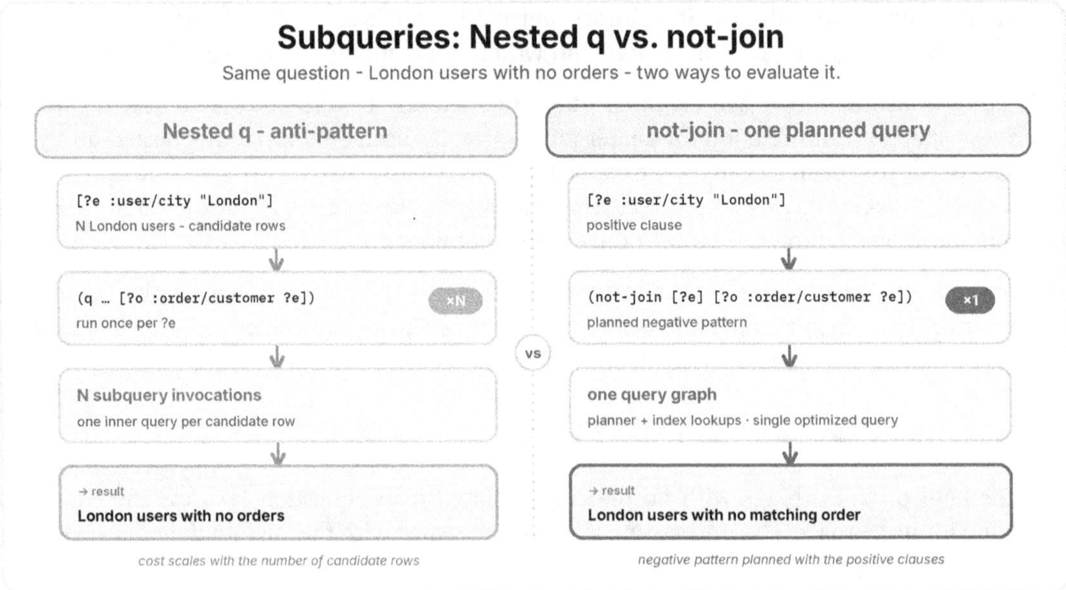

Figure 8.2: Nested q versus not-join for "London users with no orders": the nested-q anti-pattern produces N candidate rows and runs an inner subquery once per row (N invocations whose cost scales with the candidates), while not-join folds the positive clause and a planned negative pattern into one query graph evaluated by the planner with index lookups as a single optimized query

7. Declarative Power: The Query Optimizer

Because Datalog is declarative, the **order of your `:where` clauses does not matter**. Datalevin has a sophisticated **query optimizer** that analyzes your clauses and automatically determines the most efficient execution plan.

The optimizer uses the order statistics from the underlying storage layer (see Chapter 4) to make informed decisions. For example, if there are only 3 users in "London" but 10,000 users named "Alice", it will almost certainly start by finding the London users first, because it is a much smaller set to filter.

> **State constraints clearly**
>
> Do not try to manually optimize your queries by reordering `:where` clauses. State your constraints clearly and let the optimizer do its job.

8. The `:find` Specification: Shaping Your Results

The `:find` clause determines the shape of the query result. Its basic use is to return a collection of values, but it has several powerful variations for aggregation and shaping data.

8.1 Result Shapes

By default, `:find` returns a **relation**: a set of tuples, where each tuple contains the values named in the `:find` clause.

The word **set** is important. Even if two different internal rows produce the same `:find` tuple, that tuple appears only once in the final result set.

```
;; Returns: #{["Alice" 30] ["Bob" 42]}
(d/q '[:find ?name ?age
       :where [?user-eid :user/name ?name]
              [?user-eid :user/age ?age]]
     db)
```

Set semantics describe which rows belong to the relation, not the foreign language container used to carry them. Python materializes a relation as a list of row lists, and JavaScript materializes it as an Array of row Arrays. Without `order_by`/`orderBy`, the rows remain unordered even though the bridge uses an indexable sequence.

Datalevin supports these find shapes:

Shape	Syntax	Result shape
Relation	`:find ?name ?age`	A set of tuples, for example `#{["Alice" 30] ["Bob" 42]}`.
Collection	`:find [?name ...]`	A collection of scalar values, for example `["Alice" "Bob"]`.
Tuple	`:find [?name ?age]`	One tuple, for example `["Alice" 30]`.
Scalar	`:find ?name .`	One scalar value, for example `"Alice"`.

Figure 8.3 shows the differences graphically.

Use the collection find form when you only need one value from each result tuple:

```
(d/q '[:find [?name ...]
       :where [?user-eid :user/name ?name]]
     db)
```

Use tuple or scalar find only when the query logically identifies one result, for example through a unique attribute or an entity id:

```
;; One tuple
(d/q '[:find [?name ?email]
       :in $ ?user-id
       :where [?user-id :user/name ?name]
              [?user-id :user/email ?email]]
     db user-id)

;; One scalar
(d/q '[:find ?email .
       :in $ ?user-id
       :where [?user-id :user/email ?email]]
     db user-id)
```

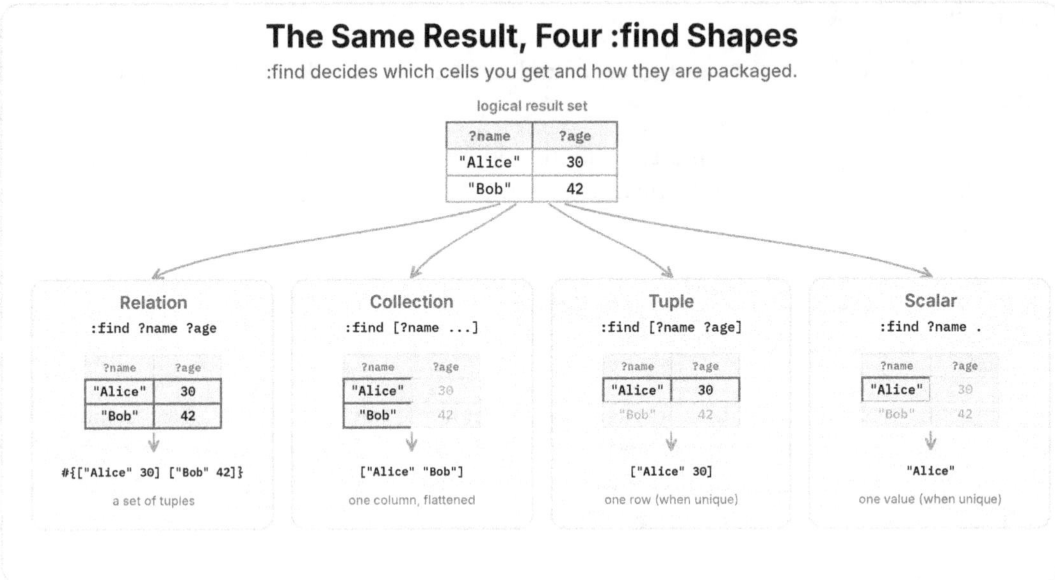

Figure 8.3: The four :find shapes applied to the same result set: relation selects all cells and returns a set of tuples, collection selects one column flattened into a vector, tuple selects one row, and scalar selects one cell

> **Tuple and scalar find need unique answers**
>
> If more than one result matches a tuple or scalar find, Datalevin returns one matching result. Do not rely on which one you get unless the query itself is unique.

For relation and tuple results, you can also ask Datalevin to return maps with `:keys`, `:strs`, or `:syms`. The number of names must match the number of `:find` elements, and return maps are not valid with collection or scalar find.

```
;; Returns: #{{:name "Alice" :age 30} {:name "Bob" :age 42}}
(d/q '[:find ?name ?age
       :keys name age
       :where [?user-eid :user/name ?name]
              [?user-eid :user/age ?age]]
     db)
```

Use `:keys` for keyword keys, `:strs` for string keys, and `:syms` for symbol keys.

8.2 Pull in Queries

When a query discovers the matching entities, `pull` can shape those entities immediately in the `:find` clause. This is the usual pattern for "find the matching entities and return application maps":

```
(d/q '[:find [(pull ?user-eid [:user/name {:user/orders [:order/id]}]) ...]
       :where [?user-eid :user/city "London"]]
     db)
```

This query finds all users in London and, for each one, pulls their name and a nested list of their orders. The important point is that `:where` discovers the entities and `pull` shapes only the entities that survived the query. Chapter 7 covers `pull` in more detail.

8.3 Aggregation

Datalog supports aggregate functions directly in the `:find` clause. Aggregates compute over the found result set and can produce summaries similar to SQL `GROUP BY`. Grouping is determined by the non-aggregate variables that remain in `:find`: `:find ?city (count ?user-eid)` groups by `?city` and returns one count per city. If `:find` contains only aggregate expressions, Datalevin treats the whole result as one group.

```
;; Find the average age of all users
(d/q '[:find (avg ?age) . ; The dot returns a single scalar result
       :where [_ :user/age ?age]]
     db)

;; Find the count of users per city
(d/q '[:find ?city (count ?user-eid)
       :where [?user-eid :user/city ?city]]
     db)
```

Common built-in aggregates include `sum`, `avg`, `count`, `min`, `max`, `median`, `distinct`, and `vec`. Use `distinct` when you want to collect the distinct values in each group into a set. Use `vec` when you want to collect the values in each group into a vector. It collects the values that reach the aggregate, including duplicates that remain in the query basis. Do not rely on the vector's order as a semantic sort order unless the aggregate function itself imposes one.

Arithmetic expressions over aggregate results can also appear in `:find`. The expression is evaluated after grouping, so each aggregate operand sees the same group it would see if it appeared as a separate `:find` column. In this query, `?customer-eid` is the grouping key, and the final column adds two aggregate results:

```
(d/q '[:find ?customer-eid
             (sum ?subtotal)
             (sum ?tax)
             (+ (sum ?subtotal) (sum ?tax))
       :where [?order-eid :order/customer ?customer-eid]
             [?order-eid :order/subtotal ?subtotal]
             [?order-eid :order/tax ?tax]]
     db)
```

Aggregate expressions support `+`, `-`, `*`, `/`, `mod`, `rem`, and `quot`. Their arguments may be aggregate calls, constants, or nested aggregate expressions, for example `(* 2 (+ (sum ?x) (sum ?y)))`.

9. Query Modifiers and Post-Processing

Beyond the core `:where` clauses, Datalog provides additional clauses to refine, filter, order, and paginate your query results.

9.1 `:with`: Preserving Rows for Aggregates

Datalog query results have set semantics. Duplicate result tuples collapse. The `:with` clause exists for cases where you need a variable to remain part of the intermediate row identity even though it should not be returned.

This most often matters with aggregates. Suppose three orders have amounts `20`, `20`, and `15`. If the query only finds `?amount`, the two `20` rows are indistinguishable and collapse before `sum` sees them:

```
;; Wrong for revenue: duplicate amount values collapse.
(d/q '[:find (sum ?amount) .
       :where [?order-eid :order/amount ?amount]]
     db)
;; => 35
```

Add `:with ?order-eid` to keep each order distinct while still returning only the aggregate:

```
;; Correct: each order contributes its amount.
(d/q '[:find (sum ?amount) .
       :with ?order-eid
       :where [?order-eid :order/amount ?amount]]
     db)
;; => 55
```

`:with` does not make a hidden column in the returned result. It only changes the set of variables that define distinct intermediate rows for aggregation. For a non-aggregate query, adding `:with` often has no observable effect because the final `:find` result is still a set.

9.2 `:having`: Filtering Aggregated Results

Like SQL's `HAVING` clause, `:having` filters results that have been aggregated in the `:find` clause by applying additional conditions.

```
;; Find customers whose total spend and largest order both pass thresholds.
(d/q '[:find ?customer-eid (sum ?total) (max ?total)
       :with ?order-eid
       :where [?order-eid :order/customer ?customer-eid]
              [?order-eid :order/total ?total]
       :having [(> (sum ?total) 1000)]
               [(>= (max ?total) 250)]]
     db)
```

Multiple `:having` predicates are conjunctive: a group is returned only if all of them are true. In the example above, `:with ?order-eid` preserves each order as a distinct contributing row, so repeated order totals are still included in the sum.

9.3 `:order-by`: Sorting Results

The `:order-by` clause sorts your results based on specified variables. You can specify ascending or descending order.

```
;; Find users, ordered by age descending, then name ascending
(d/q '[:find ?name ?age
```

```
        :where [?user-eid :user/name ?name]
               [?user-eid :user/age ?age]
        :order-by [?age :desc ?name :asc]]
      db)
```

When sorting by aggregate output, `:order-by` can also use zero-based result column indices. This is useful because aggregate expressions are result columns, not variables you can name later in `:order-by`:

```
;; Sort customers by total spend descending, then customer ascending.
(d/q '[:find ?customer-eid (sum ?total)
        :with ?order-eid
        :where [?order-eid :order/customer ?customer-eid]
               [?order-eid :order/total ?total]
        :order-by [1 :desc 0 :asc]]
      db)
```

Here column `0` is `?customer-eid`, and column `1` is `(sum ?total)`. The `:order-by` clause sorts by total spend first, then uses customer as a stable tie-breaker.

9.4 `:limit` and `:offset`: Pagination

For pagination, you can use `:limit` to restrict the number of results and `:offset` to skip a certain number of results.

```
;; Get the second page of users (limit 10, offset 10)
(d/q '[:find ?name ?age
        :where [?user-eid :user/name ?name]
               [?user-eid :user/age ?age]
        :order-by [?name :asc]
        :limit 10
        :offset 10]
      db)
```

> **Bounded sorting**
>
> Every finite ordered relation query keeps only the best `:offset + :limit` distinct tuples in a bounded priority queue. This reduces sort memory even when every candidate must still be examined.

Query-level `:offset` and `:limit` apply to distinct final tuples after the query's other clauses have run. Search functions can also have source-local bounds, such as full-text or vector `:top`; those bounds determine which candidates enter the Datalog query and are never widened to satisfy its final limit. Chapter 21, Section 1.4 explains when the optimizer can avoid unused source work and how `explain` reports that choice.

9.5 Programmatic Query Construction

The compact vector form is convenient for static EDN. Python and JavaScript use the typed `q` namespace for the same query, making runtime options ordinary function arguments or object properties:

```
[:find ?name ?age
 :where [?user-eid :user/name ?name]
        [?user-eid :user/age ?age]
 :order-by [?name :asc]
 :limit 10]
```

Datalevin also accepts a map form. This is the engine's internal query representation and is useful when application code needs to build or parameterize options such as `:limit`, `:offset`, or `:order-by`. Python's keyword arguments and JavaScript's options object provide the idiomatic binding equivalent:

```
(defn page-users
  [conn {:keys [offset limit]
         :or   {offset 0
                limit  50}}]
  (d/q {:find     '[?name ?age]
        :in       '[$]
        :where    '[[?user-eid :user/name ?name]
                    [?user-eid :user/age ?age]]
        :order-by '[?name :asc]
        :limit    limit
        :offset   offset}
       (d/db conn)))
```

In map form, the query clauses are ordinary data. Clojure code quotes static query fragments such as `:find`, `:in`, `:where`, and `:order-by`, but leaves runtime values such as `limit` and `offset` unquoted. In Java, mark query keywords and symbols with `Datalevin.kw` and `Datalevin.sym`. Python and JavaScript use the typed `q.query` builder: attribute positions accept ordinary strings, while `q.var`, `q.DB`, and helpers such as `q.asc` make grammar tokens explicit. This avoids string concatenation and keeps pagination, sorting, and optional clauses easy to assemble with normal host-language code.

The same ordering rules apply in both forms: an `:order-by` variable must appear in the query's result columns. When ordering by an aggregate result, use the appropriate result-column index in `:order-by`, as shown earlier in this section.

10. Flexible Data Sources: Beyond a Single Database

Section 5 introduced query inputs and the default source symbol $. Datalevin's Datalog engine generalizes that idea: sources are inputs too. A query can read from more than one database, or from any sequence of EAV tuples. This is why the Clojure `d/q` API takes data sources rather than a connection: queries run over data, while transactions run through connections.

10.1 Multiple Databases

You can pass multiple Datalevin databases to a single query, referenced by different symbols in `:in`:

```
;; Join across two databases: a user db and an orders db
(d/q '[:find ?name ?order-total
       :in $users $orders ?min-total
```

```
         :where [$users ?user-eid :user/name ?name]
                [$orders ?order-eid :order/customer ?user-eid]
                [$orders ?order-eid :order/total ?order-total]
                [(> ?order-total ?min-total)]]
     (d/db users-conn) (d/db orders-conn) 100)
```

`$users` and `$orders` stand for two different database sources. The triple patterns accept an extra position in front for a source symbol.

This is invaluable when:

- Keeping separate databases for different tenants or domains
- Performing migration or reconciliation between systems

10.2 EAV Sequences as Data Sources

Even more flexibly, the query engine accepts **any sequence of EAV tuples** as a data source. If you can represent your data as `[entity attribute value]` triples, you can query it with Datalog, no database required.

In each query below, `$data` is the source symbol for the tuple sequence.

```
;; Query an in-memory collection of tuples
(def my-data
  [[1 :user/name "Alice"]
   [1 :user/age 30]
   [2 :user/name "Bob"]
   [2 :user/age 25]])

(d/q '[:find ?name
       :in $data ?min-age
       :where [$data ?user-eid :user/name ?name]
              [$data ?user-eid :user/age ?age]
              [(>= ?age ?min-age)]]
     my-data 28)
;; => #{["Alice"]}
```

This capability enables:

- **Ad-hoc analysis**: Query CSV files, JSON data, or API responses by transforming them into EAV tuples
- **Testing**: Write Datalog queries against fixture data without setting up a test database
- **Data transformation**: Use Datalog's pattern matching to reshape or filter in-memory collections
- **Prototyping**: Experiment with query logic on sample data before committing to a schema

The EAV model is not only a storage format; it is also a universal data representation that makes Datalevin's query engine applicable far beyond traditional database use cases.

Summary

Datalog queries start with simple facts: match an entity, match an attribute, bind a value, and reuse variables when facts must agree. From there, the same mental model scales to joins, pull expressions, predicates, aggregates, alternatives, and multiple data sources. State the constraints clearly and let Datalevin's optimizer choose the physical path to the data.

The next chapter explores **Rules and Recursion**, which allow you to encapsulate logic for reuse and traverse complex, hierarchical, or graph-shaped data structures.

CHAPTER 9

Rules and Recursion

While simple Datalog queries are useful for finding patterns, real-world applications often require shared, reusable logic. You might find yourself repeating the same set of `:where` clauses across many queries, or needing to traverse hierarchical data of unknown depth (like a file system or an organizational chart).

This is where **rules** come in. Rules let you define named logic that can be reused, composed, and even called recursively. They are the "functions" or "views" of the Datalog world, letting you derive new knowledge from your raw facts.

1. What Are Rules?

At its simplest, a rule is a named collection of Datalog clauses. You define a rule once and then invoke it by name within your queries.

Another way to understand rules is to look back at an ordinary query. The entire `:where` clause is the body of an unnamed rule, and the `:find` variables are like the rule head: together they define a temporary relation containing the tuples that satisfy those clauses. Named rules give a piece of that logic a name, parameter list, and reusable call site.

> **Rules derive query-time relations**
>
> Rule derivation happens at query time. A rule does not transact new datoms into the database, and it does not create a stored table. It defines a virtual relation: a named set of tuples that exists while a query is being evaluated.
> If you query `(is-active? ?e)`, Datalevin computes the entities that satisfy that rule for this query. If you need the result to become durable data, your application should explicitly transact the derived facts.

1.1 The Anatomy of a Rule

A rule consists of a **head** and a **body**:

- **Head**: the rule name and its parameters, for example `(is-manager? ?e)`. The name plus the number of parameters is the rule's shape.
- **Body**: one or more Datalog clauses that must be true for the rule to match.

Read a rule as a definition of a relation. The head names the relation and its columns. The body says which tuples belong to that relation. Variables in the head are the values that callers can use; variables that appear only in the body are local intermediate variables, similar to helper bindings inside a query.

Rules are typically defined as a vector of rule definitions (a "rule set"):

```
(def user-rules
  '[[(is-active? ?e)
     [?e :user/status :active]
     [?e :user/verified? true]]

    [(is-admin? ?e)
     (is-active? ?e) ; Rules can call other rules!
     [?e :user/role :admin]]])
```

The first rule says: an entity belongs to the one-column virtual relation `is-active?` when it has status `:active` and `:user/verified?` is true. The second rule says: an entity belongs to `is-admin?` when it belongs to `is-active?` and has role `:admin`.

> **Key takeaway**: Rules are your primary tool for **abstraction** and **composition**. Use them to name common patterns and build more complex logic out of simpler, reusable components.

2. Using Rules in Queries

To use rules in a query, you must:

1. Include the `%` symbol in the `:in` clause to represent the rule set.
2. Pass the rule set as the `%` input (`db rules` in Clojure, or `rules` after the query form in `conn.query`).
3. Invoke the rule in the `:where` clause using its head syntax.

```
(let [rules '[[(is-active? ?e)
               [?e :user/status :active]
               [?e :user/verified? true]]]]
  (d/q '[:find ?name
         :in $ %
         :where [?e :user/name ?name]
                (is-active? ?e)]
       db rules))
```

Notice that unlike function calls or triple patterns, rule invocations are not wrapped in a vector.

In this example, `(is-active? ?e)` acts as a filter. For every entity `?e` that has a name, the engine checks if that entity also satisfies the clauses defined in the `is-active?` rule.

The `%` input is the rule set for this query. Different queries can pass different rule sets, which makes rules useful for application-specific policies, tenant-specific logic, tests, and experiments.

2.1 Assembling Rule Sets

A rule set is one flat collection of rule definitions. When an application grows, avoid building one giant literal rule vector in every query namespace. Keep small, named rule groups close to the domain they describe, assemble them at the application boundary, and pass the resulting flat rule set as `%`.

Assemble rules as data

- Rule groups are ordinary data. Store them in code, EDN files, or configuration records that your application controls.
- The final value passed as `%` should be a flat vector or sequence of rule definitions, not a nested vector of rule groups.
- Rules may call rules from another group as long as the called rule is present in the assembled rule set.
- Multiple definitions of the same rule are alternatives, that is, logical `or`.
- Keep the public rule names and arities stable. Treat helper rules as internal application conventions, just as you would with helper functions.

Here is a typical pattern: one group defines basic user status, another defines subscription logic, another defines feature access on top of that policy, and an optional group adds a staff override. The assembled rule set is selected per query, tenant, feature flag, or test.

In the example below, Java code uses Datalevin's keyword and symbol helpers plus `PersistentList` for EDN lists. Python and JavaScript use `q.rule_branch` / `q.ruleBranch` values directly, so they need no external EDN library or parsing step.

```
(def user-status-rules
  '[[(active-user ?u)
     [?u :user/status :user.status/active]
     [?u :user/verified? true]]])

(def subscription-rules
  '[[(paid-user ?u)
     [?sub :subscription/user ?u]
     [?sub :subscription/status :subscription.status/paid]]

    [(premium-user ?u)
     (active-user ?u)
     (paid-user ?u)]])

(def staff-override-rules
  '[[(premium-user ?u)
     (active-user ?u)
     [?u :user/role :user.role/staff]]])

(def feature-access-rules
  '[[(can-use-feature ?u ?feature)
     (premium-user ?u)
     [?feature :feature/min-tier :tier/premium]]

    [(can-use-feature ?u ?feature)
     [?u :user/feature-flag ?feature]]])

(defn assemble-rules
  [& rule-groups]
  (into [] (comp (remove nil?) cat) rule-groups))
```

```
(defn user-rules
  [{:keys [staff-override?]}]
  (assemble-rules user-status-rules
                  subscription-rules
                  feature-access-rules
                  (when staff-override?
                    staff-override-rules)))

(d/q '[:find [?feature-name ...]
       :in $ % ?email
       :where [?u :user/email ?email]
              (can-use-feature ?u ?feature)
              [?feature :feature/name ?feature-name]]
     db
     (user-rules {:staff-override? true})
     "alice@example.com")
```

Without `staff-override-rules`, `(premium-user ?u)` means "active user with a paid subscription." With that group included, the same rule also matches active staff users. The `can-use-feature` rule does not need to know which policy module made a user premium; it only depends on the assembled `premium-user` relation. Nothing about the query changes; only the rule set supplied as `%` changes. This is the main design advantage of rule assembly: query call sites can stay stable while policy modules evolve.

3. Logical OR: Multiple Rule Definitions

As with clauses in `:where`, all clauses in a rule body are joined by an implicit `and`. For reusable branch logic, define a rule with the same name and arity multiple times. The engine treats those definitions as alternatives. If *any* definition matches, the rule is satisfied.

Keep all definitions of the same rule at the same arity, meaning the same number of parameters. In the example below, every `has-access?` definition takes `?user` and `?resource`.

The Java example continues to use the EDN constructors from Section 2.1. Python and JavaScript use the typed rule builders, which also validate branch arity when `q.rules` assembles the final rule set.

```
(def access-rules
  '[[(has-access? ?user ?resource)
     [?user :user/role :admin]
     [?resource :resource/id]] ; Admins match every resource entity

    [(has-access? ?user ?resource)
     [?user :user/permissions ?resource]] ; Specific permission

    [(has-access? ?user ?resource)
     [?user :user/team ?team]
     [?team :team/permissions ?resource]]]) ; Team-based permission
```

When you query `(has-access? ?u ?r)`, Datalevin returns results when the user is an admin over a resource entity, has a direct permission, or has a team permission. This is clearer and easier to maintain than using a complex `(or ...)` block in every query.

The same mechanism can collect values for one logical relation from several attributes. In this example, `?alias` may come from a person's name, legal name, or nickname. These alternatives are not mutually exclusive: a person may have any combination of the three attributes.

```
(def alias-rules
  '[[(alias-of ?person ?alias)
     [?person :person/name ?alias]]

    [(alias-of ?person ?alias)
     [?person :person/legal-name ?alias]]

    [(alias-of ?person ?alias)
     [?person :person/nickname ?alias]]])
```

Each matching attribute value contributes a row to the derived relation `(alias-of ?person ?alias)`. A person with a name, legal name, and nickname can therefore produce three rows. If the same alias is found through more than one definition, normal Datalog set semantics remove the duplicate result. Use `or-join` for one-off branching inside a single query. Use multiple rule definitions when the expansion has a name, is reused, or is clearer as part of the domain model.

4. Rule Parameters and Binding

Rules are not limited to single variables. They can take multiple parameters, and those parameters may be bound or unbound when the rule is called.

In Datalog, "bound" means a variable already has a value from an input or an earlier clause. "Unbound" means the engine is free to discover values that make the rule true. Rules are more relational than ordinary functions: a function call usually has inputs and a return value, but a rule describes a relation.

There is one important practical nuance. Triple patterns are naturally relational: the same attribute can be used to find an entity, find a value, or test that an entity/value pair exists. Function binding clauses are directional. A clause such as `[(+ ?x ?y) ?sum]` computes and binds `?sum`; `?sum` should be unbound when that clause runs. Datalevin does not run arithmetic functions backward: `[(= ?dist2 100)]` can filter computed distances, but it does not let the engine infer which `?x` and `?y` values would produce that distance. When writing a rule that filters on a computed value, express that as a function binding for the computed variable plus a predicate over that variable. Ordinary data constraints still determine which candidate tuples the rule has to consider.

```
(def distance-rules
  '[[(distance-squared ?p1 ?p2 ?d2)
     [?p1 :point/x ?x1]
     [?p1 :point/y ?y1]
     [?p2 :point/x ?x2]
     [?p2 :point/y ?y2]
     [(- ?x2 ?x1) ?dx]
     [(- ?y2 ?y1) ?dy]
     [(* ?dx ?dx) ?dx2]
     [(* ?dy ?dy) ?dy2]
     [(+ ?dx2 ?dy2) ?d2]]])
```

This rule computes the distance between two points. You can use this rule in different ways:

- **Filter**: `(distance-squared ?a ?b ?dist2)` plus `[(= ?dist2 100)]`, compute `?dist2`, then constrain it with a predicate.
- **Calculate**: use `(distance-squared ?a ?b ?dist2)` to find pairs and *bind* the squared distance to `?dist2`.

For filtering, let the rule compute the squared distance, then test it:

```
(d/q '[:find ?name-a ?name-b
       :in $ %
       :where [?a :point/name ?name-a]
              [?b :point/name ?name-b]
              (distance-squared ?a ?b ?dist2)
              [(= ?dist2 100)]]
     db distance-rules)
```

For calculation, leave the distance variable unbound and include it in `:find`:

```
(d/q '[:find ?name ?dist2
       :in $ %
       :where [?origin :point/name "origin"]
              [?p :point/name ?name]
              (distance-squared ?origin ?p ?dist2)]
     db distance-rules)
```

The first query uses the computed third column as a filter. The second query returns it as data. The rule body is the same in both cases.

5. Recursion: Navigating Hierarchies

Recursion is one of the main strengths of rules. It lets you query data structures with arbitrary depth, which is difficult with ordinary fixed-depth joins unless you know the depth in advance.

A recursive rule follows the standard recursive pattern:

1. **Base Case**: The simplest version of the relationship.
2. **Recursive Case**: A definition that calls itself.

5.1 Example: Organizational Hierarchy

Suppose you have an `:employee/manager` attribute that points to another employee. To find everyone who reports to a manager, either directly or indirectly, you can use a recursive `reports-to` rule.

```
(def org-rules
  '[[(reports-to ?sub ?boss)
     [?sub :employee/manager ?boss]] ; Base case: direct report

    [(reports-to ?sub ?boss)
     [?sub :employee/manager ?intermediate] ; Recursive case
     (reports-to ?intermediate ?boss)]])    ; Call itself
```

With a small hierarchy (assuming `:employee/manager` is a ref attribute):

```
(d/transact! conn
  [{:db/id -1 :employee/name "Alice"}
   {:db/id -2 :employee/name "Bob"   :employee/manager -1}
   {:db/id -3 :employee/name "Cara"  :employee/manager -2}
   {:db/id -4 :employee/name "Dee"   :employee/manager -1}
   {:db/id -5 :employee/name "Eli"   :employee/manager -3}])

(d/q '[:find ?name
       :in $ % ?boss-name
       :where [?boss :employee/name ?boss-name]
              (reports-to ?sub ?boss)
              [?sub :employee/name ?name]]
     (d/db conn) org-rules "Alice")
;; => #{["Bob"] ["Cara"] ["Dee"] ["Eli"]}
```

Here is the same example worked through as Datalog evaluation. Figure 9.1 shows the steps visually.

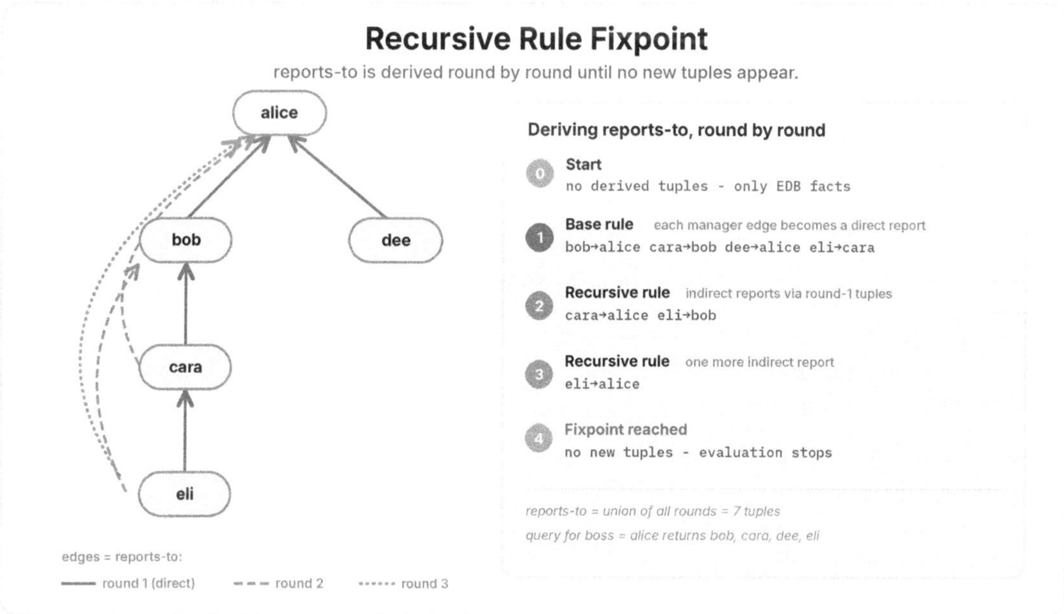

Figure 9.1: Recursive rule fixpoint: from the employee/manager facts, round 1 derives the direct reports (bob→alice, cara→bob, dee→alice, eli→cara), round 2 adds the indirect reports cara→alice and eli→bob, round 3 adds eli→alice, and round 4 produces no new tuples, reaching a fixpoint; the reports-to relation is the union of all rounds

The stored facts are the **EDB** (extensional database): facts that are directly present in Datalevin. Written as `[e a v]` datoms with names instead of entity ids, the `:employee/manager` facts are:

```
[bob  :employee/manager alice]
[cara :employee/manager bob]
[dee  :employee/manager alice]
[eli  :employee/manager cara]
```

The rule predicate `reports-to` is the **IDB** (intensional database): tuples derived by applying rules to EDB facts and to previously derived IDB tuples. Before evaluation starts, there are no derived `reports-to` tuples: `IDB round 0 = {}`.

Round 1 applies the base rule:

```
[(reports-to ?sub ?boss)
 [?sub :employee/manager ?boss]]
```

Every direct manager fact becomes a direct `reports-to` tuple:

```
(reports-to bob  alice)
(reports-to cara bob)
(reports-to dee  alice)
(reports-to eli  cara)
```

Round 2 applies the recursive rule using the tuples discovered in round 1:

```
[(reports-to ?sub ?boss)
 [?sub :employee/manager ?intermediate]
 (reports-to ?intermediate ?boss)]
```

This produces the next layer of indirect reports:

```
(reports-to cara alice)  ; [cara :employee/manager bob] + (reports-to bob alice)
(reports-to eli  bob)    ; [eli :employee/manager cara] + (reports-to cara bob)
```

Round 3 repeats the recursive step with the new round-2 tuples:

```
(reports-to eli alice)   ; [eli :employee/manager cara] + (reports-to cara
↪ alice)
```

Round 4 produces no new tuples, so evaluation has reached a **fixpoint**. The complete derived `reports-to` relation is the union of all IDB rounds:

```
(reports-to bob  alice)
(reports-to cara bob)
(reports-to dee  alice)
(reports-to eli  cara)
(reports-to cara alice)
(reports-to eli  bob)
(reports-to eli  alice)
```

The query asks only for tuples whose boss is Alice, so it returns Bob, Cara, Dee, and Eli. Chapter 21 explains how Datalevin evaluates recursive rules efficiently; the step-by-step illustration here is the conceptual model.

Termination is not validation

Datalevin's Datalog engine handles recursive expansion as a set of derived tuples, so cyclic data reaches a fixpoint instead of looping forever. Termination is not the same as data validity: if your domain requires a strict tree, your application or transaction functions should still prevent illegal management cycles.

6. Datalevin as a Reasoner: Bottom-Up Evaluation

Section 5 traced a recursive rule from direct manager facts to a fixpoint. That trace is Datalevin's rule evaluator at work: Datalevin evaluates rules bottom-up (see Chapter 21). In this mode, the database does not just store data; queries can infer new classifications or higher-level facts that were never explicitly transacted.

Bottom-up evaluation starts from base facts and repeatedly applies rules to discover derived tuples until no new tuples can be found. In rule-engine terminology, this is a forward-chaining style of reasoning: facts drive rule application, and rule application derives more facts. This still happens inside a query. Datalevin is not running a background job that stores inferred facts unless your application chooses to materialize those results.

This is different from Prolog-style backward chaining, where evaluation starts with a goal and recursively asks what subgoals could prove it. A Datalevin query names the rule relations it needs, but recursive rule evaluation itself is set-oriented and bottom-up. Chapter 21 explains how magic-set rewriting makes bottom-up evaluation goal-directed without turning it into backward chaining.

Expert systems make domain knowledge executable. Domain experts identify the concepts, relationships, constraints, and decision policies that matter, and the system applies that knowledge consistently to particular facts. An **ontology** makes the shared conceptual model explicit: what kinds of things exist in the domain, how they relate, and which distinctions the application needs to preserve. A knowledge graph supplies instances and observations in that model, while rules derive classifications and other consequences. In Datalevin, those consequences are query-time derived tuples unless the application explicitly materializes them.

This is a top-down approach to knowledge modeling: it begins with domain meaning and intended behavior, then organizes incoming evidence under that structure. It is a necessary complement to bottom-up, data-driven AI, which is excellent at discovering patterns, producing embeddings, extracting candidate facts, and making judgments under uncertainty. Data alone does not decide which vocabulary, policy, authority, or invariant an organization should adopt. In AI systems and human knowledge work alike, bottom-up methods help perceive and propose; top-down knowledge helps interpret, constrain, connect, and explain.

Here, *top-down* describes how domain knowledge enters the design, not how the query engine executes it. Datalevin still evaluates Datalog rules bottom-up, starting from facts and deriving tuples to a fixpoint. In Datalevin, ordinary facts and rules can themselves implement the application ontology: facts instantiate its concepts and relationships, while rules encode definitions and implications. An expert-authored ontology and a bottom-up rule evaluator therefore fit together naturally.

For experiments, tests, or services that do not need durable storage, an in-memory Datalevin connection can be a low-friction way to use Datalevin as a rule engine: load facts, pass rules as data, query the derived conclusions, and discard the connection when finished.

These ideas have a mature foundation in expert-system and production-system research [1,2,3]. Classic expert systems separated a knowledge base of facts and rules from an inference engine that derived conclusions. Datalevin's mechanism here is Datalog rule evaluation, not a production-system shell or a Rete implementation [2], but the modeling direction is similar: represent domain knowledge explicitly and derive higher-level facts from it. Chapter 5 discusses

RDF and OWL ontologies and explains how Datalevin can also support a lighter, application-local ontology with attributes, entities, references, and rules.

6.1 Example: Automated Classification (Expert Systems)

Imagine an e-commerce system that needs to determine whether a user qualifies for "VIP status" based on a complex set of evolving business rules. Instead of hard-coding these rules in your application, you can define them as a rule set.

```
(def reasoning-rules
  '[;; Rule 1: High Spender
    [(is-high-spender? ?u)
     [?u :user/total-spent ?amount]
     [(> ?amount 5000)]]

    ;; Rule 2: Frequent Shopper
    [(is-frequent-shopper? ?u)
     [?u :user/order-count ?count]
     [(> ?count 20)]]

    ;; Gold: both qualifications hold
    [(vip-status ?u ?tier)
     (is-high-spender? ?u)
     (is-frequent-shopper? ?u)
     [(ground :gold) ?tier]]

    ;; Silver: exactly one qualification holds
    [(vip-status ?u ?tier)
     (or (is-high-spender? ?u)
         (is-frequent-shopper? ?u))
     (not-join [?u]
       (is-high-spender? ?u)
       (is-frequent-shopper? ?u))
     [(ground :silver) ?tier]]])
```

There are three derived relations here. `is-high-spender?` and `is-frequent-shopper?` each classify a user from stored attributes. `vip-status` combines those classifications and adds a tier. These are query-time relations: evaluating them does not transact new facts.

The two definitions of `vip-status` are alternative ways to derive a tuple with that rule head. They are combined like a logical OR; their source order does not give the gold definition precedence over the silver definition. Written as Boolean conditions, the intended policy is:

```
gold   = high-spender AND frequent-shopper
silver = (high-spender OR frequent-shopper)
         AND NOT (high-spender AND frequent-shopper)
```

The `ground` function is a built-in identity function used in binding clauses. It can bind a constant into a variable, or copy an already-bound value into a new variable. The clause `[(ground :gold) ?tier]` means "bind `?tier` to `:gold`." A clause such as `[(ground ?x) ?y]`, after `?x` is bound, means "bind `?y` to the same value as `?x`." This is useful in rules because rule heads must expose variables, not raw constants.

The silver rule first uses `or` to require at least one qualification and bind `?u`. It then uses `not-join [?u]` to ask whether the same user satisfies both qualifications. The clauses inside `not-join` form one conjunction. When that conjunction has a match, `not-join` fails and rejects the silver tuple. Without this exclusion, a user satisfying both definitions would derive both `(vip-status ?u :gold)` and `(vip-status ?u :silver)`; rule order alone would not prevent that.

> **Rule order does not choose a tier**
>
> The two `vip-status` definitions form a union. The silver rule's explicit `not-join` is what makes the tiers mutually exclusive; reversing the definitions would produce the same results.

With data like this:

```
(d/transact! conn
  [{:user/id "u1" :user/total-spent 6000 :user/order-count 25}
   {:user/id "u2" :user/total-spent 6000 :user/order-count 1}
   {:user/id "u3" :user/total-spent 1    :user/order-count 25}])

(d/q '[:find ?id ?tier
       :in $ %
       :where [?u :user/id ?id]
              (vip-status ?u ?tier)]
     (d/db conn) reasoning-rules)
;; => #{["u1" :gold]
;;      ["u2" :silver]
;;      ["u3" :silver]}
```

The query declares two inputs with `:in $ %`: `$` is the database and `%` is the rule set supplied as `reasoning-rules`. The database clause binds each internal entity id to the public `:user/id`, while `(vip-status ?u ?tier)` joins that entity with the derived classification.

The evaluator reaches the result as follows:

User	High spender?	Frequent shopper?	Rule outcome
`u1`	Yes (`6000 > 5000`)	Yes (`25 > 20`)	Gold succeeds; silver is excluded by `not-join`
`u2`	Yes (`6000 > 5000`)	No (`1 <= 20`)	Gold fails; silver succeeds
`u3`	No (`1 <= 5000`)	Yes (`25 > 20`)	Gold fails; silver succeeds

A user who satisfies neither qualification derives no `vip-status` tuple and therefore does not appear in the result.

By querying `(vip-status ?u ?tier)`, you are treating Datalevin as an expert system. The evaluator starts from the stored amount and order-count facts, derives the two qualifying relations, and then derives `vip-status` tuples. Those tuples are returned by this query but are not written back to the database.

This approach is useful because:

1. **Composability**: You can build complex reasoning chains out of simple, testable rules.

2. **Performance**: Reasoning happens inside the engine, using indexes and Datalevin's semi-naive evaluator (SNE). "Semi-naive" is the standard Datalog term for an evaluation strategy that tracks newly derived tuples instead of repeatedly recomputing everything. Chapter 21, Section 2 explains the execution model.
3. **Consistency**: The "logic" of what defines a VIP is centralized in the rule set passed to queries, not scattered across microservices or application code modules.

7. Rules as an Abstraction Layer

Beyond their technical capabilities, rules serve as an **abstraction layer** for your domain logic.

Imagine a complex "Product Pricing" engine where the final price depends on user discounts, seasonal sales, and bulk quantities. Instead of putting this logic in your application code (which would require multiple round-trips to the database) or in each query, you can encode it as a rule set.

Here is a simplified rule set. It first derives candidate prices: the base price, plus one candidate for each applicable discount. The final `calculated-price` rule keeps only a candidate for which no better candidate exists.

Listing 9.1. Rule-based product pricing abstraction

```
(def pricing-rules
  '[[(price-candidate ?user ?product ?price)
     [?product :product/base-price ?price]]

    [(discount-rate ?user ?product ?rate)
     [?user :user/tier :gold]
     [(ground 0.20) ?rate]]

    [(discount-rate ?user ?product ?rate)
     [?product :product/category :clearance]
     [(ground 0.15) ?rate]]

    [(discount-rate ?user ?product ?rate)
     [?user :user/bulk-eligible? true]
     [?product :product/bulk? true]
     [(ground 0.10) ?rate]]

    [(price-candidate ?user ?product ?price)
     [?product :product/base-price ?base]
     (discount-rate ?user ?product ?rate)
     [(- 1.0 ?rate) ?multiplier]
     [(* ?base ?multiplier) ?price]]

    [(calculated-price ?user ?product ?price)
     (price-candidate ?user ?product ?price)
     (not-join [?user ?product ?price]
       (price-candidate ?user ?product ?better)
       [(< ?better ?price)])]])
```

This example uses floating-point prices to keep the arithmetic readable. In a production money model, prefer integer cents or a decimal representation and write the rule arithmetic accordingly.

Now the application query stays focused on intent:

```
(d/q '[:find ?product-id ?final-price
       :in $ % ?user-id
       :where [?user :user/id ?user-id]
              [?product :product/id ?product-id]
              (calculated-price ?user ?product ?final-price)]
     db pricing-rules "u-123")
```

The query does not need to know whether the price came from a gold-tier discount, a clearance discount, a bulk discount, or no discount at all. If the business adds a new discount rule, you add another `discount-rate` definition and keep the query shape stable.

This keeps pricing consistent, evaluates the calculation close to the data, and lets application code stay focused on query intent.

8. Benchmark Context

Datalevin's portable OpenRuleBench suite exercises transitive closure, same-generation recursion, and nonrecursive join trees across Datalevin, SQL databases, XSB, Soufflé, and Clojure rule engines [4]. It uses deterministic set-valued inputs, verifies input digests and result counts, and times query evaluation through complete result realization.

In the representative run, Datalevin had the lowest measured latency on all ten supported tasks. This is self-published workload evidence, not a universal ranking: supported tasks differ, generated-program preparation and compilation are outside the timer, and scale, cache state, and runtime settings affect the result. See [4] for the full timings and reproduction instructions.

Summary

Rules transform Datalog from a simple pattern-matching language into an expressive tool for knowledge derivation.

- **Encapsulation**: Wrap complex `:where` clauses in a named rule.
- **Reusability**: Define logic once and use it across many queries.
- **Named alternatives**: Express branching domain logic once instead of repeating `or` blocks across queries.
- **Parameters and bindings**: Use rules as relations whose arguments may be inputs, outputs, or both.
- **Recursion**: Traverse graphs and hierarchies without fixed-depth joins.
- **Derivation**: Treat rule sets as query-time reasoners that classify and derive facts from base data.
- **Abstraction layer**: Put domain calculations behind named rules so application queries can stay focused on intent.

Chapter 13 develops these patterns further for graph data.

References

[1] Stuart Russell and Peter Norvig, *Artificial Intelligence: A Modern Approach*, 4th US ed., Pearson, 2020. URL: https://aima.cs.berkeley.edu/.

[2] Charles L. Forgy, "Rete: A Fast Algorithm for the Many Pattern/Many Object Pattern Match Problem," *Artificial Intelligence* 19(1):17-37, 1982. DOI: https://doi.org/10.1016/0004-3702(82)90020-0.

[3] Peter Jackson, *Introduction to Expert Systems*, 3rd ed., Addison-Wesley, 1998.

[4] Datalevin project, "Portable OpenRuleBench Tasks," benchmark harness, methodology, representative results, and reproduction instructions. URL: https://github.com/datalevin/datalevin/tree/master/benchmarks/openrulebench.

CHAPTER 10

Key-Value Store

As explored in Chapter 4, Datalevin is built on a high-performance key-value (KV) foundation. Datalog handles complex queries well, but sometimes a direct KV interface is faster, simpler, or more appropriate for specific data shapes.

This chapter covers practical use of the Datalevin KV API, including sub-database management, public data types, transactions, and range scans.

Because Datalog is the main focus of this guide, Part II introduces transactions, read APIs, Datalog, and rules before returning to direct KV access. Treat this chapter as the direct-storage chapter you reach for when you need sorted-key access, custom indexes, list DBIs, scripting primitives, or performance-critical storage structures. For a compact list of all public KV functions, see Appendix F.

1. Opening a KV Store

Opening a KV store is distinct from opening a Datalog connection. You specify a directory location when calling `open-kv`, and Datalevin initializes the LMDB environment there. The directory does not need to exist already, but the process must have permission to create it.

Although you open a directory, the main LMDB data lives in a single memory-mapped data file inside that directory, conventionally `data.mdb`. Auxiliary files, such as `lock.mdb`, track reader/lock state, and Datalevin may create additional support files for features such as WAL. DBIs are named logical sub-databases inside the same file; opening more DBIs does not create one data file per DBI.

> **Open KV handles at application scope**
>
> Like a Datalog connection, a KV handle is a stateful resource. A normal application should open one KV handle for a local store, share it with the code that needs direct KV access, and close it during application shutdown. Do not open and close the same local KV store repeatedly for individual operations or requests.

```
(require '[datalevin.core :as d])

;; Open the store
(def kv (d/open-kv "/tmp/my-kv-store"))

;; KV ops ...

;; Close it when the application shuts down.
(d/close-kv kv)
```

The last parameter of `open-kv` can be an option map. The option map accepts many settings. Common options for `open-kv` include:

- `:mapsize`: The maximum size the database can grow to (in MiB).
- `:max-dbs`: The maximum number of named sub-databases (DBIs).
- `:max-readers`: The maximum number of concurrent reader threads. The current default is 1024.
- `:wal?`: Set to `true` to enable WAL mode that benefits from concurrent writers. WAL is **disabled by default** for both local KV stores and local embedded Datalog stores. Non-HA async read replicas require WAL on the primary; consensus-lease HA enables WAL automatically.
- `:wal-durability-profile`: Choose between `:strict` (standard `fsync`), `:relaxed` (batched syncs), and `:extra` (for example, `F_FULLFSYNC` on macOS). New WAL-enabled stores default to `:strict` when this option is omitted, including local and HA stores. Local stores can opt into `:relaxed`; consensus-lease HA accepts only `:strict` or `:extra`.
- `:wal-retention-bytes` and `:wal-retention-ms`: Set policies for how long to keep old WAL segments.
- `:flags`: A set of LMDB environment flags. The durability-relaxing write flags are `:nometasync`, which syncs data pages but not metadata pages on every commit; `:nosync`, which skips commit-time `fsync`; and `:writemap` with `:mapasync`, which uses a writable memory map and defers dirty-page writeback to the operating system. These flags can improve write throughput for rebuildable data, but they move or weaken the crash boundary. Chapter 19 covers durability tuning guidance.
- `:temp?`: Create a temporary file-backed store.
- `:inmemory?`: Create a memory-only store.
- `:spill-opts`: Control when queries spill large intermediate results to temporary disk storage. Common keys are `:spill-threshold`, a heap-pressure percentage, and `:spill-root`, the directory for temporary spill files. Set `:spill-threshold` to `100` to disable spill-to-disk.

```
;; Open an in-memory KV store (no file persistence), directory can be nil
(def mem-kv (d/open-kv nil {:inmemory? true}))
```

Both `:temp?` and `:inmemory?` are for ephemeral stores, but they are not the same mode. A `:temp?` store is still an LMDB environment backed by files in a directory; Datalevin marks it for cleanup and disables synchronous flushing. Use it when you want temporary data but still want the normal file-backed behavior during the process lifetime. An `:inmemory?` store uses the native in-memory environment mode: the environment lives in memory, may be opened with a `nil` directory, and disappears when closed. Use it for small test databases, short-lived fixtures, scratch stores, session caches, intermediate computation results, or high-speed buffers where no filesystem-backed environment is needed.

Configure `:spill-opts` when opening a KV store:

```
(def kv
  (d/open-kv "/tmp/my-kv"
             {:spill-opts {:spill-threshold 100}}))
```

For a Datalog connection, pass the same setting through `:kv-opts`, because the Datalog store owns an underlying KV store:

```
(def conn
  (d/get-conn "/tmp/my-db" schema
              {:kv-opts {:spill-opts {:spill-threshold 100}}}))
```

2. Sub-Databases (DBIs) and DUPSORT

Datalevin allows multiple KV sub-databases (DBIs) to reside in the same KV store, where they can participate in the same store-scoped transaction. Each DBI requires a unique string name. A DBI needs to be opened before use, and DBI opening is idempotent: it is OK to open a DBI multiple times. There is no need to close a DBI.

> **Keep application DBIs separate**
>
> The same file can contain Datalevin's Datalog DBIs and application-defined KV DBIs. Use distinct names for your KV DBIs and do not write directly to internal Datalog DBIs such as `datalevin/eav` or `datalevin/ave`; those are maintained by the Datalog engine.

There are two types of DBI: a regular DBI and a list DBI. Figure 10.1 shows the difference.

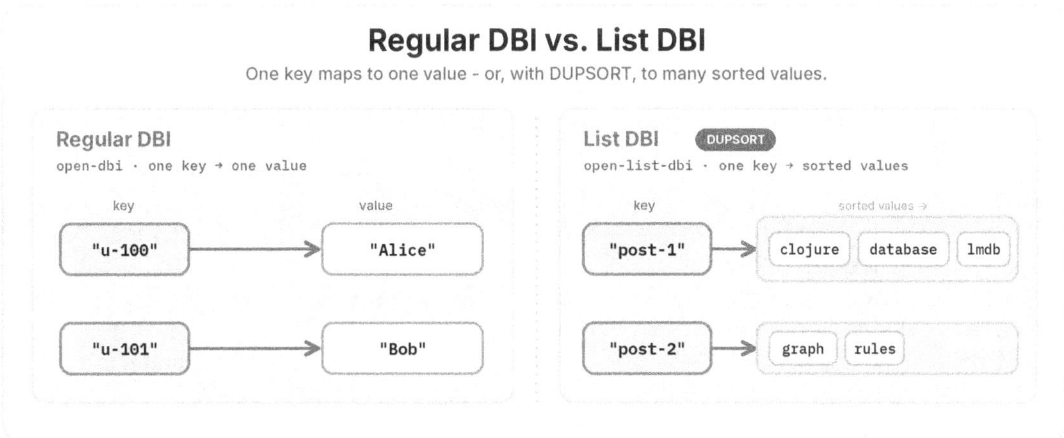

Figure 10.1: Regular DBI vs. list DBI: a regular DBI (open-dbi) maps one key to one value, such as "u-100" to "Alice"; a list DBI (open-list-dbi, using LMDB DUPSORT) maps one key to many sorted values, such as "post-1" to clojure, database, lmdb

2.1 Regular DBI

A regular DBI maps one key to exactly one value. It is opened with `open-dbi`.

```
;; Open a regular DBI called "people" in the KV store.
(d/open-dbi kv "people")
```

2.2 List DBI

A list DBI uses LMDB's `DUPSORT` feature. A single key can map to **multiple sorted values**: that is, a list. This is effectively a B+Tree of B+Trees. A list DBI is opened with `open-list-dbi`.

```
;; Open a list DBI (DUPSORT DBI) called "tags".
(d/open-list-dbi kv "tags")
```

3. Data Types

LMDB deals with raw bytes. Datalevin encodes and decodes typed values so those bytes sort correctly and store efficiently. These types can be specified as `key-type` or `value-type` parameters in KV operations.

KV type validation is separate from Datalog schema validation. In Datalog, `:db/valueType` belongs to an attribute schema. In the KV API, types are passed as operation descriptors or DBI options for keys and values. A DBI opened with `:validate-data? true` checks KV writes against those key/value types instead of checking Datalog attributes.

KV writes are more than "put this value somewhere." A write must also tell Datalevin how to encode the key, and usually how to encode the value. The encoding determines equality, ordering, range-scan behavior, and whether a value can fit in an LMDB key position.

3.1 Scalar types

The table below shows the scalar KV data types.

Type	Runtime Data	Notes
`:data`	Any EDN data	Default. Stored as opaque binary; **not sortable** for range queries.
`:string`	`String`	UTF-8 encoded, lexicographically sorted.
`:long`	`Long` (64-bit)	Numerically sorted.
`:id`	`Long` (64-bit)	Specialized id encoding used for entity ids.
`:int`	`Integer` (32-bit)	Numerically sorted.
`:float`	`Float` (32-bit)	IEEE 754 floating point.
`:double`	`Double` (64-bit)	IEEE 754 floating point.
`:bigint`	`java.math.BigInteger`	Arbitrary-precision integer within Datalevin's encoded range; numerically sorted.
`:bigdec`	`java.math.BigDecimal`	Arbitrary-precision decimal within Datalevin's encoded range; numerically sorted.
`:keyword`	`clojure.lang.Keyword`	EDN keyword, sorted by encoded namespace/name.
`:symbol`	`clojure.lang.Symbol`	EDN symbol, sorted by encoded namespace/name.
`:boolean`	`Boolean`	
`:instant`	Date/time value	Chronologically sorted.
`:uuid`	`java.util.UUID`	
`:bytes`	`byte[]`	Raw binary payloads.

If no KV type is specified, the default is `:data`, which stores an EDN binary blob.

3.2 Tuple types

Tuple types are composite KV type descriptors. They are used when a key or DUPSORT value needs to sort by more than one field, such as `[customer-id created-at]` or `[tenant-id` order-total order-id].

Do not confuse these KV type descriptors with Datalog schema composite `:db.type/tuple`. Here, the descriptor controls how a vector is encoded in the KV layer. In the KV API, a tuple type is written as a vector of scalar KV types. There are two forms:

- A **heterogeneous** descriptor gives one scalar type per position.
- A **homogeneous** descriptor gives one scalar type that repeats for every element.

Some examples:

Type Descriptor	Meaning
`[:string :instant]`	Heterogeneous tuple: exactly two elements, first a string, second an instant.
`[:keyword :long :string]`	Heterogeneous tuple: exactly three elements with the listed types.
`[:long]`	Homogeneous tuple: any number of elements, all encoded as longs.

Tuple values are ordinary vectors whose shape must match the type descriptor. Heterogeneous tuple descriptors are fixed arity: `[:string :instant]` matches `["acct-42" #inst "2026-05-31T00:00:00.000-00:00"]`. Homogeneous descriptors have one element type and can encode vectors such as `[2026 5 31]`.

Datalevin encodes typed KV keys and values into byte buffers internally. Tuple elements must use encodings that preserve the intended sort order in those buffers: `:string`, `:long`, `:float`, `:double`, `:bigint`, `:bigdec`, `:keyword`, `:symbol`, `:boolean`, `:instant`, and `:uuid`. Do not use `:data` or `:bytes` inside a tuple; arbitrary EDN data and byte arrays are not sortable tuple components. For numeric IDs inside tuples, use `:long`: tuple elements need headered encodings, while the KV-only `:id` and `:int` encodings are raw fixed-width forms without per-element type headers. If you use `:id` or `:int` as a tuple element type, Datalevin will reject the tuple descriptor rather than silently choosing a different encoding.

Tuples sort lexicographically by their encoded elements: first element first, then the second element to break ties, and so on. Put the field you most often range over or group by at the front of the tuple.

The example below uses a range expression before the full range API reference in Section 4.3. Read `[:closed start end]` as "scan from `start` through `end`, including both endpoints." Because the keys are tuples, the start and end values are tuple keys too.

```
(def event-key-type [:string :instant])

(d/open-dbi kv "events")

(d/transact-kv kv
  [[:put "events"
    ["acct-42" #inst "2026-05-31T10:00:00.000-00:00"]
    {"event/type" "login"}
    event-key-type
    :data]
   [:put "events"
```

```
    ["acct-42" #inst "2026-05-31T12:00:00.000-00:00"]
    {"event/type" "purchase"}
    event-key-type
    :data]])

;; Scan one account for a time window.
(d/get-range kv
  "events"
  [:closed
   ["acct-42" #inst "2026-05-31T00:00:00.000-00:00"]
   ["acct-42" #inst "2026-06-01T00:00:00.000-00:00"]]
  event-key-type)
```

In list DBIs, tuple values are useful for sorted secondary lists:

```
(def score-type [:double :string])

(d/open-list-dbi kv "scores-by-board")

(d/transact-kv kv
  [[:put "scores-by-board" "daily" [98.5 "alice"] :string score-type]
   [:put "scores-by-board" "daily" [87.0 "bob"] :string score-type]])

(d/get-list kv "scores-by-board" "daily" :string score-type)
;; => ([87.0 "bob"] [98.5 "alice"])
```

3.3 Important Constraints: Key Size Limit

LMDB, and therefore Datalevin, has a hard key-size limit. In Datalevin, the maximum key size is **511 bytes**. Keep KV keys compact and put large payloads in values rather than keys.

When using composite keys, the sum of the encoded tuple parts must stay within this limit. In a list DBI, values are stored in a secondary B+Tree where they essentially act as keys, so each value in a list DBI is also subject to the 511-byte limit. Regular DBI values are not subject to this key-size limit and can grow up to 2 GB.

4. KV Operations

Under the hood, LMDB uses transactions for both reads and writes. In this section, transaction means a write transaction: a batch of changes that commits atomically. Read transactions are called queries.

4.1 Transactions

Data is written to a KV store with `transact-kv`. One call applies a batch of operations atomically: either all items commit or none of them do. A single KV transaction may write to several DBIs in the same store, but it cannot span two different stores or database files.

Transaction items are vectors. Datalevin accepts two public layouts for KV transaction data.

The **self-contained shape** is used with `(d/transact-kv kv txs)`. Each item names the DBI and may carry its own key and value types:

```
[:put dbi-name key value]
[:put dbi-name key value key-type value-type]
[:put dbi-name key value key-type value-type flags]
[:del dbi-name key]
[:del dbi-name key key-type]
```

When `key-type` or `value-type` is omitted, Datalevin uses `:data`. That is fine for opaque EDN values, but it is usually the wrong choice for keys you plan to scan by range. Use explicit sortable types such as `:string`, `:long`, `:instant`, or tuple descriptors for range-oriented DBIs.

The optional `flags` slot applies to `:put` operations and is rarely needed. The common flag worth knowing is `:append`, a hint for trusted bulk loading when keys are already in ascending order. Most transactions should omit `flags`.

```
;; Add multiple values to the same key in a list DBI.
(d/transact-kv kv
  [[:put "tags" "clojure" "fast" :string :string]
   [:put "tags" "clojure" "expressive" :string :string]])
```

The **single-DBI shape** `(d/transact-kv kv dbi-name txs)` is used when every item targets the same DBI and uses the same key/value types. Pull the DBI name and types out to the function call, then each item contains only the operation, key, and value:

```
(d/transact-kv kv
  "people"
  [[:put 1 "Alice"]
   [:put 2 "Bob"]
   [:del 3]]
  :long
  :string)
```

This call is equivalent to:

```
(d/transact-kv kv
  [[:put "people" 1 "Alice" :long :string]
   [:put "people" 2 "Bob" :long :string]
   [:del "people" 3 :long]])
```

Use the self-contained shape when a batch touches multiple DBIs or mixed data types. Use the single-DBI shape when you are bulk-writing one typed DBI; it is shorter, clearer, and avoids repeating the same descriptors on every item.

For list DBIs, `:put` adds one sorted value under the key. To add or delete many list values under one key, the dedicated `put-list-items` and `del-list-items` helpers are often clearer; Appendix F lists the full list-DBI API.

4.2 Point query

The value for a key can be retrieved with `get-value`. If the DBI is a list DBI, the sorted values for a key can be retrieved with `get-list`. These are single-key point queries.

For a regular DBI, read one key:

```
(d/get-value kv "people" 1 :long :string true)
;; => "Alice"
```

For a list DBI, read all sorted values under one key:

```
;; Get all values for a key
(d/get-list kv "tags" "clojure" :string :string)
;; => ("expressive" "fast") ; sorted values
```

4.3 Range query

Because Datalevin keeps keys (and DUPSORT values) sorted, you can perform highly efficient range scans. Datalevin provides a rich set of range keywords that specify how the scan should start, end, and in which direction it should move.

A **range** is specified as a vector: `[range-type start-value end-value]`.

Range Type	Direction	Bounds
`:all` / `:all-back`	Forward / Backward	Entire keyspace.
`:at-least` / `:at-most-back`	Forward / Backward	Inclusive start.
`:at-most` / `:at-least-back`	Forward / Backward	Inclusive end.
`:closed` / `:closed-back`	Forward / Backward	Inclusive start, inclusive end.
`:closed-open` / `:closed-open-back`	Forward / Backward	Inclusive start, exclusive end.
`:open-closed` / `:open-closed-back`	Forward / Backward	Exclusive start, inclusive end.
`:open` / `:open-back`	Forward / Backward	Exclusive start, exclusive end.
`:greater-than` / `:less-than-back`	Forward / Backward	Exclusive start.
`:less-than` / `:greater-than-back`	Forward / Backward	Exclusive end.

Common range query functions are:

- `(d/get-range kv dbi range key-type)`: Returns an eager sequence of KV pairs.
- `(d/range-count kv dbi range key-type)`: Counts entries in a range without returning them.
- `(d/range-filter kv dbi pred range key-type)`: Scans a range and keeps pairs whose predicate matches.
- `(d/range-seq kv dbi range key-type)`: Returns a closeable lazy sequence of the range.

Technically, one DBI can contain keys encoded with different KV data types. Datalevin uses that capability internally in several places. For application DBIs, avoid mixed key types unless you have a deliberate encoding plan. Range queries operate on the encoded byte ordering, so a DBI that mixes strings, numbers, keywords, tuples, and raw data can produce ranges that are surprising to read and hard to maintain. In most application DBIs, pick one key type, or one tuple key type, and use it consistently.

`get-range` and `range-filter` are eager APIs: they realize the requested range as a result collection. For large ranges, Datalevin's spillable collections can move intermediate data to temporary disk storage under memory pressure. Section 1 shows how to configure `:spill-opts`. If you want to process a large range without realizing it eagerly, prefer `range-seq` and close the returned sequence when finished.

4.4 Visitor and Filter Functions

Sometimes you do not want a sequence at all: you want to stream over a range, update a counter, write to an output file, build a summary, or stop as soon as enough data has been seen. For those cases, use the visitor APIs.

The main visitor families are:

- `visit-key-range`, which visits keys only.
- `visit`, which visits key-value pairs in a regular DBI.
- `visit-list`, which visits the values stored under one list-DBI key.
- `visit-list-range`, which visits key-value pairs across a list-DBI range.

Visitors return no result from the outer call and are intended for side effects. In Clojure, a visitor callback may return `:datalevin/terminate-visit` to stop a scan early. Java, Python, and JavaScript expose visitor wrappers as side-effect callbacks in the host language. For search-style short-circuiting, use a narrow range; for list DBIs, use `list-range-some` in Clojure, `listRangeSome` in Java and JavaScript, or `list_range_some` in Python.

At the Clojure API level, visitors run in **raw mode** by default unless the final raw-mode flag is `false`. The Java, Python, and JavaScript wrappers split that choice into decoded methods such as `visit` and raw methods such as `visitRaw`/`visit_raw`; list DBIs follow the same pattern with methods such as `visitListRange`/`visit_list_range` and `visitListRangeRaw`/`visit_list_range_raw`.

In raw mode, the callback argument depends on the visitor. `visit-key-range` receives a key buffer directly. `visit-list` receives a value buffer directly. `visit` and `visit-list-range` receive an `IKV` pair, because both key and value are available. Clojure exposes raw callbacks as `ByteBuffer` values or `IKV` key/value pairs. Java exposes `RawBuffer` and `RawKV` wrappers; Python and JavaScript expose analogous `RawBuffer` and `RawKV` objects with `read`, `read_key`/`readKey`, `read_value`/`readValue`, and byte-copy helpers.

For a portable range visitor over a regular DBI, use decoded `visit` when you want application values in the callback. The examples below reuse the `events` DBI from Section 3.2, where each key is `[account instant]` and each value is an event map.

```
(def event-types (atom []))

(d/visit kv "events"
  (fn [[_account at] event]
    (swap! event-types conj [at (get event "event/type")]))
  [:closed
   ["acct-42" #inst "2026-05-31T00:00:00.000-00:00"]
   ["acct-42" #inst "2026-06-01T00:00:00.000-00:00"]]
  event-key-type
  :data
  false)
```

```
@event-types
;; => [[#inst "2026-05-31T10:00:00.000-00:00" "login"]
;;      [#inst "2026-05-31T12:00:00.000-00:00" "purchase"]]
```

`read-buffer` is the inverse of Datalevin's KV encoding. It reads a value out of the current position of a `ByteBuffer` according to the type descriptor you provide: `:string`, `:long`, `:instant`, `:data`, a tuple descriptor such as `[:string :instant]`, and so on. `put-buffer` goes the other direction: it writes an application value into a `ByteBuffer` using the same type descriptor. This is useful when a raw visitor or raw range predicate wants to compare encoded values without decoding every item.

For raw regular-DBI visits, the callback receives a raw key/value pair. Decode or copy what you need inside the callback:

```
(d/visit kv "events"
  (fn [kvp]
    (let [[_account at] (d/read-buffer (d/k kvp) event-key-type)
          event         (d/read-buffer (d/v kvp) :data)]
      (println at (get event "event/type"))))
  [:closed
   ["acct-42" #inst "2026-05-31T00:00:00.000-00:00"]
   ["acct-42" #inst "2026-06-01T00:00:00.000-00:00"]]
  event-key-type
  :data
  true)
```

Raw filters over list ranges receive a raw key/value pair. The predicate can decode only the parts it needs:

```
(d/list-range-filter kv "scores-by-board"
  (fn [kvp]
    (let [[score _user] (d/read-buffer (d/v kvp) score-type)]
      (>= score 90.0)))
  [:all]
  :string
  [:all]
  score-type
  true)
;; => [["daily" [98.5 "alice"]]]
```

> **Raw visitor buffers are callback-local**
>
> Clojure code that drops to the raw JVM buffer API can use `put-buffer` to encode a cutoff once and compare raw `ByteBuffer` values. Raw byte comparison is meaningful only when both buffers use the same type descriptor.
> Read or copy what you need inside the visitor; do not retain raw buffers outside the callback.

Raw mode avoids decoding data that the visitor may not need. If you prefer a simpler callback and are willing to decode every visited item, use decoded visitor methods. In Clojure, this means passing `false` as the final raw-mode flag; in Java, Python, and JavaScript, call the decoded wrapper rather than the Raw wrapper:

```
(d/visit-list kv "scores-by-board"
  (fn [[score user]]
    (println user score))
  "daily"
  :string
  score-type
  false)
```

Use decoded mode for clarity. Use raw mode when you are doing a large scan, need to skip many items cheaply, or need direct control over decoding.

4.5 Other Data Access Functions

Beyond point and range queries, Datalevin provides several specialized functions for the KV layer:

Function	Purpose
`get-first`	Get the first key-value pair in a DBI.
`get-first-n`	Get the first N key-value pairs in a DBI.
`get-rank`	Find the numerical rank (position) of a key in the sorted order.
`get-by-rank`	Retrieve the key at a specific numerical index.
`sample-kv`	Take a random sample of KV pairs from a DBI.

`sample-kv` is useful for quick inspection, smoke tests after import, or getting representative examples without scanning a whole DBI. By default it returns values only. Pass `false` as `ignore-key?` when you want `[key value]` pairs.

Java exposes this helper through `DatalevinInterop.kvSample`; Python and JavaScript expose it as a method on the KV handle.

```
(d/open-dbi kv "people")

(d/transact-kv kv
  [[:put "people" 1 "Alice" :long :string]
   [:put "people" 2 "Bob" :long :string]
   [:put "people" 3 "Cara" :long :string]])

(d/sample-kv kv "people" 2 :long :string false)
;; => [[2 "Bob"] [1 "Alice"]] ; example only, samples are random
```

If `n` is larger than the number of entries in the DBI, `sample-kv` returns no result (`nil`, `null`, or `None`, depending on the language).

5. Explicit Transaction `with-transaction-kv`

`transact-kv` is the standard way to write data. However, for complex logic involving reads and writes that must be atomic, use `with-transaction-kv` in Clojure or the corresponding explicit transaction helper in each host-language binding.

```
(d/open-dbi kv "counters")
```

```
(d/with-transaction-kv [tx kv]
  (let [current (or (d/get-value tx "counters" "hits" :string :long) 0)
        new-val (inc current)]
    (d/transact-kv tx
      "counters"
      [[:put "hits" new-val]]
      :string
      :long)))
```

This helper ensures that: 1. All reads and writes inside the block share the same transaction snapshot. 2. The transaction is automatically committed at the end of the block. 3. If an exception occurs, the transaction is safely aborted.

Inside the body, use the `tx` handle anywhere you would normally use `kv`. `transact-kv` accepts the same self-contained and single-DBI transaction shapes inside an explicit transaction.

Explicit KV transactions also support timeouts:

```
(d/with-transaction-kv [tx kv {:timeout-ms 5000}]
  (d/transact-kv tx "counters" [[:put "hits" 1]] :string :long))
```

KV explicit transactions use the same timeout machinery as Datalog explicit transactions, described in Chapter 6. A timeout aborts the transaction and reports `:type :transaction/timeout`; code that ignores interruption, or an already-running JavaScript async body, may continue until it returns.

Use `with-transaction-kv` for KV-only work. When one atomic operation needs to write both Datalog data and custom KV DBIs in the same store, use Datalog `with-transaction` and access the transactional KV instance from inside that block; Chapter 6 shows the pattern.

Summary

The Datalevin KV API is the direct sorted-storage layer beneath Datalog.

- **Stores and DBIs**: A KV store can contain many named DBIs. Regular DBIs map one key to one value; list DBIs use DUPSORT for one key with many sorted values.
- **Typed encoding**: Choose key and value types deliberately. Sortable scalar and tuple encodings make range scans predictable; `:data` is for opaque EDN values, not ordered access paths.
- **Point and range reads**: Use `get-value` and `get-list` for single-key reads, and `get-range`, `range-count`, `range-filter`, or `range-seq` for ordered ranges.
- **Large scans**: Eager range APIs may use spillable collections under memory pressure. Configure `:spill-opts` at open time, or use `range-seq` when you want lazy range processing.
- **Visitors**: Use visitor APIs when you want to stream through a range, stop early, or work in raw buffer mode without building a result collection.
- **Specialized access**: Rank, sample, and first-N helpers expose useful order-statistics operations over sorted DBIs.
- **Explicit transactions**: `with-transaction-kv` gives KV-only read-modify-write logic one atomic transaction, with optional timeout support.

PART III

Modeling Across Paradigms

CHAPTER 11

Schema Design

While Chapter 5 introduced the mechanics of attributes and namespaces, this chapter focuses on schema design choices. In a multi-paradigm database like Datalevin, schema is where you tell the query engine, search indexes, and storage layer how to interpret your facts.

A well-designed schema in Datalevin enables efficient joins, graph traversals, full-text search, embedding search, vector search, and path-indexed documents.

For a compact reference of every Datalog schema property and its accepted values, see Appendix C, "Datalog Schema Reference."

1. Identity: `:db.unique/identity`

One of the most important decisions in schema design is how you identify your entities. While Datalevin provides internal 64-bit integer ids, application code usually wants to refer to entities using natural keys from your domain (like an email, a SKU, a URL slug, or synthetic keys like a UUID).

> **Keep entity ids inside the database boundary**
>
> Do not expose `:db/id` as a durable application identifier. An entity id is internal, database-local state, not a stable external identity, so do not place it in URL paths or query parameters, public API payloads, external events, or bookmarks. Give the entity an explicit unique domain attribute such as a UUID, account number, or slug, resolve that value with a lookup ref at the application boundary, and use the resulting entity id only inside database operations.

1.1 Choosing Domain Identity Attributes

Chapter 7 shows how lookup refs are used when reading data. The schema-design question comes earlier: which attributes should be stable handles for logical entities?

Any attribute with `:db/unique` can be used in a lookup ref. For domain identifiers such as emails, SKUs, account numbers, URL slugs, or external ids, `:db.unique/identity` is usually the right choice. It says that the value identifies the entity, and it enables upsert behavior when a transaction map mentions an identity value that already exists.

Use `:db.unique/identity` for values that answer "which entity is this?" Use `:db.unique/value` for values where a duplicate is an error, not an instruction to merge transaction data into an existing entity.

```
(def schema
  {:user/email  {:db/valueType :db.type/string
```

```
                :db/unique    :db.unique/identity}
   :sku/id      {:db/valueType :db.type/string
                :db/unique    :db.unique/identity}
   :invite/code {:db/valueType :db.type/string
                :db/unique    :db.unique/value}})
```

Here `:invite/code` is not an upsert key for the invitation entity. If a new invitation tries to reuse an existing code, Datalevin rejects the transaction; the application can treat that as a collision or replay and generate another code. With `:db.unique/identity`, the same duplicate value would instead resolve to the existing entity and merge the rest of the transaction data into it.

If you transact a map with a unique identity that already exists in the database, Datalevin merges the new attributes into the existing entity instead of creating a duplicate. Chapter 6 covers upsert behavior in transaction context; Chapter 7 covers lookup refs as read inputs.

1.2 Composite Indexes and Identity with Tuple Attributes

When people ask for a custom index in a Datalevin Datalog database, they often mean a composite lookup over several attributes, or a repeated query that reads the same attributes together. Datalevin's usual answer is a derived tuple attribute: define an attribute with `:db/tupleAttrs`, and Datalevin maintains a composite index entry from its component attributes. `:db/unique` is optional: add it when the combination is an identity or uniqueness constraint, but omit it when the tuple is only a non-unique query access path.

This is different from the transacted tuple data type. With `:db/tupleAttrs`, the application writes the component attributes, and Datalevin maintains the derived tuple value in the normal Datalog indexes. It is a composite access path over ordinary attributes, not a tuple value that application code transacts directly. Transacted tuple values use `:db.type/tuple` with `:db/tupleType` or `:db/tupleTypes`; those forms are covered in Appendix C.

Even though application code does not write the derived tuple directly, auto-maintained composite tuples use the same tuple value encoding as transacted tuple values, so the same tuple-element limitations apply.

A line item, for example, may be identified by the pair of order id and SKU. The application writes `:line-item/order-id` and `:line-item/sku`; Datalevin makes `:line-item/order+sku` available as a derived composite lookup, as illustrated in Figure 11.1.

```
(def schema
  {:line-item/order-id   {:db/valueType :db.type/string}
   :line-item/sku        {:db/valueType :db.type/string}
   :line-item/quantity   {:db/valueType :db.type/long}
   :line-item/order+sku  {:db/tupleAttrs [:line-item/order-id :line-item/sku]
                          :db/unique     :db.unique/identity}})
```

Now the composite key can be used anywhere a lookup ref can be used:

```
;; Create or update by component attributes. The tuple is maintained.
(d/transact! conn
  [{:line-item/order-id "o-1001"
    :line-item/sku      "SKU-42"
    :line-item/quantity 2}])
```

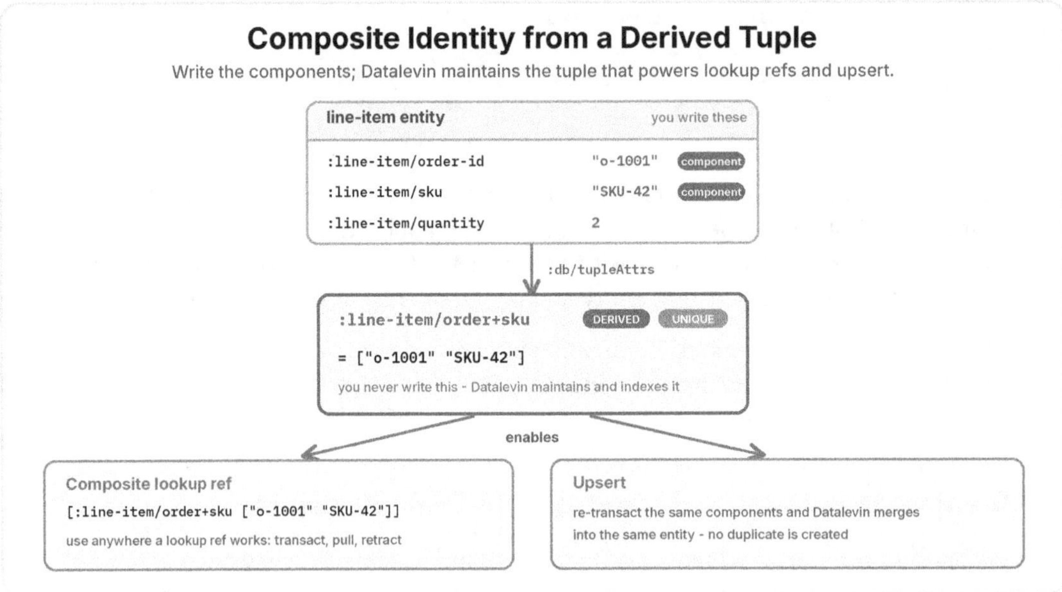

Figure 11.1: Composite identity from a derived tuple: a line-item entity stores the component attributes order-id and sku; Datalevin derives the tuple attribute :line-item/order+sku via :db/tupleAttrs, marks it unique, and never requires you to write it; the unique derived tuple then enables a composite lookup ref and upsert on the component values

```
;; Update the same line item by composite lookup ref.
(d/transact! conn
  [[:db/add [:line-item/order+sku ["o-1001" "SKU-42"]]
            :line-item/quantity
            3]])

;; Pull by the same composite identity.
(d/pull (d/db conn)
        '[:line-item/sku :line-item/quantity]
        [:line-item/order+sku ["o-1001" "SKU-42"]])
```

This pattern is useful for user-defined secondary access paths: tenant plus external id, account plus date, document plus section number, order plus SKU, or student plus course plus term. If the derived tuple attribute is also unique, it becomes a composite identity and participates in upsert.

Datalevin does not have a separate `unique-composite` schema property. A composite uniqueness constraint is expressed by combining `:db/tupleAttrs` with `:db/unique` on the derived tuple attribute. `:db/tupleAttrs` says how to derive the composite value from ordinary attributes; `:db/unique` says whether that derived value is a unique identity or unique value.

A maintained access path trades write and storage for faster reads. A derived tuple attribute materializes the joint relation among its components as one indexed value. Without it, a component plan reads the individual EAV datoms and joins them by entity, often emitting a large intermediate tuple relation. With it, a query can scan or probe already-correlated combinations directly. If the composite path emits `C` qualifying intermediate tuples while the component path emits `E`, the important saving is `C` versus `E`; for selective combinations, `C` is often much smaller.

That lower cardinality also means less work for every later join, filter, and aggregate.

The saving is paid for on the write side. Every entity carrying the components gains another logical datom indexed in both EAV and AVE, and a write that changes any component must also maintain that derived datom.

If you only need occasional ad hoc filtering by the components, ordinary Datalog joins over the component attributes may be enough. If the combination is a domain identity or upsert key, use `:db/tupleAttrs` with `:db/unique`. If a known hot query repeatedly reads the components together, `:db/tupleAttrs` without `:db/unique` can be worthwhile. Do not add uniqueness merely to obtain the query access path; doing so also changes transaction semantics.

Do not use a transacted `:db.type/tuple` as a substitute for this unless the tuple itself is the application value. A stored tuple can be unique like any other attribute, but Datalevin will not derive it from separate component attributes or keep those components in sync for you.

There are a few important rules:

- Write the component attributes, not the derived tuple attribute. Datalevin updates the derived composite index when components change.
- `:db/tupleAttrs` must name cardinality-one attributes.
- A derived tuple cannot depend on another derived tuple.
- If a component is missing, the derived tuple may contain `nil` in that position. Datalevin still compares the whole tuple value structurally, so two tuples with `nil` in the same position and the same remaining values are not distinct for uniqueness checks. This differs from SQL databases that allow multiple `NULL` values in a unique index. Decide whether that is acceptable before making the tuple unique.
- Composite lookup refs are simplest when the tuple components are scalar identity values such as tenant id, order id, SKU, date, or source-system id. If a component attribute is a reference, the tuple stores the resolved entity id; keep that in mind when constructing tuple lookup values.
- For ad hoc query-side tuple construction and destructuring, use the built-in `tuple` and `untuple` functions:

```
;; Build a tuple in the query and match it against the derived attribute.
(d/q '[:find ?line
       :where [?line :line-item/order-id ?order-id]
              [?line :line-item/sku ?sku]
              [(tuple ?order-id ?sku) ?order+sku]
              [?line :line-item/order+sku ?order+sku]]
     (d/db conn))

;; Destructure a tuple value back into its components.
(d/q '[:find ?line ?order-id ?sku
       :where [?line :line-item/order+sku ?order+sku]
              [(untuple ?order+sku) [?order-id ?sku]]]
     (d/db conn))
```

Composite Attributes for Known Hot Query Shapes

Composite attributes are also a deliberate denormalization for read performance. This works best when the workload is known: the same query repeatedly consumes several cardinality-one attributes from the same entity, and the saved read work justifies another maintained datom.

For example, the LDBC Social Network Benchmark (LDBC SNB) [3] includes the Interactive Complex 5 (IC5) query, in which a forum-membership entity carries a person, a forum, and the date the person joined. The Datalevin SF1 benchmark implementation is available with its query and bundled parameters [4]. A non-unique derived attribute can materialize those three values together:

```
{:hasMember/person+forum+date
 {:db/tupleAttrs [:hasMember/person
                  :hasMember/forum
                  :hasMember/joinDate]}}
```

Only the membership fragment changes between the two query variants. **Without the composite attribute**, IC5 reads three component patterns:

```
[?membership :hasMember/person ?person]
[?membership :hasMember/forum ?forum]
[?membership :hasMember/joinDate ?join-date]
[(> ?join-date ?min-date)]
```

With the composite attribute, the query reads and destructures the derived tuple instead:

```
[?membership :hasMember/person+forum+date ?membership-value]
[(untuple ?membership-value) [?person ?forum ?join-date]]
[(> ?join-date ?min-date)]
```

The benefit is the cardinality reduction, not merely replacing three clauses with one. The composite value has already correlated person, forum, and date. The notation `[person, forum, date > min-date]` is conceptual shorthand for that combined condition, not literal Datalog syntax.

The following comparison comes from one local diagnostic of the two membership plans against the LDBC SNB scale-factor-1 database. It used Datalevin's bundled IC5 parameter row, `person-id = 100` and `min-date = 2011-01-01T00:00:00Z`. The unit in both rows is the number of intermediate membership tuples handed to subsequent joins:

IC5 query variant	Membership access	Intermediate membership tuples
Without composite attribute	Three component patterns	1,402,224
With composite attribute	Derived `:hasMember/person+forum+date` pattern	7,759

The composite query therefore sends fewer membership tuples into subsequent joins and aggregation: about 0.55% as many—roughly 1/181. These figures are plan cardinalities, not counts of index datoms read and not a wall-clock latency comparison. They isolate why the composite query makes the later work cheaper.

The exact reduction depends on the data distribution and supplied inputs. It is largest when the joint combination is much more selective than any component attribute by itself. The component attributes remain independently queryable; the tuple is an additional access path, not a replacement for the underlying facts.

Declaring the schema entry only causes Datalevin to maintain the derived value. The query must reference the composite attribute to use it; the optimizer does not automatically replace three component patterns with a tuple pattern. Check the changed plan with `explain`, benchmark representative parameters, and keep the composite only when the read improvement pays for its write amplification, additional index space, and cache footprint.

1.3 System-Wide Identifiers: `:db/ident`

While `:db.unique/identity` is used for attributes that uniquely identify a domain entity (like a user's email), **`:db/ident`** is a built-in attribute used to assign a globally unique keyword to an entity.

This is the standard way to represent **enums** or system-wide constants. Once an entity has a `:db/ident`, you can use that keyword anywhere you would use an entity id or a lookup ref.

```
;; Schema includes a ref attribute for the enum-valued field
(def schema {:order/id     {:db/valueType :db.type/string
                             :db/unique    :db.unique/identity}
             :order/status {:db/valueType :db.type/ref}})

;; Define an "enum" entity for order status
(d/transact! conn [{:db/ident :order.status/shipped}])

;; Use the keyword directly in another transaction
(d/transact! conn [{:order/id "123" :order/status :order.status/shipped}])
```

The important point is that `:order/status` is a `:db.type/ref`, so it takes an entity id. Here a `:db/ident` serves the purpose of an entity id. Installing a `:db/ident` entity is an ordinary data write, not a schema change.

The advantage of using `:db/ident` over raw strings or keywords is that the enum itself is an ordinary entity. You can attach descriptive facts to it (like `:order.status/label "Shipped"`) without changing your order data.

For example, later you can add display details to the status entity itself:

```
(d/transact! conn
  [{:db/ident :order.status/shipped
    :order.status/label "Shipped"
    :order.status/sort 30}])

(d/pull (d/db conn)
  [:db/ident :order.status/label :order.status/sort]
  :order.status/shipped)
;;=> {:db/ident :order.status/shipped
;;     :order.status/label "Shipped"
;;     :order.status/sort 30}
```

Existing orders still point at the same status entity. You added facts about the enum; you did not rewrite the order facts that refer to it.

2. Modeling Relationships: References and Cardinality

In Datalevin, relationships are ordinary facts. They are defined using the `:db.type/ref` value type and the `:db/cardinality` property. The stored value of a reference attribute is the entity id of the target entity. In transaction data, you can still use tempids, lookup refs, or `:db/ident` keywords as convenient inputs; Datalevin resolves them to entity ids when it writes the datoms.

One of the most important decisions when modeling a relationship is its cardinality [1], also called association multiplicity: one-to-one, one-to-many, or many-to-many.

2.1 One-to-One and One-to-Many

- **One-to-One**: Use `:db.type/ref` with `:db.cardinality/one`.
  - Example: A `User` has one `Profile`.
- **One-to-Many**: Usually put a cardinality-one ref on the many side.
  - Example: A `Post` has many `Comment` entities, so each `Comment` carries a reference to its `Post` through `:comment/post`.

The one-to-many example is worth stating carefully: you do not need a `:post/comments` cardinality-many attribute only to find a post's comments. Store `:comment/post` on each comment, then query or pull the reverse direction when you need the collection. The broader performance rationale for normalized relationship facts is covered in the many-to-many discussion below.

2.2 Many-to-Many: Cardinality vs. Join Entities

For many-to-many relationships, you have two main ways to model them, as compared in Figure 11.2.

1. **`:db.cardinality/many`**: You can add an attribute with `:db.cardinality/many` to one or both entities. Use this only for small, bounded sets that are normally read with their owner. It is convenient and can produce a compact shape in `d/pull` results, but it is not the right default for large or queryable relationships.
2. **Join Entities**: You can create a third entity that "joins" the other two, similar to a join table in SQL.

While `:db.cardinality/many` is convenient, it is not a compact array stored inside one datom. Each member is a separate datom, logically shaped like `[entity attribute value]`, and it participates in the indexes. DUPSORT reduces some repeated-prefix cost at the storage layer, but a large many-valued attribute still carries real index-entry overhead and repeated entity/attribute structure. For large relationships, this shape behaves like non-normalized data and can impose a real index cost; Chapter 15 explains the EAV and AVE mechanics. Often the better model is to put a cardinality-one ref on the many side; when the relationship itself needs facts, use a join entity.

Datalevin performs best with normalized data [2]: small relationship facts whose access paths are explicit. For one-to-many relationships, this often means placing a cardinality-one reference on the many side. For many-to-many relationships, it often means creating a join entity.

This shape gives the query optimizer more granular facts to count, join, and filter. It also avoids treating a large relationship set as if it were a small property of one entity. If you expect a single

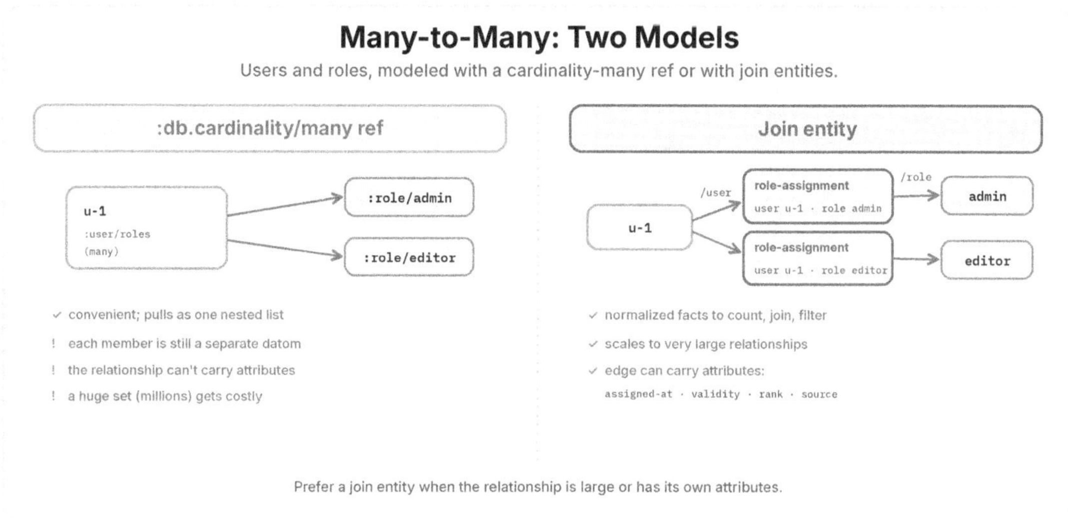

Figure 11.2: Many-to-many models compared: a :db.cardinality/many ref puts :user/roles holding role/admin and role/editor on user u-1 - convenient and pulls as one nested list, but each member is still a separate datom, the relationship cannot carry attributes, and large sets get costly; a join entity uses role-assignment entities linking u-1 to each role - normalized facts the optimizer can count, join, and filter, scaling to large relationships and able to carry edge attributes like assigned-at, validity, rank, or source

entity to have hundreds, thousands, or millions of references, a join entity is the better default for query planning and operational clarity. This is not about minimizing raw datom count; a join entity usually stores more datoms than one cardinality-many ref. It is the right model when the relationship itself has attributes, such as role assignment time, quantity, rank, validity interval, or source system.

In the example below, `:role/admin` and `:role/editor` are `:db/ident` enum entities used as reference values, as described in Section 1.3.

```
;; Instead of many references in one entity:
{:user/id "u-1" :user/roles [:role/admin :role/editor]}

;; Use join entities for large or queryable relationships:
(d/transact! conn
  [{:role-assignment/user [:user/id "u-1"]
    :role-assignment/role :role/admin}
   {:role-assignment/user [:user/id "u-1"]
    :role-assignment/role :role/editor}])
```

2.3 Reference Integrity

Datalevin represents a reference as an entity id, so a declared `:db.type/ref` always points to an entity id. This differs from a SQL foreign key: there is no required parent row whose continued existence gives the reference validity. Entities have no required attributes, and an entity id remains a valid target even when no facts currently describe that id.

Consequently, retracting all of an entity's own attributes does not retract facts on other entities that refer to its id. The reference still names the same id; its target simply has no current attributes. When you intend to remove the entity as a node in the graph, use `[:db/retractEntity <eid>]`. This operation retracts facts where the id is the entity, retracts facts where the id is the value of a declared `:db.type/ref` attribute, and recursively retracts component children.

3. Attribute Properties: Search and Ownership

Datalevin allows you to attach properties to attributes that change how the database indexes, searches, or interprets relationships. These are different from value types: `:db/valueType` says what kind of value an attribute stores, while properties such as `:db/fulltext`, `:db/embedding`, and `:db/isComponent` add behavior around that value.

3.1 Full-Text Search (`:db/fulltext`)

Setting `:db/fulltext true` tells Datalevin to maintain a specialized full-text index for that attribute, indexing values as text. This enables the `(fulltext ...)` predicate in Datalog (see Chapter 16).

3.2 Embedding Search (`:db/embedding`)

Setting `:db/embedding true` on a `:db.type/string` attribute tells Datalevin to compute text embeddings and maintain an embedding similarity index. Query it with `embedding-neighbors` using query text, not a vector (see Chapter 17).

3.3 Component Attributes (`:db/isComponent`)

When a reference attribute is marked as `:db/isComponent true`, it signals a single-owner parent-child relationship.

- **Recursive Pull**: A wildcard `d/pull` on the parent will automatically include all attributes of the child component.
- **Cascading Deletes**: When the parent entity is deleted (via the **`:db/retractEntity`** transaction function), the child component entities are also automatically deleted.

> **Components are not uniqueness constraints**
>
> `:db/isComponent` is not a uniqueness constraint. Datalevin will allow two parents to point at the same component child, but that database is not well-formed for component semantics.
>
> If a parent is retracted with `:db/retractEntity`, or the component attribute is retracted with `:db.fn/retractAttribute`, Datalevin retracts the component child as owned data. Retracting that child also retracts declared reference datoms that point to it, so another parent that pointed at the same child loses its reference too. A value-specific `[:db/retract parent component-attr child]` only removes that one reference and does not cascade.

This is ideal for modeling owned data, like line items in an invoice or segments of a document. Use ordinary reference attributes for data that can be shared by several parents.

4. Specialized Value Types

Two specialized value types have secondary indexing behavior: vector values and indexed documents. They are declared with `:db/valueType`, just like strings, longs, refs, and tuples. They appear here only to clarify where they fit in schema design.

4.1 Vector Values (`:db.type/vec`)

Use `:db.type/vec` when your application supplies vectors directly. Query these attributes with `vec-neighbors`. Vector dimensions and metric settings belong in store options such as `:vector-opts` or `:vector-domains`, not in each individual datom.

4.2 Indexed Documents (`:db.type/idoc`)

Use `:db.type/idoc` for nested maps that should be stored as one value but queried by path with `idoc-match`. This is useful for flexible metadata, JSON import, and Markdown-derived structures.

5. Best Practices: Designing for Evolution

Datalevin's flexibility makes schema maintenance an ongoing design task. These practices keep the schema understandable as the model evolves.

5.1 Namespace Everything

Always use namespaces for your attributes. A common pattern is `[domain]/[property]`, such as `:account/balance` or `:sensor/reading`. This prevents collisions if you later integrate third-party data or modularize your application.

5.2 Prefer Flat Entities

While `:db/isComponent` is useful for true ownership, do not over-nest your data. Datalevin excels at flat, normalized facts. The query engine is designed to join flat facts efficiently, so you do not need to "pre-join" data into complex documents as you might in a document store like MongoDB.

5.3 Modeling Enums with `:db/ident`

For attributes that have a fixed set of values, such as status or type, prefer enum entities with `:db/ident` over raw keyword or string values. Section 1.3 shows the schema and transaction mechanics. The modeling payoff is:

1. **Descriptive facts**: You can attach display names, descriptions, or translations directly to the enum entity, as shown in Section 1.3.
2. **Reference semantics**: If you use a `:db.type/ref` for the status attribute, Datalog joins and pulls treat the status as an entity reference rather than an arbitrary string.
3. **Discovery**: You can query for all possible statuses using Datalog.

6. Schema Evolution and Migration

Datalevin can create schema entries when new attributes appear in transaction data, but schema is still persistent operational state. A schema map supplied while opening a database is not a connection-local declaration that Datalevin checks against the store. The database stores one effective schema, and opening with a schema map can change it. Schema changes are not transacted as ordinary datoms.

6.1 Update Schema at a Glance

Use `update-schema` for deliberate changes to an open database. The smallest additive update looks like this:

```
(d/update-schema conn {:user/last-login {:db/valueType :db.type/
instant}})
```

This adds a typed attribute or brings that named property to the supplied value. The Java, Python, and JavaScript bindings expose the same operation as `updateSchema` or `update_schema`; Section 6.4 shows complete examples.

The schema-update map and each nested property map are patches:

Patch input	Effect
Omit an attribute or property	Retain its stored definition.
Supply a property value	Add or replace that property after validating the resulting complete attribute definition.
Set a property to `:db/retract`	Remove that property while retaining the attribute.
Supply an attribute through `del-attrs`	Delete its schema metadata after its datoms have been removed.
Supply an old and new name through the rename map	Rename the attribute while preserving its internal attribute id.

These patch semantics make additive and corrective changes concise. The following sections explain how they interact with stored schema, validation, existing data, and production migration policy.

> **Keep intended schema in version control**
>
> Because a schema definition is EDN data, projects in any Datalevin language binding can keep the intended schema in a checked-in EDN file and load it at startup before creating a connection or calling `update-schema`. That keeps schema changes reviewable with code.
> Treat that file as source for desired schema; the database still stores the effective schema, and live changes still go through `update-schema`.

6.2 Connection Schema and Stored Schema

When an existing store is opened with a non-`nil` schema map, Datalevin reconciles the supplied entries with the stored schema by attribute:

Attribute state	Effective result
Present only in the store	Retained. Omitting an attribute never deletes or hides it.
Present only in the supplied map	Added to the stored schema immediately and assigned an internal attribute id. No datom needs to use it yet.
Present in both and all supplied properties already match	The effective definition is unchanged; omitted stored properties remain.
Present in both with different properties	Supplied properties replace properties with the same names. Omitted stored properties and the existing internal attribute id are retained.

Both the top-level attribute map and each nested property map use patch semantics. Omitting an attribute preserves that attribute, and omitting a property preserves that property. For example, supplying only `:db/unique` for an existing typed attribute leaves its stored `:db/valueType` unchanged. To remove one property explicitly, set its value to `:db/retract`; deleting the whole attribute instead uses the `del-attrs` argument to `update-schema`.

Internal attribute ids remain owned by storage. An incoming `:db/aid` is ignored, so the result of `d/schema` is safe to reuse as schema input. A checked-in source schema will usually omit generated ids because they are not part of application intent.

> **Opening is not a schema-drift check**
>
> It is safe to omit stored attributes or properties and to declare genuinely new attributes, but do not initialize an existing production store with conflicting property values merely to see whether they match. The open can mutate the stored definitions. Open with no application schema, inspect `d/schema`, and apply intentional changes with `update-schema`; that path performs schema-mutation checks and the supported untyped-to-typed data migration.

There is one more connection-lifecycle wrinkle. `get-conn` can return the already-open, process-local connection for a path. A later `get-conn` call with a different schema may therefore reuse that connection rather than apply the new argument. Acquire the connection once at the application boundary and call `update-schema` on that connection. Chapter 22 explains connection ownership in more detail.

Use `d/schema` to inspect the current effective schema before and after a change:

```
(d/schema conn)
```

The returned map includes the schema you supplied, built-in attributes such as `:db/ident`, and Datalevin's internal attribute ids. This is useful for checking what the database actually knows about an attribute, debugging unexpected value encoding or lookup-ref behavior, and verifying that a schema migration did what you expected.

Do not confuse `d/schema` with the Java helper `Datalevin.schema()` used in multi-language examples. `d/schema` reads schema from an open database connection; `Datalevin.schema()` is a host-language builder for constructing a schema map before passing it to Datalevin.

6.3 Explicit Updates Are Atomic and Retryable

Use `update-schema` for intentional changes to an open database. It follows the patch semantics in Section 6.1, retains existing `:db/aid` values, ignores supplied ids, and validates each resulting complete attribute definition. Adding a type to an untyped populated attribute invokes the conversion described below.

The complete request is safe to retry. Reapplying the same patch, property retraction, deletion, or completed rename is a no-op. Datalevin validates a request that combines patches, value re-encoding, deletions, and renames before writing, then commits it as one serialized transaction. A validation or commit failure leaves schema and data unchanged. Concurrent schema updates are serialized, so independent patches compose without losing one another.

Idempotence does not make ambiguous requests valid. A rename fails without overwriting when both the old and new names exist, and it fails as missing when neither exists. Rename chains, cycles, and duplicate targets are rejected, as are requests that try to patch and delete the same attribute or delete an attribute participating in a rename. These conflicts are detected before any part of the request is written.

6.4 Supported Schema Changes

Use `update-schema` to add, refine, or retract properties; rename an attribute; or delete schema metadata after its facts have been removed. A rename keeps the same internal attribute id, so existing datoms become readable under the new name. The examples below show each form:

```
;; Add or refine schema for an open connection.
(d/update-schema conn
                 {:user/last-login {:db/valueType :db.type/instant}})

;; Remove one property while retaining the attribute and its other properties.
(d/update-schema conn
                 {:user/email {:db/unique :db/retract}})

;; Rename an attribute. Existing facts are read as :user/contact-email.
(d/update-schema conn nil nil {:user/email :user/contact-email})

;; Delete schema metadata only after no facts remain for the attribute.
(d/update-schema conn nil #{:user/temporary-note})
```

One especially useful evolution path is moving a prototype attribute from untyped EDN storage to a typed encoding. When you use `update-schema` to add `:db/valueType` to an existing attribute that was previously untyped, Datalevin migrates the stored values as part of the schema update:

1. It reads the existing EDN binary values for the attribute.
2. It attempts to decode each value into the new declared type.
3. It rewrites the index entries to use the typed encoding, enabling efficient typed comparisons and range scans.

If any value cannot be coerced, the schema update fails before rewriting the attribute. Datalevin raises `Cannot migrate attribute values to new type` with the attribute, target type, and the entity/value pairs that failed conversion. Fix or retract those values, then run `update-schema` again.

This path supports the common workflow of starting flexible during prototyping and tightening the schema once access patterns are known. It applies to the untyped-to-typed case; changing one declared type to another declared type is a different, incompatible migration once data exists.

6.5 Mutation Boundaries

Not every schema change is safe once data exists. Datalevin validates these boundaries against stored datoms:

- Changing `:db/valueType` is rejected when typed data already exists, except for migrating an untyped EDN attribute to a specific type.
- Changing `:db/cardinality` from many to one is rejected when the attribute has existing datoms.
- Adding `:db/unique` is allowed only when the existing values are consistent with the requested uniqueness.
- Changing embedding-related schema on a populated attribute requires an explicit secondary-index rebuild.
- Changing idoc path-indexing schema on a populated attribute also requires an explicit rebuild.

For incompatible changes, treat the change as a data migration: introduce the new attribute, backfill it from the old one, move readers and writers, retract the old facts, then delete the old schema metadata. This is safer than changing a populated attribute in place.

6.6 Production Migration Models

For **declarative schema convergence**, a simple startup policy is enough:

1. Open the database without an application schema so the stored schema is loaded unchanged.
2. Call `update-schema` with the checked-in current property patches. Repeating this step is safe, and genuinely new attributes are added even before they have datoms.
3. Read `d/schema` and verify the patch's postconditions: ordinary property values must match, properties marked `:db/retract` must be absent, and `:db/aid` can be ignored. Do not expect properties omitted from the patch to disappear. Start application traffic only after these checks pass.

This needs no separate schema-version record when releases only add attributes or apply independently safe property patches. Supplying the latest schema directly while opening also adds genuinely new attributes and patches named properties, but the explicit open-then-update sequence gives existing-attribute changes the mutation checks of `update-schema` and leaves a clear place to verify the result. Renames, deletions, data backfills, and specialized secondary-index changes are not simple declarative convergence.

For **ordered application migrations**, keep migration state in the same database unless the surrounding system has a reason to store it elsewhere:

1. Open without an application schema.

2. Use `update-schema` to ensure a small, stable migration-ledger schema, such as a unique migration id and an applied-at timestamp. This works for both a new database and an existing database whose application schema has more attributes; the extra stored attributes remain.
3. Read applied migration ids from ordinary Datalevin data, elect exactly one migrator, and run only missing migrations in order.
4. For each migration, apply its schema transition, perform any data backfill, verify the resulting schema and data, and only then transact its completion record.
5. After all migrations, verify the effective schema against the current checked-in patch postconditions before serving traffic.

An `update-schema` call is atomic and can be retried unchanged, but an application data transaction and the migration completion record are still separate operations. Make the application migration as a whole resumable after a crash: data backfills and completion-record writes must be idempotent or check their own postconditions. Test the full sequence on a restored production snapshot.

Do not open every database with a historical "v1" schema and replay all schema maps blindly. On a database already at a later version, old values can replace newer values for the properties that v1 names before the migration runner reads its ledger. Properties omitted by v1 are preserved, but that does not make a stale patch safe. Open from stored state and apply only missing transitions. A new database has no completion records, so the same runner can apply the full sequence without an out-of-band test for whether the directory is new.

7. Schema Validation and Coercion

Datalevin schema also participates in transaction preparation. Declared `:db/valueType` properties tell Datalevin how to encode values, and store options decide how strictly the transaction input is checked before those values are written.

Two store options are especially important:

- **`:validate-data? true`** checks transaction values against declared `:db/valueType` before writing. With the default `false`, Datalevin still uses declared types to canonicalize values where it can.
- **`:closed-schema? true`** rejects transactions that mention attributes not already present in the schema. The default `false` allows new attributes to be added as data is written.

Datalevin does not store `nil` values. In JavaScript and Python client code, this also means application-level `null` or `None` must not be used as a stored attribute value. A missing value is represented by the absence of a datom, not by a stored null marker. If a value becomes unknown or inapplicable, retract the existing datom or omit the attribute in new transaction data.

Derived tuple attributes are the special case described in Section 1.2: Datalevin may maintain `nil` placeholders inside a derived tuple value when component attributes are missing. Application transaction data still cannot use `nil` as the value of an attribute.

With `:validate-data? true`, a transaction that writes `"42"` to `:user/age`, a string UUID to a `:db.type/uuid` attribute, or a timestamp string to a `:db.type/instant` attribute is rejected unless the value already has the declared runtime type. This is useful at system boundaries where you want malformed input to fail early.

With the default `:validate-data? false`, Datalevin still uses the declared type to canonicalize values where possible before encoding ordinary attribute values. For example, values may be converted to strings, numeric values may be narrowed to `long`, `float`, or `double`, strings may be parsed as UUIDs, integers may be converted to instants, keywords and symbols may be normalized, and tuple values are stored as vectors. This is convenient for trusted inputs, but it is not a substitute for application validation when user input quality matters. In particular, use correctly typed values for identities and lookup-facing attributes; Datalevin may need those values before general value correction.

Use `:db.attr/preds` for checks that depend only on one attribute value. An attribute predicate may be a qualified symbol, a predicate UDF descriptor, a registered UDF keyword id, or a non-empty sequence of those. Each predicate receives the normalized attribute value and must return exactly `true`.

The descriptor form is portable across embedded languages: register the predicate in the runtime, then reference the descriptor from schema:

```
(def routing-number
  {:udf/lang :clojure
   :udf/kind :predicate
   :udf/id   :payments/routing-number?})

(udf/register! registry routing-number
  #(boolean (and (string? %) (re-matches #"\d{9}" %))))

(def schema
  {:bank/routing-number {:db/valueType  :db.type/string
                         :db.attr/preds routing-number}})
```

An attribute predicate runs when that attribute is present. It is not an entity shape rule: it does not decide that every bank account must have both a routing number and an account number, or that every Venmo account must have a handle. Use `:db/ensure` for entity invariants over the would-be transaction result.

A `:db/ensure` predicate receives the would-be database after applying the transaction followed by its resolved arguments, and should return truthy on success. It runs after transaction functions, tempid resolution, upserts, map expansion, retractions, and other transaction data have produced the would-be final datoms. Chapter 6 covers the full transaction mechanics; from a schema design point of view, the important distinction is simple: `:db.attr/preds` checks one value, while `:db/ensure` checks the would-be entity or transaction result.

The transaction shape is portable. Assume `bank-account/bankAccount/ bank_account` is a registered predicate descriptor that checks the would-be entity:

```
(d/transact! conn
  [{:db/id               "acct"
    :account/identity    "acct-1"
    :account/type        :account.type/bank
    :bank/routing-number "021000021"
    :bank/account-number "123456789"}
```

```
[:db/ensure bank-account "acct"]])
```

If a failing test wrote `nil`, `null`, or `None` for `:bank/account-number`, Datalevin's nil-value rejection would fail first. To demonstrate that the entity-shape invariant rejects a missing required fact, omit the datom and let `:db/ensure` inspect the would-be result before commit.

Summary: The Schema Checklist

When adding a new attribute to your Datalevin database, ask yourself:

1. **What is the type?** Use a specific type (`:db.type/long`, `:db.type/string`, etc.) for performance and range queries.
2. **Is it a reference?** Use `:db.type/ref` to enable joins and graph traversals.
3. **Is it a stable application identity?** Use `:db.unique/identity` for natural or synthetic keys that you will use to look up or upsert entities. Do not expose the database-local `:db/id` in URLs, APIs, or other external references.
4. **Is it a system-wide constant?** Use **`:db/ident`** to give a unique, globally namespaced keyword to an entity, well suited for enums and static system data.
5. **Is it many-valued?** Use `:db.cardinality/many` only for small, bounded sets of values. Before adding it, ask whether the ref belongs on the many side as `:db.cardinality/one`; use join entities when the relationship needs its own facts.
6. **Does it need keyword search?** Use `:db/fulltext` for attributes you want to search as text.
7. **Does it need semantic text search?** Use `:db/embedding` for string attributes Datalevin should embed.
8. **Do you already have vectors?** Use `:db.type/vec` and configure vector domains.
9. **Is it flexible nested data?** Use `:db.type/idoc` for path-indexed documents.
10. **Is it a component?** Use `:db/isComponent` for single-owner nested entities.
11. **Does a known hot query read several attributes together?** Consider a non-unique `:db/tupleAttrs` access path after measuring the query, and budget for its additional write work and index storage.
12. **How will it evolve?** Use `update-schema` property patches for explicit changes, `:db/retract` for property removal, and planned migrations for incompatible changes on populated attributes.
13. **How strict should writes be?** Use `:validate-data?` and `:closed-schema?` when transaction input should be checked against the declared schema.
14. **Where do invariants belong?** Use `:db.attr/preds` for value-local checks, and `:db/ensure` for invariants over the would-be transaction result.
15. **Can it be absent?** Represent absence by omitting or retracting datoms, not by storing `nil`, `null`, or sentinel strings.

Applied consistently, these choices keep the schema flexible during development and stable enough for complex applications.

References

[1] Peter Pin-Shan Chen, "The Entity-Relationship Model: Toward a Unified View of Data," *ACM Transactions on Database Systems* 1(1):9-36, 1976. DOI: https://doi.org/10.1145/320434.320

440.

[2] E. F. Codd, “Further Normalization of the Data Base Relational Model,” IBM Research Report RJ909, August 31, 1971. Republished in Randall J. Rustin, ed., *Data Base Systems: Courant Computer Science Symposia Series 6*, Prentice-Hall, 1972.

[3] LDBC, “LDBC Social Network Benchmark (LDBC SNB),” official benchmark overview and links to the Interactive workload, datasets, and specification. URL: https://ldbcouncil.org/benchmarks/snb/.

[4] Datalevin project, “LDBC SNB Benchmark,” SF1 benchmark implementation, queries, bundled parameters, and reproduction instructions. URL: https://github.com/datalevin/datalevin/tree/master/benchmarks/LDBC-SNB-bench.

CHAPTER 12

Relational Modeling

Because Datalevin works well with normalized data [1], industry-standard **Entity-Relationship (ER) modeling** [2] is a useful starting point for designing your database.

SQL applications often use object-relational mapping (ORM) to bridge the gap between application objects and tables. Datalevin starts from facts instead: entities, attributes, and relationships are explicit datoms that can be queried directly. This chapter shows how to apply classic relational modeling patterns to a Datalog-powered triplestore.

1. The Grammar of Data: Nouns, Adjectives, and Verbs

When modeling a domain, think of your schema as a language.

1.1 Nouns (Entities)

Nouns are the "things" in your system: users, products, orders, invoices, courses, enrollments, and so on. In Datalevin, an entity is not represented by a table row, but by a set of facts that share one internal entity id. You usually describe that entity with a set of namespaced attributes such as `:user/name`, `:user/gender`, or `:user/email`.

> **Use singular attribute namespaces**
>
> Use singular nouns for attribute namespaces: `:user/name`, not `:users/name`; `:product/price`, not `:products/price`.

The namespace suggests which domain concept the attribute primarily describes. It is a modeling **convention**, not a table declaration or a class tag. If an entity has `:user/email` and `:user/name`, readers can recognize it as a user; Datalevin still stores ordinary datoms.

Singular namespacing makes Datalog queries feel like natural sentences: `[?user-eid :user/name ?name]` reads as "the entity `?user-eid` has a user name `?name`."

1.2 Adjectives (Attributes)

Adjectives are properties that describe an entity. These are the attributes already shown above. Attributes have value types: strings, longs, booleans, instants, keywords, and other types.

- **Example**: `:product/color "Red"`, `:order/status :order.status/pending`.

1.3 Verbs (Relationships)

Verbs describe how entities interact. In Datalevin, relationships are ordinary attributes whose value type is `:db.type/ref`.

- **Example:** `:user/follows`, `:order/customer`, `:line-item/product`.

2. ER Design Decisions in Datalevin

ER modeling is useful because it forces you to make a few decisions explicitly before you write schema. Datalevin does not require tables, but the same questions still produce better facts. Normalize by default: store durable relationships as refs, and promote a relationship into its own entity when the relationship needs facts of its own.

2.1 Choose Stable Keys

Every important entity type should have a stable domain identifier when the domain has one. Use Datalevin's internal entity ids for storage and joins, but put natural identifiers in unique attributes so transactions, imports, and APIs can use lookup refs.

```
{:user/id      {:db/valueType :db.type/string
                :db/unique    :db.unique/identity}
 :product/sku  {:db/valueType :db.type/string
                :db/unique    :db.unique/identity}
 :invoice/uuid {:db/valueType :db.type/uuid
                :db/unique    :db.unique/identity}}
```

Prefer stable identifiers such as account IDs, SKUs, slugs, or UUIDs. Avoid using mutable labels such as display names as identities unless the domain treats them as immutable.

2.2 Record Cardinality and Participation

ER diagrams distinguish one-to-one, one-to-many, and many-to-many relationships. Datalevin represents these with reference attributes and cardinality, but it does not enforce all ER participation constraints for you.

- **One-to-many**: Put a cardinality-one ref on the many side, such as `:order/user` or `:comment/post`.
- **Many-to-many**: Use a join entity when the relationship is queried directly, may grow large, or has attributes of its own.
- **Optional participation**: Omit the datom when the relationship is unknown or not applicable; do not invent sentinel values like `"N/A"`.
- **Required participation**: Enforce it in transaction construction, application validation, or import checks. The schema documents the model, but a missing datom is still possible unless your write path rejects it.

> **Absence is a normal fact shape**
>
> A query for users without an active subscription is a query over missing relationship facts. `nil/null` is not a valid value in Datalevin. Optional relationships are omitted; required relationships need write-path validation or import checks.

2.3 Promote Relationships That Have Attributes

In ER modeling, a relationship can have attributes. In Datalevin, that is your signal to make the relationship an entity.

```
;; A thin model: the relationship has nowhere to put its own facts.
{:order/products [[:product/sku "SKU-1"] [:product/sku "SKU-2"]]}

;; A stronger model: each line item is a join entity.
{:line-item/order    [:order/id "ord-1001"]
 :line-item/product  [:product/sku "SKU-1"]
 :line-item/quantity 2
 :line-item/price    1999}

;; The same pattern handles group membership.
{:membership/user      [:user/id "u-101"]
 :membership/group     [:group/id "g-202"]
 :membership/role      :role/admin
 :membership/joined-at #inst "2026-01-15T09:30:00.000-00:00"}
```

Join entities handle classic ER cases such as order line items, memberships, assignments, reservations, approvals, and permissions. They also handle ternary relationships cleanly: if a supplier sells a product to a store under a contract, model the contract as an entity with three refs, not as three separate binary links that lose the original meaning.

This is a modeling choice, not a raw datom-count optimization. Promote the relationship when it needs identity, attributes, lifecycle, uniqueness, direct queries, or selective filtering. Chapter 11, Section 2.2 covers the performance trade-offs between small `:db.cardinality/many` refs and join entities.

2.4 Model Weak Entities and Ownership

ER weak entities depend on an owner for identity or lifecycle. In Datalevin, model them as ordinary entities with a reference to the owner, and use `:db/isComponent` only when the child is truly owned by the parent and should be deleted with it.

Good component candidates include invoice line items, document sections, or profile settings that have no useful life outside the parent. Poor component candidates include users, products, organizations, or tags that can be shared or referenced independently.

Datalevin allows multiple parents to point at the same component child, but cascading retraction treats the child as owned data; use an ordinary reference for shared targets.

2.5 Represent Types Without Inheritance Tables

ER specialization maps naturally to facts. For a small closed set of subtypes, use an enum-style ref or `:db/ident` value:

```
{:account/id   "acct-1"
 :account/type :account.type/business
 :account/name "Acme Corp"}
```

Use shared attributes in the common namespace and subtype-specific attributes only where they apply, such as `:business-account/tax-id` or `:personal-account/birth-date`. This keeps the model queryable without forcing an inheritance-table pattern into a fact database.

2.6 Be Deliberate About Derived Facts

ER analysis often identifies derived attributes, such as order totals, account balances, or comment counts. Prefer deriving these with Datalog when they are cheap and needed inside a transactionally current view. Store them only when they are expensive, part of an audit record, or must preserve a historical snapshot.

When you do store a derived fact, document the source of truth and update rule with `:db/doc` (Section 5). Future readers should be able to tell whether `:order/total` is authoritative, cached, or a historical snapshot.

3. Choosing Between Similar Shapes

Several Datalevin features can look like solutions to the same modeling problem. The right choice depends on whether you are modeling a relationship, an identity, or a value.

Modeling problem	Prefer	Why
One entity points to one parent	Cardinality-one `:db.type/ref` on the child	Normalized, simple to join, and available through reverse pull.
A small bounded owned set of refs or values	`:db.cardinality/many`	Convenient only when the set stays small and is normally read with its owner; otherwise put a cardinality-one ref on the many side or use a join entity.
A relationship has attributes, lifecycle, uniqueness, or direct queries	Join entity	The relationship itself is a thing the domain talks about.
A combination of attributes identifies an entity	Composite unique identity with `:db/tupleAttrs` and `:db/unique`	Datalevin maintains a derived lookup/index entry from ordinary attributes; this is the composite-uniqueness pattern.
A known hot query repeatedly reads several cardinality-one attributes from one entity	Non-unique composite attribute with `:db/tupleAttrs`	Treat the combination as a deliberate, measured access path; Chapter 11, Section 1.2 gives the selection criteria and IC5 evidence.
A small vector-like value whose elements all have one type	Stored homogeneous tuple with `:db/tupleType`	The tuple is one value, such as a coordinate or numeric interval.
A small record-like value with fixed positions of different types	Stored heterogeneous tuple with `:db/tupleTypes`	The tuple is one value with typed positions, such as image bounds `[x y width height]`.

Modeling problem	Prefer	Why
Arbitrary nested imported data	`:db.type/idoc`	Use document modeling when the structure is open-ended or path-oriented; Chapter 14 covers `idoc`.

3.1 Composite Tuple Attributes Are Access Paths

Use `:db/tupleAttrs` when the component attributes remain meaningful facts on their own but their combination is either a domain identity or a measured hot-query access path. Add `:db/unique` only when the combination is genuinely unique. Chapter 11, Section 1.2 is the canonical treatment of derivation mechanics, composite lookup refs, and the IC5 performance evidence.

3.2 Stored Tuple Values Are Values

Stored tuple values are different. Use them when the tuple itself is a compact value object and its parts usually do not have useful meaning as standalone facts.

```
{:place/latlon {:db/valueType :db.type/tuple
                :db/tupleType :db.type/double}

 :image/bounds
 {:db/valueType  :db.type/tuple
  :db/tupleTypes [:db.type/double
                  :db.type/double
                  :db.type/long
                  :db.type/long]}}
```

`:place/latlon` is homogeneous: every element is a double. `:image/bounds` is heterogeneous: the first two positions are doubles and the last two are longs.

If you frequently ask "which places are north of this latitude?" or "which images are wider than this?", model those parts as ordinary attributes instead:

```
{:place/lat {:db/valueType :db.type/double}
 :place/lon {:db/valueType :db.type/double}

 :image/x      {:db/valueType :db.type/double}
 :image/y      {:db/valueType :db.type/double}
 :image/width  {:db/valueType :db.type/long}
 :image/height {:db/valueType :db.type/long}}
```

Separate attributes give the query planner direct facts and indexes for each part. A stored tuple is best when the application normally treats the whole vector as one value and the tuple elements do not usually matter outside that value.

3.3 A Practical Decision Checklist

Ask these questions before reaching for a tuple:

1. Is this a relationship between entities? Use refs, and use a join entity if the relationship has its own facts.
2. Is the combination of several existing facts a domain identity or a measured hot-query access path? Consider `:db/tupleAttrs`, adding `:db/unique` only for real identity or uniqueness; see Chapter 11, Section 1.2 before adding a non-unique access path.
3. Is this one compact value whose components do not have useful meaning as standalone facts? Use a stored tuple.
4. Are all stored tuple elements the same type? Use `:db/tupleType`.
5. Do the stored tuple positions have different types and fixed meanings? Use `:db/tupleTypes`.
6. Do you need to query, validate, document, or evolve the parts separately? Prefer separate attributes.

The short version: relationships should remain facts, composite identities and measured composite query paths should be derived from facts, and stored tuples should be reserved for values.

4. A Worked ER Example: Course Enrollment

Consider a small school registration system. In an ER diagram you would likely draw several stable domain entities, one course offering, and one join entity:

- `Student`: identified by a student id.
- `Course`: identified by a course code.
- `Term`: identified by an academic term id.
- `Instructor`: identified by an instructor id.
- `CourseOffering`: a particular section of a course in a term, taught by an instructor.
- `Enrollment`: the fact that a student is enrolled in a course offering.

Here, a course offering is a scheduled section: a reusable catalog course such as `CS101`, offered in a particular term, with a section number, instructor, and capacity. It is not a special Datalevin construct. It is a contextual entity, because students enroll in that specific offering rather than directly in the catalog course.

Figure 12.1 is an ER diagram showing the relationships among the entities. The figure labels `CourseOffering` as a scheduled section to make that modeling role visible.

The important modeling move is the same one the figure highlights: `Enrollment` points to the course offering, not directly to the catalog course. The enrollment has attributes of its own: status, grade, enrollment time, and perhaps the source system that created the record. That makes it a join entity in Datalevin. Listing 12.1 shows the actual schema.

Listing 12.1. Course enrollment schema with scheduled offerings

```
(def school-schema
  {:student/id   {:db/valueType :db.type/string
                  :db/unique    :db.unique/identity}
   :student/name {:db/valueType :db.type/string}

   :course/code  {:db/valueType :db.type/string
                  :db/unique    :db.unique/identity}
   :course/title {:db/valueType :db.type/string}
```

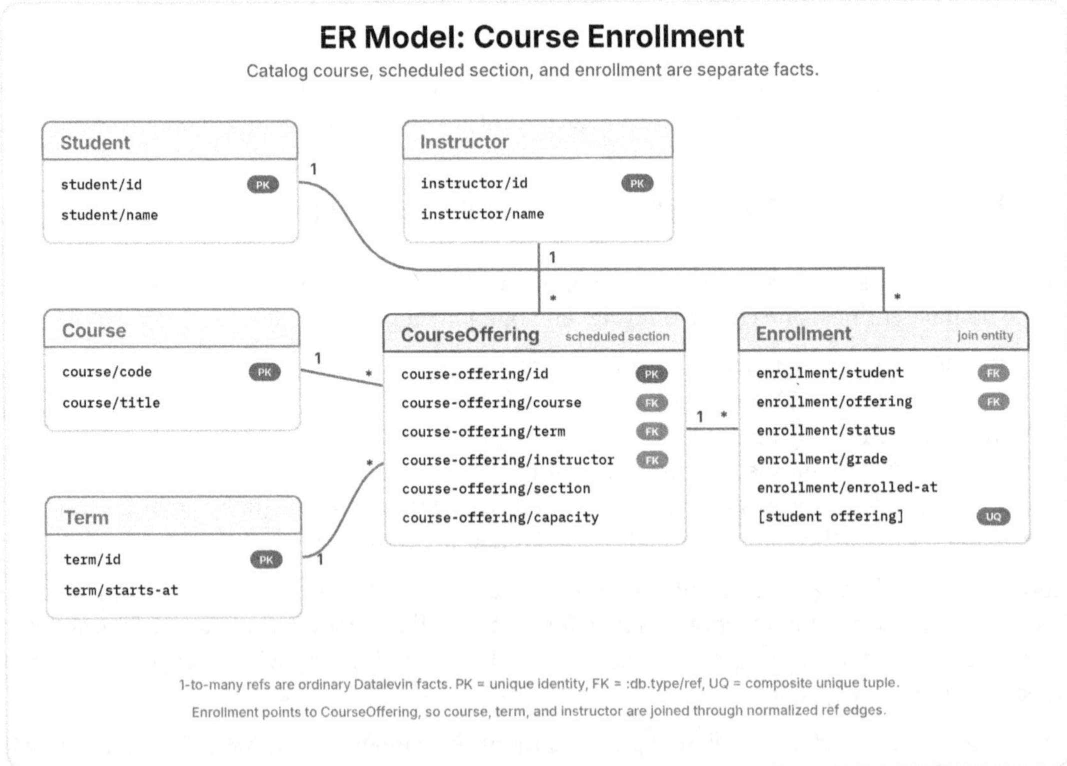

Figure 12.1: ER model for course enrollment: Student, Course, Term, and Instructor are strong entities; CourseOffering represents a scheduled section by referencing Course, Term, and Instructor; Enrollment joins Student to CourseOffering and carries status, grade, enrollment time, and a composite unique key over student and offering

```
:term/id         {:db/valueType :db.type/string
                  :db/unique    :db.unique/identity}
:term/starts-at {:db/valueType :db.type/instant}

:instructor/id   {:db/valueType :db.type/string
                  :db/unique    :db.unique/identity}
:instructor/name {:db/valueType :db.type/string}

;; A scheduled section of a catalog course.
:course-offering/id          {:db/valueType :db.type/string
                              :db/unique    :db.unique/identity}
:course-offering/course      {:db/valueType :db.type/ref}
:course-offering/term        {:db/valueType :db.type/ref}
:course-offering/instructor {:db/valueType :db.type/ref}
:course-offering/section     {:db/valueType :db.type/string}
:course-offering/capacity    {:db/valueType :db.type/long}

;; Optional: enforce one offering per course/term/section.
:course-offering/key         {:db/tupleAttrs [:course-offering/course
                                              :course-offering/term
                                              :course-offering/section]
                              :db/unique     :db.unique/identity
                              :db/doc        "Uniqueness key for one scheduled
                               ↪  course section."}

;; Enrollment is the relationship between a student and an offering.
:enrollment/student      {:db/valueType :db.type/ref}
:enrollment/offering     {:db/valueType :db.type/ref}

;; Optional: enforce one enrollment per student/offering.
:enrollment/key          {:db/tupleAttrs [:enrollment/student
                                          :enrollment/offering]
                          :db/unique     :db.unique/identity
                          :db/doc        "Uniqueness key for one student's
                           ↪  enrollment in one offering."}

:enrollment/status       {:db/valueType :db.type/keyword
                          :db/doc       "Current lifecycle state of the
                           ↪  enrollment."}
:enrollment/grade        {:db/valueType :db.type/string}
:enrollment/enrolled-at {:db/valueType :db.type/instant}})
```

The refs are the relationship model. A course offering points to the course, term, and instructor entities. An enrollment points to the student and offering entities. Ordinary joins recover the corresponding domain identifiers when needed. There is no need to copy `:course/code`, `:term/id`, or `:instructor/id` onto the enrollment only to make later queries convenient. If you need uniqueness for either relationship, derive a composite tuple from the refs themselves.

```
(d/transact! conn
  [{:student/id "s-100" :student/name "Ada"}
   {:course/code "CS101" :course/title "Intro to Databases"}
```

```
{:term/id "2026-fall"}
{:instructor/id "i-10" :instructor/name "Grace Hopper"}

{:course-offering/id         "2026-fall-CS101-001"
 :course-offering/course     [:course/code "CS101"]
 :course-offering/term       [:term/id "2026-fall"]
 :course-offering/instructor [:instructor/id "i-10"]
 :course-offering/section    "001"
 :course-offering/capacity   30}

{:enrollment/student     [:student/id "s-100"]
 :enrollment/offering    [:course-offering/id "2026-fall-CS101-001"]
 :enrollment/status      :enrollment.status/active}])
```

To update that enrollment later, find the join entity by joining through the domain identifiers. The composite `:enrollment/key` still enforces one student/offering pair, but its tuple stores the resolved entity ids of those two refs. Code that starts with domain identifiers such as `"s-100"` and `"CS101"` therefore joins through those identifiers first instead of using `[:enrollment/key ...]` directly.

The query moves from course and term to the offering, then from offering and student to enrollment:

```
(let [enrollment
      (d/q '[:find ?enrollment .
             :in $ ?student-id ?course-code ?term-id ?section
             :where
             [?student :student/id ?student-id]
             [?course :course/code ?course-code]
             [?term :term/id ?term-id]
             [?offering :course-offering/course ?course]
             [?offering :course-offering/term ?term]
             [?offering :course-offering/section ?section]
             [?enrollment :enrollment/student ?student]
             [?enrollment :enrollment/offering ?offering]]
           (d/db conn)
           "s-100"
           "CS101"
           "2026-fall"
           "001")]
  (d/transact! conn
    [[:db/add enrollment :enrollment/grade "A"]]))
```

The read query uses the same ref edges:

```
(d/q '[:find ?student-name ?course-title ?section ?instructor-name ?status
       :in $ ?term-id
       :where
       [?term :term/id ?term-id]
       [?offering :course-offering/term ?term]
       [?offering :course-offering/course ?course]
       [?offering :course-offering/section ?section]
       [?offering :course-offering/instructor ?instructor]
```

```
    [?enrollment :enrollment/offering ?offering]
    [?enrollment :enrollment/student ?student]
    [?enrollment :enrollment/status ?status]
    [?student :student/name ?student-name]
    [?course :course/title ?course-title]
    [?instructor :instructor/name ?instructor-name]]
  (d/db conn)
  "2026-fall")
```

This example illustrates the layers that often appear together in a real model:

1. **Domain entities** such as students, courses, terms, and instructors.
2. **Scheduled or contextual entities** such as course offerings, reservations, shipments, or price lists. These are real things in the domain, not merely attributes copied onto another entity.
3. **Join entities** such as enrollments, line items, memberships, and assignments.
4. **Composite identities** for join entities that need upsert, lookup, or import stability.

Use composite identities as constraints and access paths, not as a reason to duplicate facts. When the relationship is already expressed by refs, Datalevin's joins are the natural way to move from external domain keys to the relationship entity.

5. Documenting the Schema: `:db/doc`

> **Schema is maintenance documentation**
>
> A schema is for both the database engine and the developers who will maintain the system for years to come. Important attributes should be documented, especially when their meaning is not obvious from the name alone.

Datalevin supports a `:db/doc` property in the schema map. Use it to explain the purpose and constraints of an attribute.

```
(def schema
  {:user/email {:db/valueType :db.type/string
                :db/unique    :db.unique/identity
                :db/doc       "The primary unique identifier for a user
                 ↪ account."}
   :order/total {:db/valueType :db.type/bigdec
                 :db/doc       "The total price of the order in USD, including
                  ↪ tax."}})
```

Think of `:db/doc` as "comments that live in the database." They can be queried and used to generate documentation automatically.

This matters even more when LLMs or other tools help write queries. SQL databases usually attach comments to table and column objects, and the exact mechanism differs by vendor. Datalevin attaches documentation to the attribute itself. The same attribute name that appears in a Datalog clause can carry its meaning, constraints, and domain intent in schema metadata. A query assistant can inspect `:db/doc` before choosing between attributes such as `:order/total`,

`:order/subtotal`, and `:order/tax`, instead of guessing from names alone. Good `:db/doc` strings therefore improve both human maintenance and machine-assisted query generation.

6. From Tables to Facts: Migrating SQL Models

Migrating from SQL to Datalevin is not a rejection of relational thinking. Datalevin keeps the relational principle that facts can be joined by shared values, but represents data as explicit datoms that can be joined, traversed, pulled, counted, sampled, and indexed in several ways.

When a SQL database adds JSON, text search, vector indexes, or graph-like extensions, it may reduce the number of services you run, but the application is still written against tables and rows as the primary interface. Datalevin starts from facts and uses Datalog as the surface language, so migration is not merely a change of syntax; it is a change in what the database exposes as its basic unit of reasoning.

Figure 12.2 shows the conceptual steps that are often involved in such a migration.

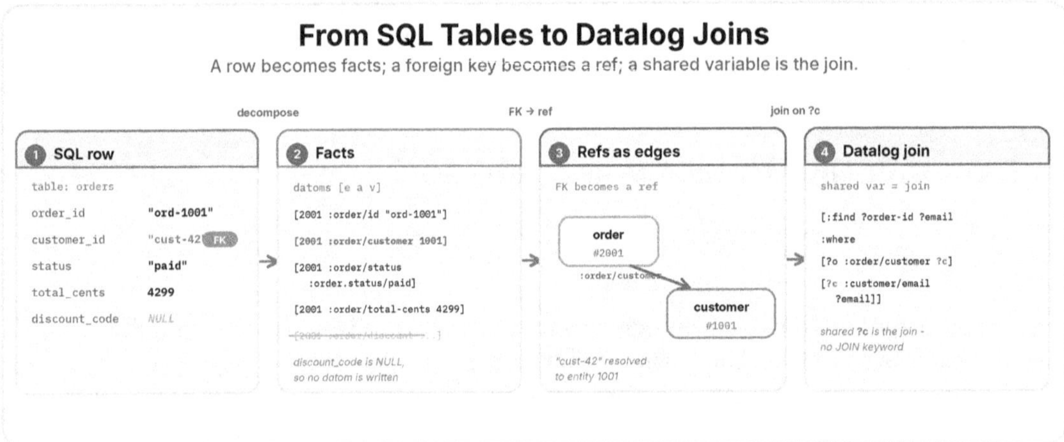

Figure 12.2: From SQL tables to Datalog joins: a SQL orders row decomposes into datoms on entity 2001, where its NULL discount_code column produces no datom at all; its customer_id foreign key becomes an :order/customer ref edge to the customer entity; and a Datalog query joins by sharing the variable ?c between the order and customer patterns, with no JOIN keyword

6.1 Translate the Vocabulary

The familiar SQL concepts still have Datalevin equivalents, but the boundaries move from tables to attributes and facts:

SQL Concept	Datalevin Equivalent	Migration Note
Table	**Namespace** such as `:order/`	A namespace groups attributes by domain meaning; it is not a physical table.
Row	**entity id**	An entity is the set of facts that share the same entity id.
Column	**Attribute**	Attributes are schema objects with their own value type, cardinality, uniqueness, and index settings.

SQL Concept	Datalevin Equivalent	Migration Note
Primary Key	**Unique identity attribute**	Use stable domain IDs with `:db.unique/identity`; keep Datalevin entity ids internal.
Foreign Key	**Ref attribute**	A `:db.type/ref` stores another entity id and joins naturally in Datalog.
Join Table	**Join entity**	Model the join as its own entity when the relationship is large, queried directly, or has attributes.
NULL	**No datom**	Datalevin does not store `nil/null`; absence means the fact is not present.

6.2 Decompose Rows into Facts

In SQL, a row is the unit that receives a full set of column values:

In Datalevin, the transaction can still be written as an entity map, but the stored representation is a set of separate facts. The unknown `discount_code` is left out of the map rather than set to SQL `NULL`:

```
{:order/id          "ord-1001"
 :order/customer    [:customer/id "cust-42"]
 :order/status      :order.status/paid
 :order/total-cents 4299}
```

Conceptually, after lookup refs are resolved, the database contains datoms like:

```
[2001 :order/id "ord-1001"]
[2001 :order/customer 1001]
[2001 :order/status :order.status/paid]
[2001 :order/total-cents 4299]
```

Notice there is no `:order/discount-code` datom at all. Where SQL stores a `NULL` in the `discount_code` cell, Datalevin stores nothing: the fact is simply absent. This is why sparse and evolving domains work well in Datalevin. Optional fields do not require nullable columns. If a discount code, shipment, or cancellation reason is unknown, the corresponding datom is absent.

6.3 Recast Joins as Shared Variables

The most visible change for SQL developers is query syntax. SQL names tables and join conditions explicitly:

Datalog describes the facts that must be true. The join happens because `?customer-eid` is shared by the order and customer patterns:

```
[:find ?order-id ?email
 :where [?order-eid :order/id ?order-id]
        [?order-eid :order/status :order.status/paid]
        [?order-eid :order/customer ?customer-eid]
        [?customer-eid :customer/email ?email]]
```

This is the same relational idea expressed declaratively. There is no `JOIN` keyword because unification over shared variables is the join.

Aggregations follow the same principle. SQL uses `GROUP BY`; Datalog groups by the non-aggregate variables in `:find`:

```
[:find ?status (count ?order-eid)
 :where [?order-eid :order/status ?status]]
```

6.4 Model Foreign Keys and Join Tables Deliberately

One-to-many relationships map directly to a reference on the "many" side, such as `:order/customer` or `:comment/post`.

Many-to-many relationships require a choice:

1. Use a cardinality-many ref only when the set is small, bounded, directly owned by the entity, and has no attributes of its own.
2. Put a cardinality-one ref on the many side when that represents the relationship directly.
3. Use a join entity when the relationship can grow large, needs facts, lifecycle, uniqueness, or direct queries.

For SQL migrations, join tables usually become join entities. A `user_groups` table with `user_id`, `group_id`, `role`, and `joined_at` is more than a link; it is a membership entity.

6.5 Use a Practical Migration Checklist

For an existing SQL application, migrate in this order:

1. Map each table to a namespace and each column to an attribute.
2. Preserve stable primary keys as `:db.unique/identity` attributes.
3. Convert foreign keys to `:db.type/ref` attributes that point to lookup refs.
4. Convert join tables with payload columns into join entities.
5. Treat SQL `NULL` as an omitted datom by default; revisit only sentinel values or columns where `NULL` carried domain-specific meaning.
6. Add `:db/doc` to attributes whose meaning came from SQL constraints, comments, or application conventions.
7. Validate imports by comparing row counts, relationship counts, uniqueness constraints, and representative business queries.

6.6 Example: A Normalized E-commerce Schema

Suppose the SQL source has four familiar tables: `products`, `customers`, `orders`, and `line_items`. A direct migration might be tempted to make an order entity contain a nested list of products, quantities, and prices. That would recreate an object graph, not a normalized relational model.

In Datalevin, keep the same conceptual separation that made the SQL model useful. Products and customers are stable domain entities. Orders are events or business records that point to customers. Line items are join entities: each one connects one order to one product and carries facts about that relationship, such as quantity. The result is still relational, but the joins are expressed as Datalog variable sharing over refs rather than SQL `JOIN` clauses over table names.

```
(def ecommerce-schema
  {;; Noun: Product
```

```
 :product/sku   {:db/unique    :db.unique/identity
                 :db/valueType :db.type/string}
 :product/title {:db/fulltext  true
                 :db/valueType :db.type/string}
 :product/price {:db/valueType :db.type/long}

 ;; Noun: Customer
 :customer/id    {:db/unique    :db.unique/identity
                  :db/valueType :db.type/string}
 :customer/email {:db/unique    :db.unique/identity
                  :db/valueType :db.type/string}

 ;; Noun: Order
 :order/id       {:db/unique    :db.unique/identity
                  :db/valueType :db.type/string}
 :order/customer {:db/valueType :db.type/ref}

 ;; Join entity: Line Item (joins Order and Product)
 :line-item/order    {:db/valueType :db.type/ref}
 :line-item/product  {:db/valueType :db.type/ref}
 :line-item/quantity {:db/valueType :db.type/long}})
```

The key decision is the `:line-item/` namespace. A line item is not a product attribute and not an order attribute; it is the relationship between an order and a product. Modeling it as its own entity lets the application ask precise questions: which products appeared in paid orders, which customers bought a given SKU, how many units of each product were sold, or which orders contain backordered items. Those queries can start from whichever fact is selective and join through the two refs.

The schema also shows how migration preserves stable identifiers without exposing Datalevin entity ids. `:product/sku`, `:customer/id`, and `:order/id` are unique identity attributes, so imported rows and application commands can use lookup refs. `:order/customer`, `:line-item/order`, and `:line-item/product` store entity ids internally, but callers can transact them using the domain keys they already have.

This is intentionally only the starting point. A production schema might add order status, timestamps, shipment entities, payment attempts, discounts, inventory reservations, or a composite `:line-item/key` over order and product when the domain allows only one line per product per order. The modeling rule is the same: keep durable nouns as entities, represent foreign keys as refs, and promote relationships with their own facts into join entities.

Summary: Relational Best Practices

1. **Think in singular namespaces**: `:user/email`, not `:users/emails`.
2. **Choose stable keys**: Use `:db.unique/identity` for domain identifiers you will use in imports, APIs, and lookup refs.
3. **Normalize relationships**: Prefer many-side refs and join entities for large or directly queried relationships; see Chapter 11, Section 2.2.
4. **Model ownership carefully**: Use `:db/isComponent` for true single-owner lifecycle ownership, not for convenient nesting alone.

5. **Represent subtypes as facts**: Use enum entities or subtype-specific namespaces rather than inheritance-table patterns.
6. **Distinguish composite attributes from tuple values**: Use `:db/tupleAttrs` for a domain identity or measured access path, and stored tuple types for compact value objects; Chapter 11, Section 1.2 covers the composite-index mechanics.
7. **Document as you go**: Use `:db/doc` to encode the "why" behind every attribute, especially derived or denormalized facts.
8. **Migrate SQL incrementally**: Preserve stable keys, convert foreign keys to refs, and validate migrated facts with the business queries the application already depends on.

By applying these ER principles, you keep your Datalevin database explicit, queryable, and understandable as your domain grows.

References

[1] E. F. Codd, "Further Normalization of the Data Base Relational Model," IBM Research Report RJ909, August 31, 1971. Republished in Randall J. Rustin, ed., *Data Base Systems: Courant Computer Science Symposia Series* 6, Prentice-Hall, 1972.

[2] Peter Pin-Shan Chen, "The Entity-Relationship Model: Toward a Unified View of Data," *ACM Transactions on Database Systems* 1(1):9-36, 1976. DOI: https://doi.org/10.1145/320434.320440.

CHAPTER 13

Graph Modeling

In many databases, "graph" is a specialized mode or a separate engine. In Datalevin, graph-shaped data is part of the normal fact model. Every `:db.type/ref` you define is a directed edge between two nodes (entities).

> **Graph cost follows fan-out**
>
> Because Datalevin maintains EAV and AVE indexes, both forward and reverse navigation have indexed access paths. Actual cost still depends on fan-out, selectivity, and result size: an indexed edge that reaches millions of neighbors still returns millions of candidates.

This chapter shows how to use Datalevin for graph-shaped data, focusing on navigation, recursive patterns, and modeling complex relationships.

1. Entities as Nodes, Refs as Edges

Figure 13.1 depicts a graph model as a network of interconnected entities.

- **Nodes**: Entities with their properties (attributes).
- **Edges**: Reference attributes (`:db.type/ref`) that point from one entity to another.

1.1 Forward and Reverse Navigation

Datalevin allows you to traverse relationships in both directions without any additional indexing. The query planner chooses the right index for navigation:

- **Forward**: Starting from the known entity position and following a reference attribute to the value position using the EAV index.
  - Pattern: `[?bound-eid :user/follows ?unknown-eid]`
- **Reverse**: Starting from a known value position and finding entities whose reference attribute has that value using the AVE index.
  - Pattern: `[?unknown-eid :user/follows ?bound-eid]`

This works well because most queries in a well-designed data model know the attribute they are traversing.

2. Modeling Complex Edges (Join Entities)

In a simple graph, an edge can be a direct reference such as `:user/follows`. When the edge itself has facts, promote the edge to an entity. Property-graph literature models this as an edge

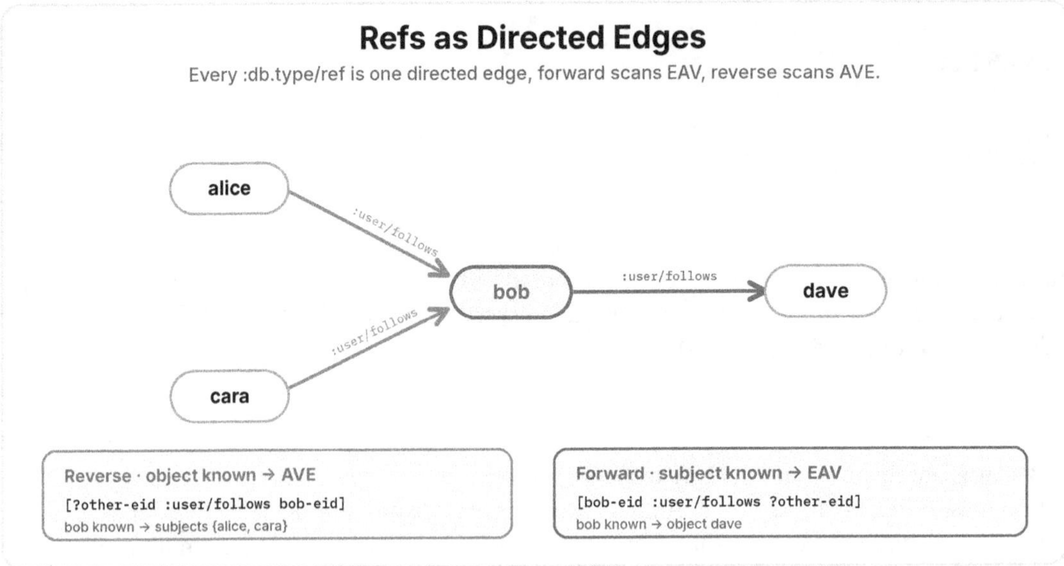

Figure 13.1: Refs as directed edges: alice and cara follow bob, and bob follows dave, via :user/follows. Forward navigation starts from a known entity position and follows the reference to the value position using the EAV index; reverse navigation starts from a known value position and finds matching entities using the AVE index

with properties. In Datalevin, model it as a join entity: ordinary entities connected by ordinary ref attributes.

In Figure 13.2, the direct edge `A -> B` becomes a `Follow` entity with two refs, `:follow/follower` and `:follow/following`, plus facts such as `:follow/created-at` or `:follow/strength`. This gives you property-graph-style modeling [1] while keeping Datalevin's normalized fact representation [2]. Chapter 12 shows the same move in ER examples such as memberships, line items, and enrollments.

3. Graph Traversal Patterns

Graph queries range from fixed-depth joins to recursive reachability, branching path shapes, and exclusions. Datalevin uses ordinary Datalog joins for fixed-depth traversals, recursive rules for paths of arbitrary length, and `or-join` or `not-join` when graph shapes branch or exclude one another.

Fixed-depth traversals are ordinary joins. To find people followed by the people you follow:

```
(d/q '[:find ?suggestion-eid
       :in $ ?user-eid
       :where
       [?user-eid :user/follows ?intermediate-eid]
       [?intermediate-eid :user/follows ?suggestion-eid]
       [(not= ?user-eid ?suggestion-eid)]]
     db user-eid)
```

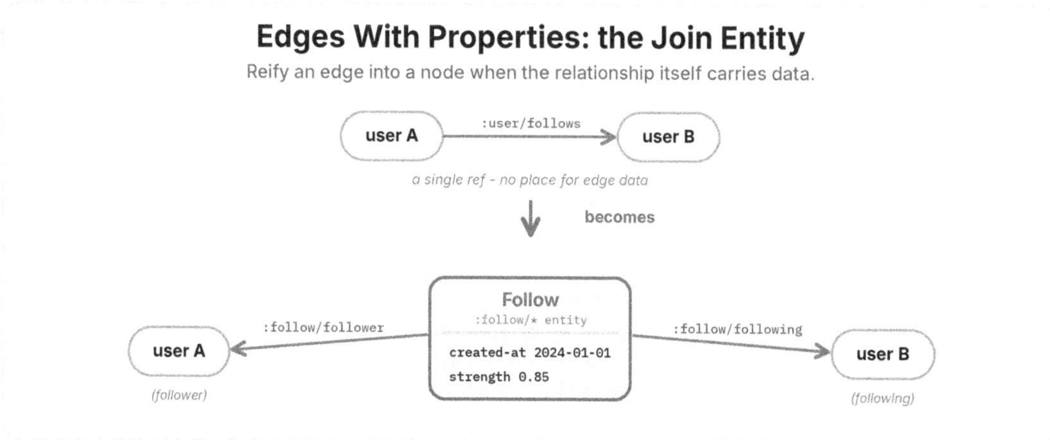

Figure 13.2: Reifying an edge into a join entity: a simple user A :user/follows user B edge becomes a Follow entity that references user A as :follow/follower and user B as :follow/following, and carries properties such as created-at and strength on the relationship itself

3.1 Deep Traversal: Transitive Closure

Transitive closure means the full set of nodes reachable by following the same edge relation one or more times. Add a reflexive base case when the start node should also count as reachable from itself. In practical terms, transitive closure is how you find connections at any distance, such as all reachable nodes in a network. Use recursive rules for transitive closure:

```
(def reachability-rules
  '[[(reachable ?start-eid ?end-eid)
     [?start-eid :link/to ?end-eid]]

    [(reachable ?start-eid ?end-eid)
     [?start-eid :link/to ?mid-eid]
     (reachable ?mid-eid ?end-eid)]])
```

Datalevin evaluates recursive rules to a fixpoint: the point where another round of rule evaluation produces no new result tuples. That means cyclic data does not create a recursive call-stack loop. A fixpoint can still contain many derived tuples, so constrain broad traversals with selective starting facts, bounded rules, or application-level limits when the graph can fan out widely.

3.2 Hierarchies and Taxonomies

Recursive rules are also a natural fit for tree-like graph structures. The same pattern works for place hierarchies, product categories, organization charts, and tag taxonomies. A city belongs to a country through `:place/isPartOf`, and a tag class belongs to a broader class through `:tagclass/isSubclassOf`.

```
(def place-country-rules
  '[[(place-country ?place-eid ?country-eid)
     [?place-eid :place/type "Country"]
     [(ground ?place-eid) ?country-eid]]

    [(place-country ?place-eid ?country-eid)
```

```
    [?place-eid :place/isPartOf ?parent-eid]
    (place-country ?parent-eid ?country-eid)]])

(d/q '[:find ?country-name
       :in $ % ?place-name
       :where
       [?place-eid :place/name ?place-name]
       (place-country ?place-eid ?country-eid)
       [?country-eid :place/name ?country-name]]
     db place-country-rules "Paris")
```

The base case uses `[(ground ?place-eid) ?country-eid]` as an identity binding: after `?place-eid` is known to be a country, the rule copies that entity id into `?country-eid`. In other words, a country is its own country for the purpose of this rule, while non-country places recurse through `:place/isPartOf`.

This rule works for both direct and indirect containment. If `Paris` is part of `France`, and `France` is part of `Europe`, the rule stops at the first ancestor whose `:place/type` is `"Country"`.

3.3 Bounded Neighborhoods with Distance

Some graph queries should not traverse arbitrarily far. They ask for neighbors within a fixed radius, while preserving the distance so the final result can be ordered or aggregated.

Listing 13.1 shows one bounded-neighborhood pattern: unroll the fixed distances into separate rule bodies and bind the distance explicitly.

In this example, friendship is a join entity with two reference attributes, `:knows/person1` and `:knows/person2`, following the Section 2 pattern. This is also how LDBC SNB models friendship. The queries treat the edge as directional from `:knows/person1` to `:knows/person2`; for mutual friendship, store both orientations or query both orientations.

Listing 13.1. Bounded graph neighborhood rules with distance

```
(def friends-3-rules
  '[[(friends-3 ?start-eid ?friend-eid ?dist)
     [?knows-eid :knows/person1 ?start-eid]
     [?knows-eid :knows/person2 ?friend-eid]
     [(ground 1) ?dist]]

    [(friends-3 ?start-eid ?friend-eid ?dist)
     [?knows1-eid :knows/person1 ?start-eid]
     [?knows1-eid :knows/person2 ?mid-eid]
     [?knows2-eid :knows/person1 ?mid-eid]
     [?knows2-eid :knows/person2 ?friend-eid]
     [(not= ?friend-eid ?start-eid)]
     [(ground 2) ?dist]]

    [(friends-3 ?start-eid ?friend-eid ?dist)
     [?knows1-eid :knows/person1 ?start-eid]
     [?knows1-eid :knows/person2 ?mid1-eid]
     [?knows2-eid :knows/person1 ?mid1-eid]
     [?knows2-eid :knows/person2 ?mid2-eid]
```

```
    [(not= ?mid2-eid ?start-eid)]
    [?knows3-eid :knows/person1 ?mid2-eid]
    [?knows3-eid :knows/person2 ?friend-eid]
    [(not= ?friend-eid ?start-eid)]
    [(not= ?friend-eid ?mid1-eid)]
    [(ground 3) ?dist]]])

(d/q '[:find (min ?dist) ?friend-id
       :in $ % ?person-id
       :where
       [?start-eid :person/id ?person-id]
       (friends-3 ?start-eid ?friend-eid ?dist)
       [?friend-eid :person/id ?friend-id]]
     db friends-3-rules person-id)
```

The rule emits one row for each path shape. The query uses `(min ?dist)` to keep the shortest distance for each reachable person.

3.4 Mutual Recursion: Alternating Edge Types

Some graph traversals are not a single repeated edge. Datalevin's rule engine also supports mutually recursive rules, where two or more rules call each other until the result reaches a fixpoint. This pattern is useful for alternating path semantics.

For example, suppose a graph has two edge types, `:edge/a` and `:edge/b`, and you want to find paths that end with an `:edge/a` step after alternating between the two edge types:

```
(def alternating-rules
  '[[(path-a ?from-eid ?to-eid)
     [?from-eid :edge/a ?to-eid]]

    [(path-a ?from-eid ?to-eid)
     [?mid-eid :edge/a ?to-eid]
     (path-b ?from-eid ?mid-eid)]

    [(path-b ?from-eid ?to-eid)
     [?from-eid :edge/b ?to-eid]]

    [(path-b ?from-eid ?to-eid)
     [?mid-eid :edge/b ?to-eid]
     (path-a ?from-eid ?mid-eid)]])

(d/q '[:find ?to-eid
       :in $ % ?from-eid
       :where
       (path-a ?from-eid ?to-eid)]
     db alternating-rules start-eid)
```

`path-a` depends on `path-b`, and `path-b` depends on `path-a`. Datalevin evaluates these rules together, rather than expanding one call stack top-down. That is what allows mutually recursive graph definitions to remain practical.

3.5 Branching Traversals with `or-join`

Complex queries often need to combine several graph shapes into one logical binding. For graph neighborhoods such as "people within two hops" from a start person, where a person can be reached either directly or through one intermediate friend, `or-join` can be used to cover the alternatives:

```
(d/q '[:find ?person-eid
       :in $ ?person-id
       :where
       [?start-eid :person/id ?person-id]
       (or-join [?start-eid ?person-eid]
                ;; one hop
                (and [?knows-eid :knows/person1 ?start-eid]
                     [?knows-eid :knows/person2 ?person-eid])

                ;; two hops
                (and [?knows1-eid :knows/person1 ?start-eid]
                     [?knows1-eid :knows/person2 ?mid-eid]
                     [?knows2-eid :knows/person1 ?mid-eid]
                     [?knows2-eid :knows/person2 ?person-eid]
                     [(not= ?person-eid ?start-eid)]))]
     db person-id)
```

Use `or-join` when the alternatives bind the same logical result through different graph patterns. The join variables, here `?start-eid` and `?person-eid`, tell the query engine which values must be shared between the branch and the rest of the query. Each `or-join` branch that contains more than one clause is wrapped in `(and ...)`, because the branch itself must be one syntactic alternative. This is especially useful before aggregating over messages, forums, tags, or memberships reached through several allowed path shapes.

3.6 Excluding Graph Shapes with `not-join`

Recommendation queries often need an **anti-join**: keep candidates from one pattern only when no matching fact exists in another pattern. In graph terms, this means finding nodes reachable by one path, but excluding nodes reachable by another path. A common example is friends-of-friends who are not already direct friends.

```
(d/q '[:find ?person-id
       :in $ ?start-person-id
       :where
       [?start-eid :person/id ?start-person-id]

       ;; exactly two hops
       [?knows1-eid :knows/person1 ?start-eid]
       [?knows1-eid :knows/person2 ?mid-eid]
       [?knows2-eid :knows/person1 ?mid-eid]
       [?knows2-eid :knows/person2 ?person-eid]
       [(not= ?person-eid ?start-eid)]

       ;; exclude direct friends
       (not-join [?start-eid ?person-eid]
```

```
                [?knows-eid :knows/person1 ?start-eid]
                [?knows-eid :knows/person2 ?person-eid])

       [?person-eid :person/id ?person-id]]
     db start-person-id)
```

Use `not-join` when the negated graph pattern depends on a specific set of outer variables. Here the exclusion is about the pair `?start-eid` and `?person-eid`; it should not depend on unrelated variables that happen to appear elsewhere in the query.

4. More Graph Problem Shapes

The previous examples cover social networks and hierarchies, but the same modeling tools apply to many graph problems. The important step is to identify what the nodes are, what the edges mean, and whether the edge needs its own facts.

4.1 Co-Appearance and Degrees of Separation

The "six degrees of Kevin Bacon" problem, depicted in Figure 13.3, is a useful graph example because it is not a tree. It is a dense bipartite graph: two kinds of nodes, people and movies, with many connections between the two sets. People connect to movies through credits, and two people are adjacent if they worked on the same movie. Andrew Dennis used this problem to show both Datalog joins and application-level breadth-first search over database facts [3]. The same idea applies naturally to Datalevin.

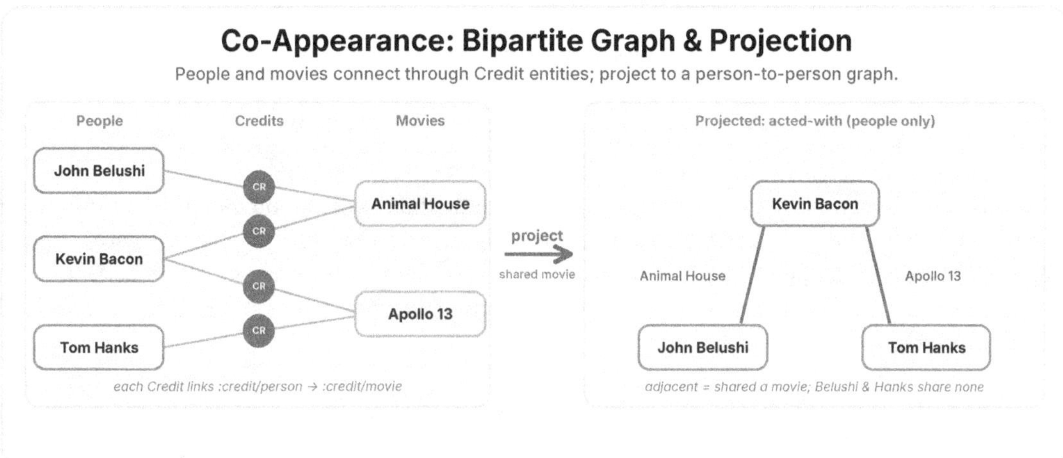

Figure 13.3: Co-appearance as a bipartite graph: people (Kevin Bacon, John Belushi, Tom Hanks) and movies (Animal House, Apollo 13) connected through Credit entities, projected to a person-to-person acted-with graph where Bacon links to Belushi via Animal House and to Hanks via Apollo 13, while Belushi and Hanks share no movie

> **Model co-appearance through relationship nodes**
>
> Model the credit as an entity, not as redundant lists on both actors and movies. The credit is the durable fact that connects a person to a movie and can carry role, billing, source, or time facts later.

```
(def movie-schema
  {:person/name    {:db/valueType :db.type/string
                    :db/unique    :db.unique/identity}
   :movie/title    {:db/valueType :db.type/string
                    :db/unique    :db.unique/identity}
   :movie/year     {:db/valueType :db.type/long}
   :credit/person  {:db/valueType :db.type/ref}
   :credit/movie   {:db/valueType :db.type/ref}
   :credit/role    {:db/valueType :db.type/keyword}})
```

A direct co-appearance query is a join through `:credit/movie`:

```
(d/q '[:find ?movie-title
       :in $ ?name-1 ?name-2
       :where
       [?person1-eid :person/name ?name-1]
       [?person2-eid :person/name ?name-2]
       [?credit1-eid :credit/person ?person1-eid]
       [?credit1-eid :credit/movie ?movie-eid]
       [?credit2-eid :credit/person ?person2-eid]
       [?credit2-eid :credit/movie ?movie-eid]
       [(not= ?person1-eid ?person2-eid)]
       [?movie-eid :movie/title ?movie-title]]
     db
     "Kevin Bacon"
     "John Belushi")
```

That answers "Bacon number 1" questions. For a fixed small radius, you can add more joins or generate bounded rules, as shown earlier in this chapter. For an unknown shortest path, use the database to supply neighbors and run a standard graph algorithm in the application. Datalevin makes this practical because the same current database snapshot can be queried repeatedly and refs are indexed.

Listing 13.2. Shortest-path helper over Datalevin graph facts

```
(defn acted-with
  "Return [other-eid movie-eid] pairs adjacent to actor-eid."
  [db actor-eid]
  (d/q '[:find ?other-eid ?movie-eid
         :in $ ?actor-eid
         :where
         [?credit1-eid :credit/person ?actor-eid]
         [?credit1-eid :credit/movie ?movie-eid]
         [?credit2-eid :credit/movie ?movie-eid]
         [?credit2-eid :credit/person ?other-eid]
         [(not= ?actor-eid ?other-eid)]]
```

```
      db actor-eid))

(defn entity-label [db eid]
  (let [m (d/pull db '[:person/name :movie/title] eid)]
    (or (:person/name m)
        (:movie/title m))))

(defn bacon-path
  "Return one shortest alternating [actor movie actor ...] path."
  [db start-name target-name]
  (let [start-eid  (d/q '[:find ?person-eid .
                           :in $ ?name
                           :where [?person-eid :person/name ?name]]
                         db start-name)
        target-eid (d/q '[:find ?person-eid .
                           :in $ ?name
                           :where [?person-eid :person/name ?name]]
                         db target-name)
        edges   (memoize #(acted-with db %))]
    (loop [queue (conj clojure.lang.PersistentQueue/EMPTY [start-eid])
           seen  #{start-eid}]
      (when-let [path (peek queue)]
        (let [actor-eid (peek path)]
          (if (= actor-eid target-eid)
            (mapv (partial entity-label db) path)
            (let [next-edges (remove #(seen (first %)) (edges actor-eid))]
              (recur
                (into (pop queue)
                      (map (fn [[other-eid movie-eid]]
                             (conj path movie-eid other-eid))
                           next-edges))
                (into seen (map first next-edges))))))))))
```

Listing 13.2 treats Datalevin as the graph store and keeps shortest-path policy in ordinary code. That is often a practical split: use Datalog for the local neighborhood relation, and use a graph algorithm when the query needs a global search strategy.

4.2 Dependency and Impact Analysis

Package managers, build systems, data pipelines, and microservice deployments often need the reverse of a dependency graph question. If `core` changes, what must be rebuilt? If a table changes, which reports are affected?

Model each dependency as an edge from the dependent item to the thing it uses:

```
{:component/name {:db/valueType :db.type/string
                  :db/unique    :db.unique/identity}
 :component/uses {:db/valueType :db.type/ref}}
```

Then define reachability over `:component/uses`:

```
(def dependency-rules
  '[[(depends-on ?component-eid ?dependency-eid)
```

```
  [?component-eid :component/uses ?dependency-eid]]

 [(depends-on ?component-eid ?dependency-eid)
  [?component-eid :component/uses ?direct-eid]
  (depends-on ?direct-eid ?dependency-eid)]])
```

To find everything impacted by a changed component, bind the dependency side and ask which components depend on it:

```
(d/q '[:find ?impacted-name
       :in $ % ?changed-name
       :where
       [?changed-eid :component/name ?changed-name]
       (depends-on ?impacted-eid ?changed-eid)
       [?impacted-eid :component/name ?impacted-name]]
     db dependency-rules "core")
```

The same pattern works for "which policies inherit this rule?", "which dashboards read from this dataset?", and "which generated files depend on this source file?"

4.3 Weighted Paths and Route Search

Some graph questions are about reachability; others are about optimal paths. Route planning, network latency, risk propagation, and cost-based dependency planning all attach weights to edges.

Model a weighted edge as an entity:

```
{:route/from {:db/valueType :db.type/ref}
 :route/to   {:db/valueType :db.type/ref}
 :route/km   {:db/valueType :db.type/double}}
```

Datalog can enumerate bounded path shapes and sum their weights when the bound is small:

```
(d/q '[:find ?mid-name ?km
       :in $ ?from-name ?to-name
       :where
       [?from-eid :place/name ?from-name]
       [?to-eid :place/name ?to-name]
       [?route1-eid :route/from ?from-eid]
       [?route1-eid :route/to ?mid-eid]
       [?route1-eid :route/km ?km1]
       [?route2-eid :route/from ?mid-eid]
       [?route2-eid :route/to ?to-eid]
       [?route2-eid :route/km ?km2]
       [(+ ?km1 ?km2) ?km]
       [?mid-eid :place/name ?mid-name]]
     db "Paris" "Berlin")
```

Use algorithms for global path search

For unbounded weighted shortest path, use Dijkstra, A*, or another application algorithm. Let Datalevin answer the indexed neighbor query, but let the algorithm own the priority queue and stopping condition. That keeps the database model clear and avoids forcing every graph algorithm into one Datalog query.

4.4 Lineage Through Relationship Nodes

Some graphs are not stored as one obvious edge. For example, doctoral student advisor lineage is traced through dissertations: a person advises a dissertation, and the dissertation points to the student who authored it [4]. The logical edge "advisor advised student" is therefore a derived relation, not a stored attribute.

```
{:person/name          {:db/valueType :db.type/string}
 :person/advised       {:db/valueType   :db.type/ref
                        :db/cardinality :db.cardinality/many}
 :dissertation/cid     {:db/valueType :db.type/ref}
 :dissertation/univ    {:db/valueType :db.type/string}
 :dissertation/area    {:db/valueType :db.type/string}
 :dissertation/title   {:db/valueType :db.type/string}}
```

Rules give names to the useful graph relations:

```
(def academic-rules
  '[[(author ?dissertation-eid ?person-eid)
     [?dissertation-eid :dissertation/cid ?person-eid]]

    [(advised ?advisor-eid ?student-eid)
     [?advisor-eid :person/advised ?dissertation-eid]
     (author ?dissertation-eid ?student-eid)]

    [(academic-ancestor ?ancestor-eid ?student-eid)
     (advised ?ancestor-eid ?student-eid)]

    [(academic-ancestor ?ancestor-eid ?student-eid)
     (advised ?ancestor-eid ?middle-eid)
     (academic-ancestor ?middle-eid ?student-eid)]])
```

Now "find all academic ancestors of a person" is a graph traversal over a relation that was assembled from several attributes:

```
(d/q '[:find ?ancestor-name
       :in $ % ?student-name
       :where
       [?student-eid :person/name ?student-name]
       (academic-ancestor ?ancestor-eid ?student-eid)
       [?ancestor-eid :person/name ?ancestor-name]]
     db academic-rules "David Scott Warren")
```

This is a good pattern when the edge has a first-class domain object behind it. The dissertation is not merely a join artifact; it has title, institution, area, and year facts of its own. Keeping that node visible preserves domain meaning while rules recover the convenient graph relation.

4.5 Type Graphs and Class Derivation

Graph reasoning is not limited to social links or dependency chains. In some cases, entities have concrete `:type` facts, and rules derive membership in broader classes such as `:Student`, `:Faculty`, and `:Person` [5].

The attributes in this section and the next follow the compact OpenRuleBench/LUBM-style benchmark schemas, so they use names such as `:type`, `:advisor`, `:par`, and `:sib`. In application schemas, prefer the namespacing convention from Chapter 5, for example `:university/type`, `:student/advisor`, `:person/parent`, or `:person/sibling`.

```
(def university-type-rules
  '[[(is-a ?x-eid ?class)
     [?x-eid :type :GraduateStudent]
     [(ground :Student) ?class]]
    [(is-a ?x-eid ?class)
     [?x-eid :type :UndergraduateStudent]
     [(ground :Student) ?class]]
    [(is-a ?x-eid ?class)
     (is-a ?x-eid ?student-class)
     [(= ?student-class :Student)]
     [(ground :Person) ?class]]

    [(is-a ?x-eid ?class)
     [?x-eid :type :FullProfessor]
     [(ground :Professor) ?class]]
    [(is-a ?x-eid ?class)
     [?x-eid :type :AssociateProfessor]
     [(ground :Professor) ?class]]
    [(is-a ?x-eid ?class)
     (is-a ?x-eid ?professor-class)
     [(= ?professor-class :Professor)]
     [(ground :Faculty) ?class]]
    [(is-a ?x-eid ?class)
     (is-a ?x-eid ?faculty-class)
     [(= ?faculty-class :Faculty)]
     [(ground :Person) ?class]]])
```

A query can then ask at the semantic level it cares about:

```
(d/q '[:find ?student-eid ?advisor-eid
       :in $ %
       :where
       (is-a ?student-eid ?class)
       [(= ?class :Student)]
       [?student-eid :advisor ?advisor-eid]]
     db university-type-rules)
```

This does not require Datalevin attributes themselves to become an ontology. The class graph is ordinary application data, and the rules describe how that application wants to interpret it.

4.6 Same-Generation Relations

A second OpenRuleBench-style derived relation is same generation over `:par` and `:sib` base relations. This is different from reachability: it derives a relation between pairs of nodes that occupy the same structural level.

```
(def same-generation-rules
  '[[(same-generation ?x-eid ?y-eid)
     [?x-eid :sib ?y-eid]]
    [(same-generation ?x-eid ?y-eid)
     [?x-eid :par ?z-eid]
     (same-generation ?z-eid ?z1-eid)
     [?y-eid :par ?z1-eid]]])
```

The base case says that sibling pairs are same-generation. The recursive case says that if `?x-eid` has a `:par` edge to `?z-eid`, `?y-eid` has a `:par` edge to `?z1-eid`, and `?z-eid` and `?z1-eid` are same-generation, then `?x-eid` and `?y-eid` are same-generation. This kind of query is useful for structural comparisons, lineage analysis, and rule-engine tests because the derived relation is about two moving positions in the graph, not one source-to-target path.

5. Example: Finding the Forum for a Message (LDBC SNB IS6)

The LDBC Social Network Benchmark (LDBC SNB) is an industry standard for evaluating graph databases. It models a social network with people, friendships, forums, posts, comments, likes, tags, organizations, and places [6]. This section examines one real Interactive Short query from the benchmark to show graph traversal in practice.

Query IS6: Given a message, find the forum that contains it and the person who moderates that forum. Because comments are not directly contained in forums, the query traverses to the original post in the thread.

Define a recursive rule to find the root post of any message:

```
(def root-post-rule
  '[;; base case
    [(root-post ?message-eid ?post-eid)
     [?message-eid :message/isContainedIn _]
     [(ground ?message-eid) ?post-eid]] ; copy the root message eid into
↪    ?post-eid

    ;; recursive case
    [(root-post ?message-eid ?post-eid)
     [?message-eid :message/replyOf ?parent-eid]
     (root-post ?parent-eid ?post-eid)]])
```

The rule has two branches:

1. If the message has `:message/isContainedIn`, it is itself the root post.
2. Otherwise, follow `:message/replyOf` recursively until the query reaches the root.

Now the query to find the forum and moderator:

Listing 13.3. Recursive graph query for LDBC SNB forum lookup

```
(d/q '[:find ?forum-id ?forum-title ?moderator-id
              ?moderator-first-name ?moderator-last-name
       :in $ % ?message-id          ; $=db, %=rules, ?message-id=input
       :where
       ;; locate the message
       [?message-eid :message/id ?message-id]

       ;; traverse to root post (recursive rule)
       (root-post ?message-eid ?post-eid)

       ;; get forum containing the post
       [?post-eid :message/isContainedIn ?forum-eid]
       [?forum-eid :forum/id ?forum-id]
       [?forum-eid :forum/title ?forum-title]

       ;; get forum moderator
       [?forum-eid :forum/hasModerator ?moderator-eid]
       [?moderator-eid :person/id ?moderator-id]
       [?moderator-eid :person/firstName ?moderator-first-name]
       [?moderator-eid :person/lastName ?moderator-last-name]]
     db root-post-rule message-id)
```

This query uses several Datalevin features together:

- **Recursive traversal**: The `root-post` rule follows reply chains of arbitrary depth.
- **Graph navigation**: Multiple hops from message to post to forum to moderator.
- **Rule composition**: The `%` parameter passes rules into the query.
- **Unified schema**: All entities use the same attribute-based model.

The Datalevin project includes an unofficial LDBC SNB SF1 comparison with Neo4j [7]. In the August 23, 2026 snapshot, run on an Apple M3 Pro with 36 GB RAM, Datalevin was faster on 13 of 14 Interactive Complex reads and all seven Interactive Short reads; Neo4j was faster on IC10. Both embedded runners used the same dataset and parameter schedule, all 21 queries matched the expected result counts and digests, and timing covered planning through full result realization.

This is a reproducible, self-published snapshot, not an official LDBC result or a general performance guarantee. See [7] for versions, the full timing table, harness, and reproduction instructions. Hardware, scale, indexes, parameters, and concurrency can change the result.

5.1 Other LDBC SNB Graph Questions

LDBC SNB contains many graph problems beyond plain traversal:

- **Temporal neighborhoods**: find friends or friends-of-friends who produced activity inside a time window, then exclude activity before that window.
- **Co-occurrence**: find tags, topics, or forums that co-occur in posts written by a social neighborhood.
- **Recommendation**: find friends-of-friends who are not direct friends, then rank them by shared interests or interaction evidence.

- **Referral and expert search**: combine social paths, organizations, places, replies, and tag-class descendants into one query.
- **Trusted paths**: enumerate bounded social paths and score them by message interactions along each edge.

These examples all use the same Datalevin building blocks: refs for graph edges, join entities when an edge has facts, rules for reusable derived relations, and `not-join` when one graph shape must exclude another. The important modeling lesson is that "graph query" often means a mix of topology, time, text, taxonomy, and ranking, and these are all readily represented in Datalevin's native data model.

Summary: Graph Design Principles

1. **Edges are Refs**: Use `:db.type/ref` for relationships between Datalevin entities.
2. **Think Navigation in Both Directions**: Forward and reverse lookups both use indexed access paths. The optimizer can choose between them automatically.
3. **Use Rules for Paths**: Encapsulate reusable traversal logic in recursive rules.
4. **Preserve Distance When Needed**: Bounded rules can return path length as data for ranking, grouping, or filtering.
5. **Use `or-join` for Branch Shapes**: Combine alternative graph patterns when several paths bind the same result.
6. **Use `not-join` for Exclusion**: Express "reachable this way but not that way" with explicit anti-joins.
7. **Use Nodes for Edge Metadata**: If an edge needs properties, make the edge an entity.
8. **Normalize for Depth**: Deep graph queries are easier to plan when nodes are small and relationships are explicit.
9. **Name Derived Graphs**: Use rules when the useful graph relation is assembled from several lower-level facts.
10. **Use Algorithms When Needed**: For arbitrary shortest paths, weighted routing, or other global graph algorithms, use Datalevin for indexed neighbor lookup and keep the search policy in application code.

By treating graph shape as part of the normal data model, you can combine navigation, rules, search, and ordinary facts without moving data into a separate graph-only system.

References

[1] Renzo Angles, "The Property Graph Database Model," *Proceedings of the 12th Alberto Mendelzon International Workshop on Foundations of Data Management*, CEUR Workshop Proceedings 2100, 2018. URL: https://ceur-ws.org/Vol-2100/paper26.pdf.

[2] E. F. Codd, "Further Normalization of the Data Base Relational Model," IBM Research Report RJ909, August 31, 1971. Republished in Randall J. Rustin, ed., *Data Base Systems: Courant Computer Science Symposia Series 6*, Prentice-Hall, 1972.

[3] Andrew Dennis, "Using Datomic as a Graph Database," Hashrocket, April 10, 2014. URL: https://hashrocket.com/blog/posts/using-datomic-as-a-graph-database.

[4] Datalevin project, "Math Bench," benchmark implementation based on the Mathematics Genealogy Project. URL: https://github.com/datalevin/datalevin/tree/master/benchmarks/math-bench.

[5] Datalevin project, "OpenRuleBench," benchmark implementation including transitive closure, same-generation, and LUBM-style type inference rules. URL: https://github.com/datalevin/datalevin/tree/master/benchmarks/openrulebench.

[6] LDBC, "LDBC Social Network Benchmark (LDBC SNB)," official benchmark overview and links to the SNB workloads, datasets, and specification. URL: https://ldbcouncil.org/benchmarks/snb/.

[7] Datalevin project, "LDBC SNB Benchmark," August 23, 2026 comparison using the build intended for Datalevin 1.1.0, methodology, and implementation, commit `0ac42023`. URL: https://github.com/datalevin/datalevin/blob/0ac42023b330ceaf478e0a31b046f5a251f50dc2/benchmarks/LDBC-SNB-bench/README.md#current-local-datalevinneo4j-comparison.

CHAPTER 14

Document Modeling

Datalevin gives you two ways to model document-shaped data: **logical documents**, built from normal datoms and component references, and **indexed documents** (`idoc`), stored as single values with path-level indexes. By default an `idoc` indexes every scalar leaf path, and schema or domain options can restrict indexing to selected path prefixes.

Figure 14.1 contrasts the two representations.

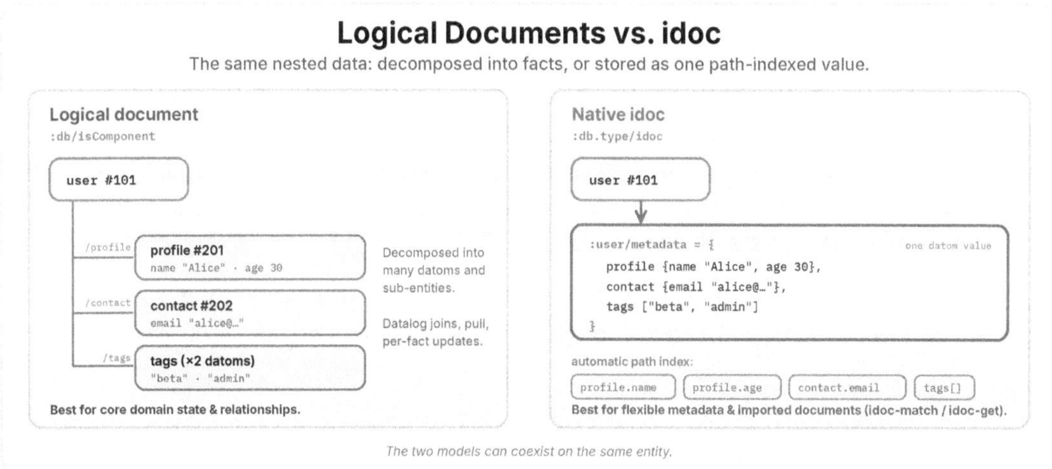

Figure 14.1: Logical documents vs. idoc, modeling the same nested user data: the logical document (:db/isComponent) decomposes it into a user entity that owns separate profile and contact component entities plus tag datoms - many facts that join, pull, and update independently; the native idoc (:db.type/idoc) stores the whole nested map as one datom value on :user/metadata with a path index over configured leaf paths, all leaf paths by default, such as profile.name, profile.age, contact.email, and tags. The two models can coexist on the same entity

Use logical documents when the nested structure is part of your core domain model. Use `idoc` when the nested structure is flexible metadata, imported JSON, Markdown-derived structure, or application-defined data that should be searched by path without forcing every field into the schema. Section 6 shows a boundary technique for querying nested data without storing it at all.

1. Logical Documents with `:db/isComponent`

The most common way to model durable domain documents in Datalevin is to use component attributes. A component reference stores a nested entity as ordinary datoms, but treats that entity as owned by its parent.

> **Components require single ownership**
>
> Use component attributes for single-owner logical documents. If a nested record can be shared by more than one parent, model it with ordinary references or as an `idoc` value instead.

1.1 Example: A Nested Blog Post

```
;; Schema: :post/comments owns comment entities.
{:post/slug      {:db/valueType :db.type/string
                  :db/unique    :db.unique/identity}
 :post/comments {:db/valueType   :db.type/ref
                  :db/cardinality :db.cardinality/many
                  :db/isComponent true}
 :user/email     {:db/valueType :db.type/string
                  :db/unique    :db.unique/identity}
 :comment/author {:db/valueType :db.type/ref}}
```

Logical documents are still normal facts. Every nested field can be typed, indexed, joined, pulled, and updated independently. This is useful when nested data has identity, relationships, constraints, or a lifecycle that should remain visible to Datalog.

For example, comments can be modeled as components of a post while their authors remain separate top-level entities, addressed through a unique email lookup ref:

```
(d/transact! conn
  [{:user/name "Ada"
    :user/email "ada@example.com"}
   {:post/slug "indexes-as-capabilities"
    :post/title "Indexes as Capabilities"
    :post/comments [{:comment/body "This clarifies the mental model."
                     :comment/author [:user/email "ada@example.com"]}]}])
```

For logical documents, use pull patterns to navigate nested paths:

```
(d/pull db
        [:post/title {:post/comments [:comment/body {:comment/author
 ↪  [:user/name]}]}]
        [:post/slug "indexes-as-capabilities"])
```

2. Native Indexed Documents with `idoc`

An `idoc` stores a nested document as one datom value, while Datalevin maintains an index for paths inside the document, as shown in Figure 14.2. This gives you document-database-style path queries while still using Datalog joins and transactions.

`idoc` indexes are orthogonal to full-text and vector indexes. The same entity can have structured facts, full-text fields, embeddings, and `idoc` metadata, and a single Datalog query can combine all of those constraints.

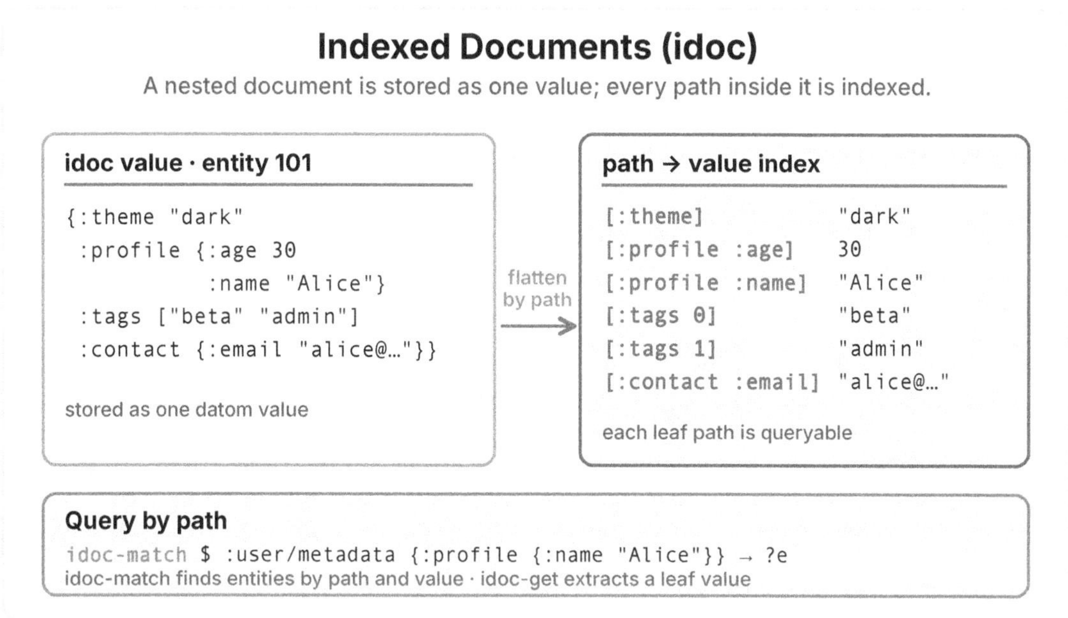

Figure 14.2: Indexed documents: a nested indexed-document value is flattened to a path-to-value index whose indexed leaf paths are queried with idoc-match and idoc-get

2.1 Defining an `idoc` Attribute

Declare an `idoc` attribute with `:db/valueType :db.type/idoc`. The optional `:db/idocFormat` controls how payloads are interpreted, the optional `:db/domain` gives the backing index a stable domain name, and the optional `:db.idoc/indexedPaths` / `:db.idoc/excludedPaths` keys control which path prefixes are indexed. Section 2.3 explains path selectors.

```
(def schema
  {:user/metadata {:db/valueType :db.type/idoc
                   :db/domain    "user_metadata"
                   :db.idoc/indexedPaths [:theme :profile :tags :contact]
                   :db.idoc/excludedPaths [[:profile :secret]]}
   :user/raw-json {:db/valueType  :db.type/idoc
                   :db/idocFormat :json}
   :doc/title     {:db/valueType :db.type/string}
   :doc/markdown  {:db/valueType  :db.type/idoc
                   :db/idocFormat :markdown}})
```

Supported `:db/idocFormat` values are `:edn` (the default), `:json`, and `:markdown`. If `:db/domain` is omitted, Datalevin derives a domain from the attribute name without the leading colon; for example, `:user/metadata` becomes `"user/metadata"`. If no path controls are supplied, Datalevin indexes all scalar leaf paths in the document.

2.2 `idoc` Domains

An `idoc` domain names the backing path index for one `idoc` attribute. If `:db/domain` is omitted, Datalevin derives the domain from the attribute name. Most schemas can treat this as an internal index name and query documents by attribute.

This is related to, but not the same as, full-text, vector, or embedding domains. Those indexes use domain lists or store-level domain maps to configure retrieval and indexing behavior, and some support default or automatic domains (Chapters 16 and 17).

`idoc` has no `:db.idoc/domains` list or `autoDomain` flag. An `idoc` attribute has one `idoc` domain, selected by `:db/domain` or derived from the attribute name. Several `idoc` attributes may intentionally share the same domain, but that is an index design choice, not a ranking or retrieval policy.

Store-level `:idoc-opts` and `:idoc-domains` can configure which paths a domain indexes (Section 2.3), but they do not make one attribute participate in multiple domains. Markdown `idoc` values normalize heading text into path segments, so they should usually live in their own domains instead of sharing a domain with EDN or JSON `idoc` values. Section 3 focuses on attribute-scoped `idoc-match` queries; Chapter 31 lists the domain-scoped and full-DB forms.

2.3 Selective Path Indexing

By default, an `idoc` domain indexes every scalar leaf path in each document. For large or user-defined documents, this can index more paths than the application needs. Selective path indexing lets you keep only the path prefixes your queries use.

Path selectors may be a top-level keyword, a top-level string, or a vector of keyword/string path segments. They are prefix selectors: `:profile` includes leaf paths such as `[:profile :age]` and `[:profile :name]`. `:db.idoc/indexedPaths` is the include list for an attribute's domain. When it is absent, all paths are included. `:db.idoc/excludedPaths` is the omit list and wins over included paths.

The original document is still stored as the datom value. Selective indexing only controls which leaf paths enter the path dictionary and inverted index. A query cannot use `idoc-match` to find documents by an omitted path, but once a document has been found by another indexed path, `idoc-get` can still extract any value from the stored document.

Use schema-level path controls when the indexing policy belongs to one attribute. The same controls can also be supplied as store options: `:idoc-opts` applies defaults to every `idoc` domain, schema keys configure the domain used by an `idoc` attribute, and `:idoc-domains` configures a named domain directly.

For a single attribute and domain, later layers override earlier ones: `:idoc-opts`, then schema keys, then the matching `:idoc-domains` entry. If several `idoc` attributes share the same domain, Datalevin combines their include and exclude selectors. Excluded paths are still applied after included paths. The example below expresses the same policy as the schema keys in Section 2.1 at the store level; use one mechanism or the other for ordinary schemas.

```
(d/create-conn
  "/tmp/idoc-db"
  schema
  {:idoc-domains
   {"user_metadata" {:indexed-paths [:theme :profile :tags :contact]
                     :excluded-paths [[:profile :secret]]}}})
```

Path indexing choices are part of index configuration. Changing `:db.idoc/indexedPaths`, `:db.idoc/excludedPaths`, `:idoc-opts`, or `:idoc-domains` after data has been indexed requires an explicit rebuild, because the existing path dictionary and inverted index were built under the old selection policy.

2.4 Transacting Documents

`idoc` values must be document-like maps. Nested vectors are treated as arrays.

```
(d/transact! conn
  [{:db/id 101
    :user/metadata {:theme "dark"
                    :profile {:age 30 :name "Alice" :secret "internal"}
                    :tags ["beta" "admin"]
                    :contact {:email "alice@example.com"}}
    :user/raw-json "{\"middle\":null,\"theme\":\"dark\"}"}])
```

Because `[:profile :secret]` is excluded, `(idoc-match $ :user/metadata {:profile {:secret "internal"}})` does not find this document. A query that finds the document by another indexed path, such as `{:theme "dark"}`, can still bind `?doc` and read `(idoc-get ?doc :profile :secret)`.

idoc input rules

- The top-level value must be a map.
- Keys may be keywords or strings, depending on the format and client language.
- Lists are not allowed; use vectors for arrays.
- `nil` values are normalized to `:json/null`.
- `:json/null` is reserved and rejected on input.
- For `:edn` format, string payloads are read as EDN and must yield a map.
- JSON and Markdown attributes accept string payloads in those formats.

The Python and JavaScript transaction above stores native string keys, so the matching query examples below also use ordinary strings for paths such as `"profile"` and `"age"`. Use `q.kw` instead when the stored document actually contains EDN keyword keys; query operators such as `:or`, `:?`, and `:*` remain keywords regardless of the document's key type.

2.5 Markdown Documents

Markdown `idoc` values are useful when the source data is authored as prose but you still want to query by section. Datalevin parses Markdown headings into nested map paths and stores the section text as values.

```
(d/transact! conn
  [{:db/id 201
    :doc/title "Search Guide"
    :doc/markdown
    "# Guide

## Install
```

```
Run `clj -M:dev`.

## Configure Search

Set `:index-position? true` for phrase search."}])
```

The parsed `idoc` value is equivalent to the following nested structure:

```
{:guide {:install "Run clj -M:dev."
         :configure-search "Set :index-position? true for phrase search."}}
```

Heading text is normalized to keyword-like paths: case is folded, leading numbering is removed, punctuation is dropped, and spaces become hyphens. Inline Markdown markup is stripped from the stored text. Content before the first heading is rejected, because Datalevin needs a heading path for the text.

You can match a section and extract its text in the same query:

```
(d/q '[:find ?title ?install
       :where
       [?e :doc/title ?title]
       [(idoc-match $ :doc/markdown
                    {:guide {:install "Run clj -M:dev."}})
        [[?e _ ?doc]]]
       [(idoc-get ?doc :guide :install) ?install]]
     db)
```

For Markdown `idoc` values, query paths are normalized too. This means string paths that look like headings work:

```
(d/q '[:find ?e
       :where
       [(idoc-match $ :doc/markdown
                    {"Guide" {"Configure Search"
                              "Set :index-position? true for phrase search."}})
        [[?e _ _]]]]
     db)
```

Use Markdown `idoc` values for section-level lookup over authored documents. If a section needs independent identity, permissions, relationships, or versioning, model it as normal component entities instead.

2.6 Patching Documents

Use `:db.fn/patchIdoc` for partial updates when you do not want to resend the whole document.

```
(d/transact! conn
  [[:db.fn/patchIdoc 101 :user/metadata
    [[:set    [:profile :display-name] "Alice A."]
     [:unset  [:profile :middle]]
     [:update [:tags] :conj "active"]]]])
```

Patch paths are vectors of map keys and vector indices. Supported operations are:

- `:set` sets the value at a path.

- `:unset` removes a map key or vector element.
- `:update` applies these operations:
  - `:conj` appends one or more values to a vector. If the path is absent, it starts from an empty vector.
  - `:merge` merges one or more maps into the map at the path. If the path is absent, it starts from an empty map.
  - `:assoc` sets one or more key/value pairs in the map at the path. If the path is absent, it starts from an empty map.
  - `:dissoc` removes one or more keys from the map at the path. If the path is absent, it starts from an empty map.
  - `:inc` increments an integer counter at the path. If the path is absent, it starts from `0`.
  - `:dec` decrements an integer counter at the path. If the path is absent, it starts from `0`.

> **Patching cardinality-many documents needs the old value**
>
> For cardinality-many `idoc` attributes, provide the old document value so Datalevin can identify which value should be patched.

3. Querying Path-Indexed Documents

Use `idoc-match` inside Datalog to find entities with matching nested document content. It returns matching datoms as `[e a v]` triples, so the result can be joined with ordinary facts.

```
;; Find users with a dark theme in their metadata document.
(d/q '[:find ?e
       :where [(idoc-match $ :user/metadata {:theme "dark"}) [[?e _ _]]]
       :limit 20]
     db)
```

`idoc-match` provides no ranked order, so a limited result without `:order-by` has no ordering guarantee and an offset does not define stable pagination. Query-level `:offset` and `:limit` apply after surrounding joins and filters, not to raw document matches. Chapter 21, Section 1.4 explains when the optimizer can avoid reading the entire matching relation.

Nested maps express nested path matches:

```
(d/q '[:find ?e
       :where [(idoc-match $ :user/metadata
                           {:profile {:age 30 :name "Alice"}})
               [[?e _ _]]]]
     db)
```

Vectors are treated as arrays. A match succeeds if any element matches:

```
(d/q '[:find ?e
       :where [(idoc-match $ :user/metadata {:tags "admin"}) [[?e _ _]]]]
     db)
```

3.1 Logical Combinators

Use `[:and ...]`, `[:or ...]`, and `[:not ...]` inside query maps:

```
(d/q '[:find ?e
       :where [(idoc-match $ :user/metadata
                           {:profile [:or {:age 30} {:age 40}]})
               [[?e _ _]]]]
     db)
```

3.2 Predicates and Ranges

Predicates can appear inline in query maps, or as path predicates supplied as query inputs:

```
;; Inline predicate.
(d/q '[:find ?e
       :where [(idoc-match $ :user/metadata {:profile {:age (> 21)}})
               [[?e _ _]]]]
     db)

;; Path predicate supplied as data.
(d/q '[:find ?e
       :in $ ?q
       :where [(idoc-match $ :user/metadata ?q) [[?e _ _]]]]
     db
     '(>= [:profile :age] 30))

;; Range via multi-arity comparison.
(d/q '[:find ?e
       :in $ ?q
       :where [(idoc-match $ :user/metadata ?q) [[?e _ _]]]]
     db
     '(< 20 [:profile :age] 40))
```

Supported predicates are `nil?`, `>`, `>=`, `<`, and `<=`. These predicate forms are EDN lists, not vectors. In Python and JavaScript, `q.edn_list`/`q.ednList` preserves that distinction without parsing an EDN string; use `q.sym` for the predicate operator. `q.expression` is for expressions in Datalog query grammar.

3.3 Wildcard Paths

Use wildcard path segments when the exact nested key is not known:

```
;; Match any single key under :profile with value >= 30.
(d/q '[:find ?e
       :in $ ?q
       :where [(idoc-match $ :user/metadata ?q) [[?e _ _]]]]
     db
     '(>= [:profile :?] 30))

;; Match a name field at any depth.
(d/q '[:find ?e
       :where [(idoc-match $ :user/metadata {:* {:name "Alice"}})
```

```
                [[?e _ _]]]]
    db)
```

`:?` matches exactly one path segment. `:*` matches any depth, including zero segments.

3.4 Null Matching

To match null values, use `(nil?)`:

```
(d/q '[:find ?e
       :where [(idoc-match $ :user/raw-json {"middle" (nil?)}) [[?e _ _]]]]
    db)
```

> **Match JSON keys as strings**
>
> JSON keys are strings, so JSON documents should be matched with `"middle"`, not `:middle`.

3.5 Extracting Values with `idoc-get`

`idoc-get` extracts a nested value from an `idoc` document inside a Datalog query:

```
(d/q '[:find ?e ?email
       :where [(idoc-match $ :user/metadata {:theme "dark"}) [[?e _ ?doc]]]
              [(idoc-get ?doc :contact :email) ?email]]
    db)
```

When a path traverses arrays, `idoc-get` returns a vector of matching values.

4. How `idoc` Indexes Work

Datalevin stores the `idoc` value itself as the datom value. The path index records enough path and value information to narrow candidates without duplicating full documents. Each domain has a separate path index, and indexing is synchronous during transactions. Unlike full-text and vector indexes, `idoc` has no async indexing mode. After a transaction commits, `idoc-match` sees the updated document index immediately.

> **idoc path indexes do not split the fact**
>
> `idoc` gives you indexed path filters, but the document remains one value from the perspective of ordinary datoms. If a nested field deserves identity, constraints, independent history, or joins as a first-class fact, model that part with normal attributes instead.

4.1 Benchmark Snapshot

The September 2026 indexed-document benchmark (https://github.com/datalevin/datalevin/tree/master/benchmarks/idoc-bench) compares Datalevin `idoc` with PostgreSQL JSONB, SQLite JSON1, and MongoDB. It uses 10,000 synthetic documents, 10,000 measured operations, a fixed mix of nested equality, range, wildcard, and array queries, and aligned strict durability settings. Correctness is checked against an independent oracle, and the reported pass runs in a separate JVM from the full warmup pass. Measurements were collected on a 12-core Apple Silicon Mac with Datalevin 1.1.0, PostgreSQL 18.4, SQLite 3.51.1, and MongoDB 7.0.35.

The single-thread results below are completed operations per second. Workload A mixes reads and updates, C is read-only, and F uses read-modify-write.

Workload	Datalevin	PostgreSQL	SQLite	MongoDB
A	3,358/s	2,270/s	438/s	309/s
C	11,454/s	2,039/s	437/s	2,868/s
F	2,680/s	1,789/s	416/s	192/s

Datalevin led all three single-thread workloads. On the C run, Datalevin's median latency across the five document-query shapes ranged from 0.051 to 0.328 ms.

These results show the benefit of path indexes for this document and query mix; they are not a comparison between `idoc` and normalized Datalevin datoms, and they do not replace application-specific measurement. Choose between the two Datalevin models by the semantic trade-offs in the next section.

5. Choosing Between Logical Documents and `idoc`

Feature	Logical document (`:db/isComponent`)	Native `idoc`
Storage	Decomposed into many datoms	Stored as one datom value
Schema	Attributes are declared explicitly	Nested fields can evolve freely
Indexing	Standard EAV/AVE and attribute indexes	Path-level indexing, all leaf paths by default or selected prefixes
Querying	Datalog joins, pull, entity API	`idoc-match` and `idoc-get`
Updates	Fine-grained datom updates	`patchIdoc` or full replacement
Best for	Core domain state and relationships	Flexible metadata and imported documents

Use logical documents when nested data needs type checking, referential integrity, joins to other entities, or independent updates as facts. Use `idoc` when nested data is sparse, user-defined, imported from another document store, or expected to change shape frequently.

These models can coexist on the same entity. A product can have structured facts for price and inventory, component entities for variants, full-text fields for descriptions, and an `idoc` attribute for merchant-specific metadata.

6. Ad Hoc Queries over Nested Data

Sometimes you do not want to persist a nested value at all. You have a map from an API response, a fixture, or a one-off data export, and you want Datalog's joins and predicates for exploration. Chapter 8 showed that Datalevin can query an in-memory sequence of tuples. A small boundary helper can turn a nested value into such a relation.

The following helpers walk maps/objects and vectors/arrays and emit one tuple for each leaf value. By default, the tuple is the path segments followed by the leaf value.

Listing 14.1. Turning nested data leaves into Datalog tuples

```
(defn leaf-paths
  "Returns paths to scalar leaves in a nested map/vector."
  ([root]
   {:pre [(or (map? root) (vector? root))]}
   (leaf-paths [] root))
  ([parent x]
   (cond
     (map? x)
     (mapcat (fn [[k v]]
               (leaf-paths (conj parent k) v))
             x)

     (vector? x)
     (mapcat (fn [[i v]]
               (leaf-paths (conj parent i) v))
             (map-indexed vector x))

     :else
     [parent])))

(defn nested->tuples
  "Converts nested maps/vectors into tuples queryable by Datalog."
  ([x]
   (nested->tuples x nil))
  ([x {:keys [paths?] :or {paths? false}}]
   (with-meta
     (mapv (fn [path]
             (let [tuple (conj path (get-in x path))]
               (if paths?
                 (into [path] tuple)
                 tuple)))
           (leaf-paths x))
     {:datalevin.docs/original x})))
```

For example:

```
(def order-doc
  {:order/id "o-1001"
   :customer {:email "ada@example.com"
              :name  "Ada"}
   :lines    [{:sku "book" :qty 2}
              {:sku "pen"  :qty 5}]})

(def tuples (nested->tuples order-doc))
;; => [[:order/id "o-1001"]
;;     [:customer :email "ada@example.com"]
;;     [:customer :name "Ada"]
;;     [:lines 0 :sku "book"]
```

```
;;      [:lines 0 :qty 2]
;;      [:lines 1 :sku "pen"]
;;      [:lines 1 :qty 5]]
```

Now the vector index in the path can be used as a join key:

```
(d/q '[:find ?sku ?qty
       :where
       [:lines ?i :sku ?sku]
       [:lines ?i :qty ?qty]]
     tuples)
;; => #{["book" 2] ["pen" 5]}
```

When the paths option is true, the full path is also carried as the first element of each tuple. This is useful when a query result should drive a later path lookup or update:

```
(def tuples-with-paths (nested->tuples order-doc {:paths? true}))

(d/q '[:find ?path ?sku
       :where
       [?path :lines ?i :sku ?sku]]
     tuples-with-paths)
;; => #{[[:lines 0 :sku] "book"]
;;      [[:lines 1 :sku] "pen"]}
```

> **Ad hoc tuple views are not persistence**
>
> This technique is for ad hoc analysis and local transformation. It does not create attributes, schema, indexes, lookup refs, or durable entities. If the nested data is part of the application's long-lived domain model, turn it into ordinary Datalevin facts or component entities. If it is flexible document data that should be stored and searched by path, use `:db.type/idoc`.

Summary

Datalevin lets you choose a representation for each part of your data. Use logical documents for structured domain state where facts, relationships, and constraints matter. Use `idoc` for flexible document-shaped data that should be stored as one value but queried by path. By default, `idoc` indexes every scalar leaf path; use selective path indexing when only a few prefixes should be searchable or when large user-defined documents would otherwise create more index surface area than the application needs. For one-off nested values, convert them to tuples and query them in memory instead of creating durable schema or `idoc` indexes. Both stored models participate in Datalog, so you can join document filters with graph, relational, full-text, and vector queries in one transactionally consistent database. For eligible finite relation queries, the optimizer can page `idoc-match` incrementally and stop after enough final rows survive the surrounding logic.

PART IV

Indexes as Capabilities

CHAPTER 15

Index Architecture

Earlier chapters introduced the logical fact shape: a datom written as `[entity attribute value]`. This chapter asks the storage question: how should those facts be ordered so queries can find the relevant facts efficiently?

In Datalevin, the core indexes are not optional side structures added after the data. They are the database's primary representations of facts. Every datom is stored in sorted orders that make entity-local reads, attribute/value lookups, range scans, joins, and graph traversal efficient. The two workhorse indexes are **EAV** and **AVE**.

1. The Logic of Sorted Triples

To make searching efficient, Datalevin stores each datom in two different sorted orders within the underlying KV store (DLMDB).

1.1 The Full Datom Object

Most chapters write datoms as `[e a v]` because those are the three fields used by Datalog patterns and by the EAV/AVE index keys. That is the logical shape of a fact:

```
[101 :user/email "alice@example.com"]
```

The implementation of a datom in memory is a richer Datom object. A full Datom has five fields:

Field	Meaning
`:e`	Entity id. This is Datalevin's internal numeric entity id.
`:a`	Attribute. Usually a namespaced keyword such as `:user/email`.
`:v`	Value. The value asserted for that entity and attribute.
`:tx`	Transaction id associated with the datom in transaction-report contexts.
`:added`	`true` for an assertion, `false` for a retraction.

The `:tx` value is Datalevin's internal numeric transaction id for the transaction report. Datalevin maintains the latest transaction counter as store metadata (`max-tx`), but per-datom transaction ids are not stored in the persistent EAV/AVE indexes. A transaction id is also not a queryable transaction entity with automatically stored attributes. The optional `tx-meta` argument to `transact!` is returned in the transaction report and sent to listeners, but it is not stored as queryable transaction metadata unless your application explicitly transacts its own audit facts.

In a transaction report, the transaction fields are available on `:tx-data`.

```
(let [report (d/transact! conn [{:user/email "alice@example.com"}])
      datom  (first (:tx-data report))]
  {:e     (:e datom)
```

```
   :a      (:a datom)
   :v      (:v datom)
   :tx     (:tx datom)
   :added (:added datom)})
;=> {:e 101, :a :user/email, :v "alice@example.com", :tx 42, :added true}
```

Datalevin's EAV/AVE indexes expose the facts present in the database state you are reading, so datoms returned from `datoms`, `search-datoms`, and `index-range` (see below) should be treated as `[e a v]` facts. Transaction reports are where `:tx` and `:added false` most often matter: a report's `:tx-data` can include datoms that were retracted by the transaction.

This distinction explains a recurring convention in this guide: `[e a v]` means "the logical fact used by a query or index key"; the full Datom object carries the extra transaction metadata needed by lower-level APIs and transaction reports.

1.2 Why Datalevin Is Not a History Database

Datalevin's persistent indexes are not history indexes. One way to support historical database values is to store the transaction dimension with every datom: keep each assertion and retraction with its transaction id, then answer historical queries by selecting the datoms whose transaction ids are visible at a chosen point in time. Datalevin deliberately does not make that per-datom transaction dimension part of the persistent EAV/AVE indexes.

> **Indexes describe current state**
>
> EAV and AVE answer questions about the latest committed facts. If an application needs audit history, provenance, or temporal semantics, model those facts explicitly instead of assuming the ordinary indexes retain every past datom.

Why not database-as-value history? Datalevin is designed first as an OLTP database: an operational store for current application state. For that workload, the ordinary developer mental model is that a database represents mutable state: users update profiles, orders change status, jobs move through queues, and the application asks questions about the current world. Automatically retaining every historical datom would make that model harder to explain and would make every application pay for a history feature that many operational systems do not need.

Keeping full transaction history is not free. It increases storage, widens the indexes the system must maintain, and makes reads and writes account for old states as well as current state. That tradeoff can be right for domains such as financial ledgers, audit stores, event logs, or analytical systems where history is the product. It is a poor default for an embedded or operational database whose common path is "store the latest state and query it efficiently."

This point matters especially for the agent-memory use cases in Part VI. A useful agent memory is not an append-only transcript. It needs deletion, expiry, consolidation, summarization, and retraction. Human memory research treats forgetting as an adaptive feature of cognition: transience can reduce interference, support abstraction, and help a system stay responsive when the world changes [1]. An agent that can only accumulate memories will eventually retrieve stale, irrelevant, or harmful context. Datalevin's current-state model makes forgetting a normal database operation rather than an exception to the model.

When applications need temporal or provenance-aware data, the requirements are usually richer than "keep every transaction forever." A transaction id can tell you when a fact was asserted or retracted in the database, but it does not tell you why a rule fired, which source passages supported a RAG answer, how confidence should propagate through a derivation, which explanation justifies a query result, or how to maintain an annotated derived view incrementally.

Datalevin's direction is therefore not to make transaction history the one built-in answer. The more general plan is to implement support for the Green-Karvounarakis-Tannen semiring provenance framework [2]. In that model, facts carry annotations from a chosen semiring; joins combine annotations with the semiring's multiplication, and alternative derivations combine them with addition. Temporal history becomes one concrete instance, for example with an interval semiring, but the point is broader than as-of history. The same framework can also support count provenance, why-provenance, confidence or probability propagation, min-cost explanations, rule-application tracing, RAG source attribution, debugging, and incremental view maintenance [2,3].

This does not require changing Datalevin's datom storage format. The core fact remains `[e a v]`; provenance annotations are represented as additional datoms: source entities, derivation entities, intervals, confidence values, costs, rule identifiers, and links between derived facts and their supporting facts. Datalog can then query and propagate those annotations explicitly. In the meantime, model audit and provenance facts explicitly when the domain needs them, and keep the core EAV and AVE indexes optimized for current OLTP state.

1.3 Why Sorted Indexes Speed Up Queries

Sorted indexes are efficient because they avoid looking at unrelated facts. The basic idea is the same reason binary search is efficient compared with a scan: when data is ordered, the engine can jump toward the place where a key must be instead of checking every item.

LMDB stores each Datalevin index as a B+Tree [4]. A point lookup, such as "find the datom for entity `101` and attribute `:user/email`", starts at the root page, chooses the child page whose key range could contain the target, repeats that step through internal pages, and lands on the leaf page where the key is stored or would be stored. Because B+Tree pages have high fanout, this is usually only a few page hops even for a large database.

Range queries use the sorted order in a second way. To answer "find all users with age between `21` and `65`", the engine first seeks to the lower bound in the AVE index. From there, it walks forward through neighboring leaf entries in sorted order until the upper bound is reached. It does not restart from the root for every matching datom, and it does not scan unrelated attributes. This is why the order of an index matters.

Datalevin's two core sorted orders are EAV and AVE.

EAV (Entity-Attribute-Value)

EAV makes facts for one entity contiguous.

- **Primary Use Case**: "Give me everything about Entity 101."
- **Query Role**: This index powers the Pull API and pivot scans in queries (see Chapter 21). In a query, the EAV index is scanned for attribute values when the entity is already

known. Because all attributes for a single entity are stored contiguously in the B+Tree, retrieving a complete record for an entity is a single localized range scan.

AVE (Attribute-Value-Entity)

AVE makes facts for one attribute/value range contiguous.

- **Primary Use Case**: "Find all entities where the `:user/age` is `30`."
- **Query Role**: This is the "search" index. It powers `:where` clauses that filter by a known attribute and value, or by a known attribute and value range.
- **Universal Indexing**: In Datalevin, **every attribute is indexed in AVE.**

DLMDB adds order statistics (Chapter 4): Datalevin can efficiently count how many entries fall in a value range without materializing the datoms. That inexpensive range count matters to the query optimizer. When a predicate such as `[(> ?age 21)]` can be pushed down to an AVE range, the engine can estimate the size of that candidate set and choose a better join order (Chapter 21). AVE is therefore both a scan path and a source of cardinality information.

Together, EAV and AVE make common Datalog patterns line up with sorted access paths. Figure 15.1 shows the mapping: entity-known patterns favor EAV, while attribute/value patterns and value ranges favor AVE.

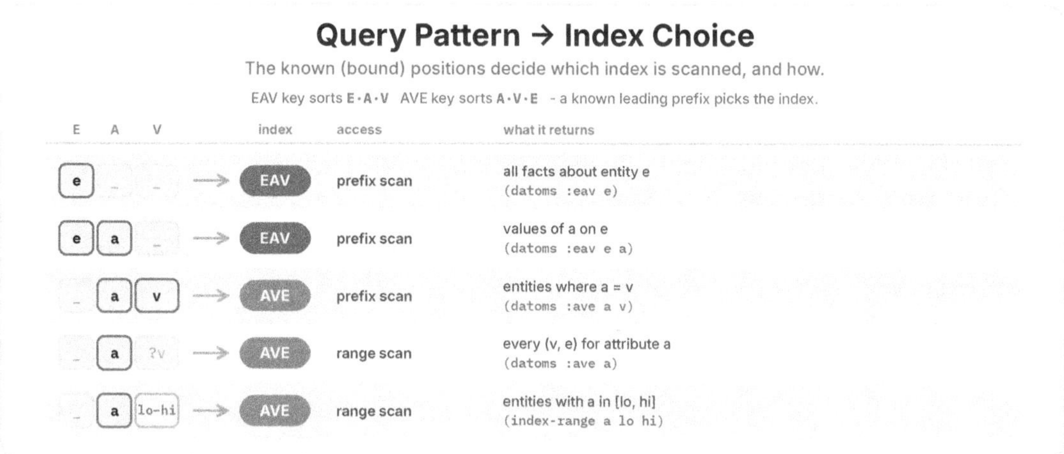

Figure 15.1: Query pattern to index choice: a pattern's known positions, when they form a leading prefix of an index sort key, pick the index. Entity-known patterns use EAV prefix scans; attribute-and-value patterns use AVE prefix scans; an attribute with a value range uses an AVE range scan via index-range

2. Index-Level APIs: Direct Access

Datalevin exposes these indexes directly to application code. You are not limited to the Datalog query engine; you can choose a direct index API when the access path is already known.

Direct index APIs return datoms in index order. They are best for simple, known access paths where you want the index itself, not the query planner, to be the API boundary.

2.1 Prefix Scans with datoms

Use `datoms` when you already know whether the lookup is entity-local (`:eav`) or value-local (`:ave`).

```
;; All datoms for entity 101, ordered by attribute and value.
(d/datoms db :eav 101)

;; A single attribute on entity 101.
(d/datoms db :eav 101 :user/email)

;; All entities whose :user/age value is 30.
(d/datoms db :ave :user/age 30)
```

2.2 Wildcard Lookup with search-datoms

Use `search-datoms` when you want to describe the logical datom pattern as `(e, a, v)` and let Datalevin choose the index. A `nil` / `None` / `null` component is a wildcard. In JavaScript calls, omit a wildcard key from the argument object or leave it `null`.

```
;; All attributes and values for entity 101.
(d/search-datoms db 101 nil nil)

;; All entities whose :user/age value is 30.
(d/search-datoms db nil :user/age 30)

;; The exact datom, if present.
(d/search-datoms db 101 :user/email "ada@example.com")
```

2.3 Counting with count-datoms

Use `count-datoms` when a query path needs selectivity, pagination totals, or an inexpensive existence check.

```
;; How many datoms are attached to entity 101?
(d/count-datoms db 101 nil nil)

;; How many users have age 30?
(d/count-datoms db nil :user/age 30)
```

`count-datoms` is faster than `(count (search-datoms ...))` because it leverages DLMDB's order statistics and does not have to allocate the matching datoms.

2.4 cardinality and analyze

Use `cardinality` when you need the number of distinct values currently present for one attribute. Do not confuse this with the schema property `:db/cardinality`, which says whether an attribute is single-valued or multi-valued. `d/cardinality` is a data statistic.

```
;; How many distinct ages exist?
(d/cardinality db :user/age)

;; How many datoms use :user/age at all?
```

```
(d/count-datoms db nil :user/age nil)

;; How many datoms say :user/age is 30?
(d/count-datoms db nil :user/age 30)
```

Datalevin periodically refreshes eid samples and attribute cardinality counts in the background for the query planner. After a bulk load or substantial reshape, `analyze` can refresh these statistics for one attribute or all attributes:

> **Refresh statistics after reshaping data**
>
> Background collection keeps ordinary workloads moving, but a large import, migration, or data-shape change can leave the planner working from stale estimates. Run `analyze` when the data shape changes faster than the background refresh can catch up.

```
(d/analyze db :user/age)
(d/analyze db)
```

2.5 Cursor Scans with `seek-datoms` and `rseek-datoms`

Use `seek-datoms` when you want to start at an index key and continue forward. If no datom exactly matches the supplied components, Datalevin starts at the first datom that is greater in that index order. `rseek-datoms` uses the same idea in reverse.

The arguments after `db`/connection are the index name, up to three components in that index order, and an optional result limit:

Index	`c1`	`c2`	`c3`
`:eav`	entity id	attribute	value
`:ave`	attribute	value	entity id

For example, `(seek-datoms db :ave :user/age 30 nil 10)` starts in AVE order at attribute `:user/age`, value `30`, leaves the entity-id component open, and returns at most 10 datoms. In Clojure and Java, pass `nil/null` for skipped later components when you need to supply the final limit. Python and JavaScript use named `limit` options instead.

```
;; Start a forward scan at :user/age 30 in AVE order.
(d/seek-datoms db :ave :user/age 30 nil 10)

;; Start a reverse scan near the high end of :user/created-at.
(d/rseek-datoms db :ave :user/created-at 4102444800000 nil 5)

;; Start at the :user/email position inside entity 101.
(d/seek-datoms db :eav 101 :user/email)
```

Without the result limit, the cursor keeps walking to the end of that index direction. That is occasionally useful for low-level scans, but it is often broader than application code intends.

> **Bound cursor scans deliberately**
>
> `seek-datoms` and `rseek-datoms` are cursor-style APIs. Unless the caller passes a limit or stops consumption itself, the scan continues in index order until the end of that direction.

2.6 Range Scans with `index-range`

Use `index-range` for the most common AVE range pattern: one attribute, lower bound, upper bound.

```
;; Find users with age between 20 and 30, inclusive.
(d/index-range db :user/age 20 30)

;; Return just the matching entity ids.
(map :e (d/index-range db :user/age 20 30))
```

These functions are intentionally low-level. Prefer Datalog for joins, rules, result shaping, and complex predicates. Reach for direct index access when the shape is simple enough that the index order itself is the query plan.

> **Use direct indexes at API boundaries**
>
> Direct index calls are best when the application surface is itself an index scan, range lookup, count, or cursor. When the work involves joins, rules, predicate logic, or shaped results, let Datalog compose the same indexes for you.

2.7 Choosing an Index Access Function

After seeing the individual APIs, the choice can be summarized as a question about which parts of the datom are known and whether you need values or only a count. Figure 15.2 lays this out as a decision chart; it is about API choice, not query planning.

Most rows in the table below are direct read paths. `analyze` is the exception: it refreshes planner statistics after data shape changes.

Function	Use it when	Index behavior
`datoms`	You know the target index and a prefix of its sort key.	Scans `:eav` or `:ave` in that index's natural order.
`search-datoms`	You have an `(e, a, v)` pattern with wildcards.	Chooses an efficient index for the supplied components.
`count-datoms`	You only need the number of matching datoms.	Counts the same wildcard pattern without materializing results.
`cardinality`	You need the number of distinct values currently present for one attribute.	Counts unique values for that attribute through the AVE index.

Function	Use it when	Index behavior
`analyze`	You bulk-loaded or substantially reshaped data and want fresh planner statistics.	Refreshes statistics for one attribute or all attributes.
`seek-datoms`	You need a forward cursor starting at a lower bound.	Starts at the first datom greater than or equal to the supplied index key.
`rseek-datoms`	You need a reverse cursor starting near an upper bound.	Same as `seek-datoms`, but walks backward.
`index-range`	You need all values of one attribute between two bounds.	Scans the AVE range for one attribute.

The two attribute-scoped value lookups are both range scans for the same reason. Fixing the attribute pins the leading bytes of the AVE key, so LMDB has a known lower and upper byte bound to scan between. Reading every value of an attribute walks from that attribute's minimum encoded value to its maximum; a bounded predicate such as `[(< 20 ?age 30)]` just tightens those bounds. Neither case scans the whole index.

3. Physical Representation and DUPSORT

Physically, these indexes are implemented using LMDB's `DUPSORT` feature (see Chapter 10). This allows Datalevin to store many values for a single key efficiently.

- **In EAV**: The key is `E`, and the values are `(A, V)` pairs.
- **In AVE**: The key is `(A, V)`, and the values are `E` (entity ids).

Figures 15.3 and 15.4 show this nesting for each index.

Thinking in terms of traditional database storage models:

- **EAV is a row store**: Each entity id (key) maps to a list of attribute-value pairs, analogous to a row where all column values are stored together. Retrieving an entity is a single key lookup.
- **AVE is a column store**: Each `(A, V)` combination (key) maps to a tightly packed list of entity ids, the "row IDs" that share that column value. This is ideal for analytical queries that scan a column.

This nested storage eliminates redundant prefixes. In EAV, an entity with 10 attributes stores the entity id once as the key, with 10 `(A, V)` pairs as values. In AVE, each `(A, V)` combination is stored once as the key, with all matching entity ids as values, so finding all entities where `:user/age` is `30` is a single key lookup.

This storage shape also explains the performance tax of using `:db.cardinality/many` for large relationship sets. In EAV, the owner entity carries one `(A, V)` duplicate value for every member, so scanning that attribute under the entity has to walk a long group instead of reading one cardinality-one value. In AVE, the cost is often larger. A normalized many-side reference such as `[?comment :comment/post ?post]` groups all comments for one post under the single `(A, V)` key `(:comment/post, ?post)`, so the lookup is one contiguous duplicate scan. If the same relationship is stored as `[:post/comments ?comment]` on the parent, each comment value has

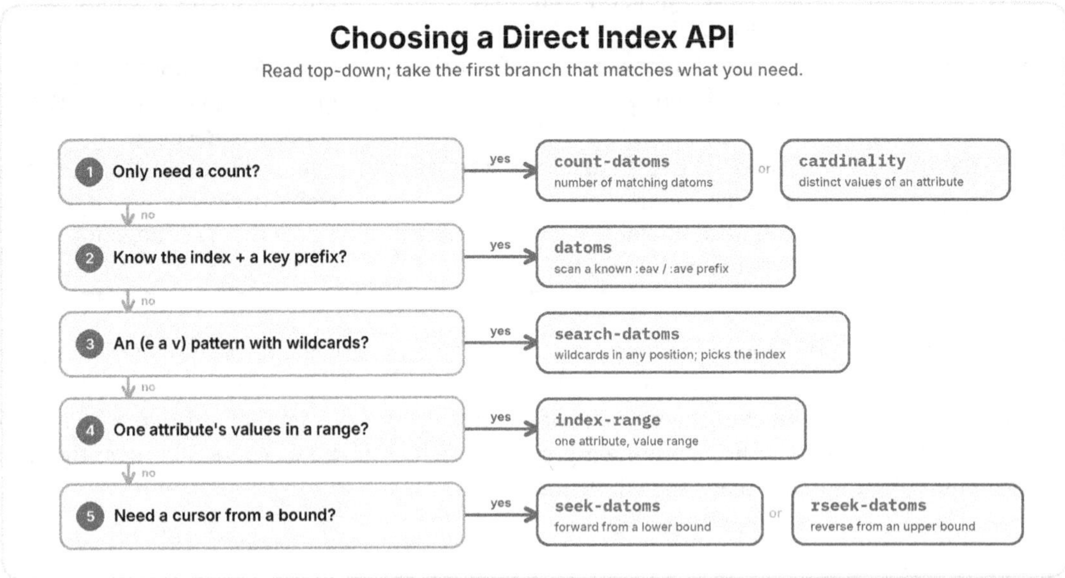

Figure 15.2: Choosing a direct index API as a top-down decision flow: if you only need a count use count-datoms or cardinality; else if you know the index and a key prefix use datoms; else if you have an (e a v) wildcard pattern use search-datoms; else if you want one attribute's values in a range use index-range; else if you need a cursor from a bound use seek-datoms or rseek-datoms

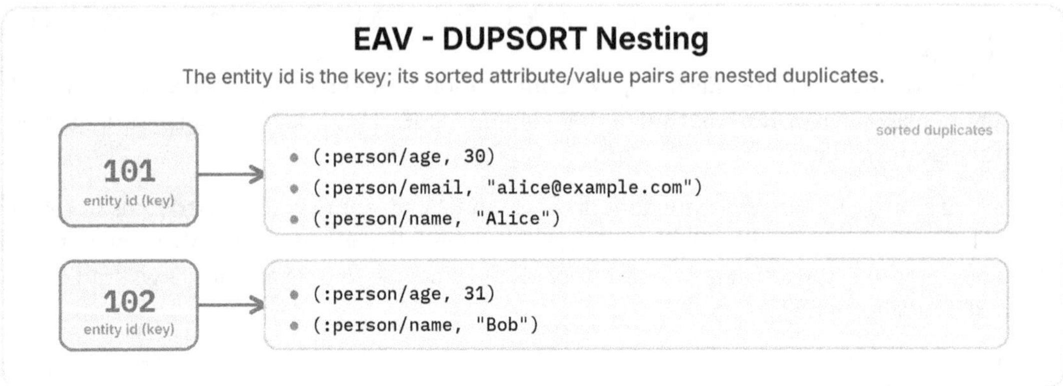

Figure 15.3: EAV DUPSORT nesting: the entity id is the key, and its sorted attribute/value pairs are stored as nested duplicate values, so the entity id is not repeated for every datom

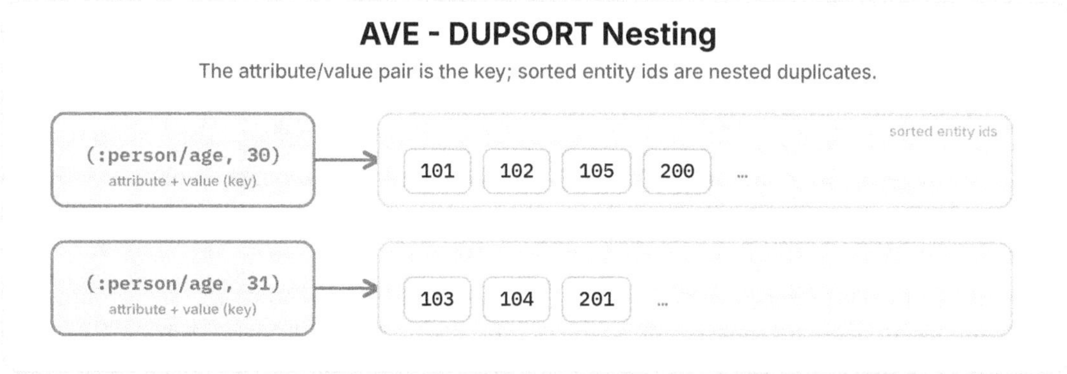

Figure 15.4: AVE DUPSORT nesting: the attribute/value pair is the key, and the matching entity ids are stored as a sorted list of nested duplicate values

a different `(A, V)` key; value-oriented access over the collection becomes many seeks or short scans that jump around AVE. That is why Chapter 11 recommends cardinality-many only for small, bounded sets read with their owner, and recommends a cardinality-one ref on the many side or a join entity for larger relationships.

In the Join Order Benchmark database, this `DUPSORT` nesting reduced index space by roughly 20%. Treat that number as an anecdotal workload observation, not a general guarantee: the savings depend on the number of repeated entity ids in EAV and repeated `(A, V)` prefixes in AVE. This is separate from DLMDB's page-level prefix compression, which can provide additional savings depending on how much of neighboring encoded keys is actually shared.

A `:db/tupleAttrs` declaration materializes one additional logical datom represented in both EAV and AVE. Every component update maintains that datom, so the physical cost is additional index space and write work. Chapter 11, Section 1.2 covers the modeling decision, explicit query use, derivation mechanics, and IC5 cardinality evidence.

Summary: Indexes as Capabilities

By indexing every attribute by default and exposing direct index APIs, Datalevin lets application code use the same access paths the query engine plans against.

- **EAV** provides locality for entities and documents.
- **AVE** provides indexed lookups and range scans for values.
- **Derived composite attributes** add explicit multi-attribute datoms to EAV and AVE; Chapter 11, Section 1.2 explains when to define them.
- **Optimizer-selected access paths** are physical alternatives to normal complete execution. They let an eligible AVE, idoc, full-text, or vector source drive a finite query when the cost model predicts a benefit; Chapter 21, Section 1.4 explains the comparison.
- **Direct Access** gives you `datoms`, `search-datoms`, `count-datoms`, `cardinality`, `seek-datoms`, `rseek-datoms`, `index-range`, and `analyze` for custom traversal and statistics logic when the access pattern is already known.

Understanding these indexes makes the later chapters on full-text search, vector similarity, and query optimization easier to reason about.

References

[1] Blake A. Richards and Paul W. Frankland, "The Persistence and Transience of Memory," *Neuron* 94(6):1071-1084, 2017. DOI: https://doi.org/10.1016/j.neuron.2017.04.037.

[2] Todd J. Green, Grigoris Karvounarakis, and Val Tannen, "Provenance Semirings," in *Proceedings of the Twenty-Sixth ACM SIGMOD-SIGACT-SIGART Symposium on Principles of Database Systems (PODS '07)*, 2007, pp. 31-40. DOI: https://doi.org/10.1145/1265530.1265535.

[3] Camille Bourgaux, Pierre Bourhis, Liat Peterfreund, and Michael Thomazo, "Revisiting Semiring Provenance for Datalog," arXiv:2202.10766, 2022. URL: https://arxiv.org/abs/2202.10766.

[4] Douglas Comer, "The Ubiquitous B-Tree," *ACM Computing Surveys* 11(2):121-137, 1979. DOI: https://doi.org/10.1145/356770.356776.

CHAPTER 16

Full-Text Search

Many database applications add a separate engine like Lucene, Elasticsearch, or Solr for full-text search. Datalevin includes an integrated search engine built directly on top of its KV substrate.

This is a deliberate design choice. Lucene is an excellent general-purpose information-retrieval library, and systems such as Elasticsearch and Solr build large search platforms around it. Datalevin has a narrower goal: make search a native secondary index of an embedded logical database. For that goal, an integrated engine has practical advantages. Search updates can participate in the same transaction boundary as datom updates, backup and restore stay within one LMDB environment, query results can be datoms that join directly with Datalog clauses, and the application does not need to operate or reconcile a separate search index beside the database.

The tradeoff is intentional. Datalevin is not trying to replace the Lucene ecosystem for every search problem. If an application needs a large distributed search cluster, a specialized Lucene analyzer stack, or search features outside Datalevin's scope, using a dedicated search system can be the right choice. Datalevin's built-in engine is for the common case where search should be a transactional capability of the database itself.

This chapter explains how to enable full-text search, write boolean search queries, and understand the underlying ranking algorithm.

1. Enabling Full-Text Search

Full-text search is search over text rather than exact database values. Instead of asking whether one stored value equals another, a full-text engine breaks text documents into searchable terms, normalizes those terms, records where they occur, and ranks matching documents by relevance. This is what lets a user search for words, phrases, prefixes, or boolean combinations across longer fields such as titles, comments, articles, documentation pages, and messages. Full-text search is also used for searching logs and other textual information.

To enable full-text search for an attribute, declare `:db/fulltext true` in your schema:

```
(def schema
  {:post/title   {:db/valueType :db.type/string
                  :db/fulltext  true
                  :db.fulltext/autoDomain true}
   :post/content {:db/valueType :db.type/string
                  :db/fulltext  true}})
```

With this schema, normal transactions update two things: the ordinary Datalog datoms and the full-text index entries for the marked attributes. When a value is added to `:post/title` or `:post/content`, Datalevin converts the value to text, runs the configured analyzer, records the resulting searchable terms, and links those terms back to the matching datom. When such a value is retracted or replaced, the corresponding index entries are updated as part of transaction processing. In the default synchronous indexing mode, a committed transaction is immediately visible to `fulltext` queries.

The examples use `:db.type/string` because full-text search is usually applied to string. It is not a hard requirement of `:db/fulltext`: during transaction processing, Datalevin converts the value to string. `:db.fulltext/autoDomain` will be explained in Section 5.

2. The Default Analyzer Pipeline

An analyzer is the function that turns source text into search terms. During indexing, Datalevin stores the analyzer output in the **inverted index**, a lookup structure that maps each term to the datoms that contain it. During querying, Datalevin analyzes the query text in the same term space, so the stored terms and query terms can be compared.

That "same term space" requirement is important. For ordinary full-text search, the index analyzer and the query analyzer should be the same analyzer. If documents are stemmed, lowercased, or filtered one way at index time but queries are analyzed another way at search time, the query may look for terms that were never stored, or may miss phrase and proximity matches because positions no longer mean the same thing.

The default English analyzer performs three steps:

1. **Tokenization**: split text on whitespace and English punctuation.
2. **Normalization**: lowercase token text.
3. **Stop-word removal**: remove common English words such as `"the"` and `"in"`.

The analyzer output is a sequence of `[term position offset]`. The **term** is the key in the inverted index. The **position** is the token number used for phrase and proximity matching. The **offset** is the character offset in the original text, used for highlighting.

For example, the default English analyzer treats this text:

```
Running runners in databases
```

conceptually like this:

Original token	Indexed term	Position	Offset
`Running`	`running`	0	0
`runners`	`runners`	1	8
`in`	filtered stop word	2	16
`databases`	`databases`	3	19

Notice that stop-word removal does not renumber the surrounding terms. The removed token leaves a position gap. That matters for phrase search: phrases match analyzed terms at consecutive positions, so analyzer choices affect both what terms are searchable and which phrase queries can match.

3. Customizing Analyzers

The default analyzer is intentionally conservative. Applications often need different behavior: stemming for morphology, prefix terms for autocomplete, ngrams for substring-style matching, or different token filters for a specific language or corpus.

Custom analyzers can be provided when creating a standalone search engine or configuring a Datalog full-text domain. The `:analyzer` option is the **index analyzer**: Datalevin applies it to document text before storing terms in the inverted index. The `:query-analyzer` option is the **query analyzer**: Datalevin applies it to the user's search text before matching it against the index. If `:query-analyzer` is omitted, the query side uses the same analyzer as the index side.

The usual rule is simple: use the same analyzer for indexing and querying. A separate `:query-analyzer` is an advanced option. It should produce terms that are intentionally compatible with the indexed terms, not an unrelated analysis pipeline.

> **Keep analyzer term spaces compatible**
>
> An analyzer choice is part of the index contract. If indexing stems, lowercases, filters, prefixes, or stores positions one way, the query analyzer must produce terms and positions that make sense against that stored index.

As shown in Figure 16.1, a custom analyzer is still a pipeline from text to a `[term position offset]` collection. The example pipeline lowercases terms, removes the stop word `"in"`, and stems the remaining terms.

Custom analyzers can be direct functions or runtime UDF descriptors. For the UDF route, register a function with kind `:analyzer` or `:query-analyzer`, pass the UDF registry in `:runtime-opts :udf-registry`, and put the descriptor in the full-text domain options. The analyzer callable receives one string and returns a sequence of `[term position offset]` triples.

3.1 Stemming and Stop Words

Datalevin provides tokenizer and token-filter helpers for building analyzers: `datalevin.search-utils` in Clojure, `SearchUtils` in Java, and matching top-level helpers in Python and JavaScript. Utility functions for stemming, ngrams, stop-word removal, and regular-expression tokenization are included.

The next example creates an analyzer that splits on non-word characters, lowercases tokens, removes English stop words, and applies English stemming. Across languages, the search-utils helpers return Datalevin analyzer, tokenizer, and token-filter objects that are passed back into search options:

```
(require '[datalevin.search-utils :as su])

(def stem-analyzer
  (su/create-analyzer
    {:tokenizer (su/create-regexp-tokenizer #"\W+")
     :token-filters [su/lower-case-token-filter
                     su/en-stop-words-token-filter
                     (su/create-stemming-token-filter "english")]}))
```

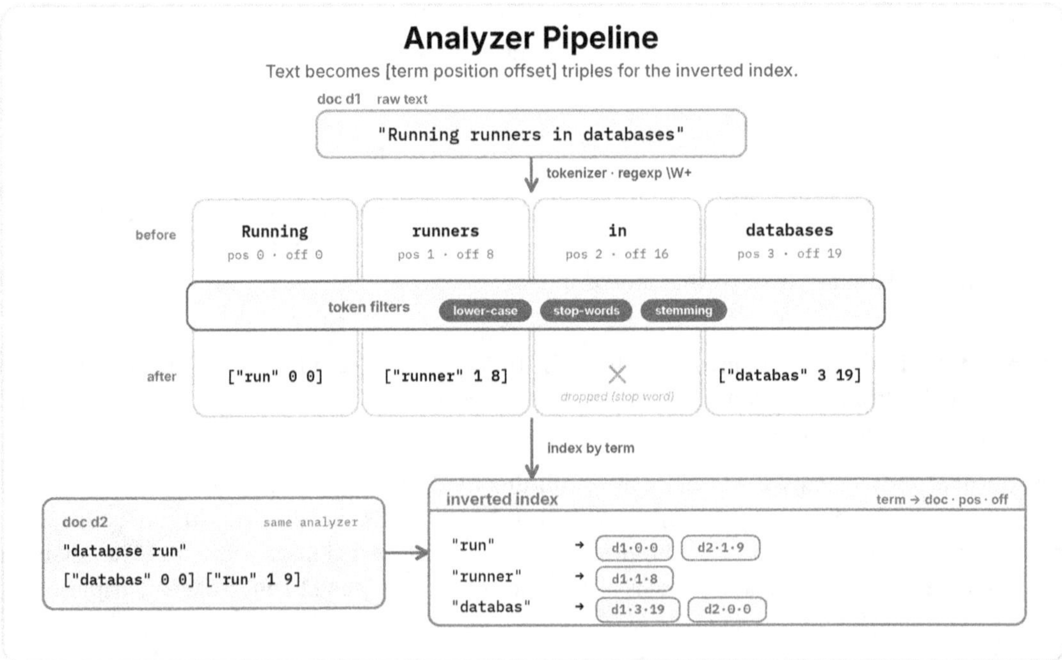

Figure 16.1: Analyzer pipeline: document d1 "Running runners in databases" is split by a regexp tokenizer into tokens with positions and offsets; token filters lower-case, drop the stop word "in", and stem the rest, preserving positions to yield `["run" 0 0]`, `["runner" 1 8]`, `["databas" 3 19]`; a second document d2 "database run" runs through the same analyzer, and both are indexed by term in the inverted index, which maps each term to the documents and positions where it appears (run to d1·0 and d2·1, runner to d1·1, databas to d1·3 and d2·0)

```
(mapv vec (stem-analyzer "Running runners in databases"))
;; => [["run" 0 0] ["runner" 1 8] ["databas" 3 19]]
```

Datalevin provides about 30 named Snowball stemmers, selected by name when creating a stemming token filter. Snowball is a family of small rule-based stemming algorithms; a stemmer reduces related word forms to a common stem, so "running" can index as "run". The English analyzer above uses one of them.

Use the analyzer in Datalog full-text options when the search domain should use the same normalization at index time and query time. `:search-opts` configures the default search domain; `:search-domains` configures named domains. Section 5 covers domains in detail:

```
(def conn
  (d/create-conn
    "/tmp/posts"
    {:post/body {:db/valueType :db.type/string
                 :db/fulltext  true}}
    {:search-opts {:analyzer stem-analyzer}}))

(d/transact! conn
  [{:db/id 10 :post/body "Runners are running fast"}
   {:db/id 11 :post/body "Databases store data"}])

(d/q '[:find [?e ...]
       :where [(fulltext $ "run") [[?e _ _]]]]
     (d/db conn))
;; => [10]
```

3.2 Prefix and N-Gram Indexing

Autocomplete is a controlled exception to the "same analyzer" rule. Apply `prefix-token-filter` at index time, but use a normal query analyzer that still produces compatible lowercase terms. This indexes "search" as "s", "se", "sea", and so on, while a query for "sea" remains a single token that can match one of the indexed prefixes.

```
(def prefix-index-analyzer
  (su/create-analyzer
    {:tokenizer (su/create-regexp-tokenizer #"\W+")
     :token-filters [su/lower-case-token-filter
                     su/prefix-token-filter]}))

(def query-analyzer
  (su/create-analyzer
    {:tokenizer (su/create-regexp-tokenizer #"\W+")
     :token-filters [su/lower-case-token-filter]}))

(def engine
  (d/new-search-engine
    (d/open-kv "/tmp/search-autocomplete")
    {:analyzer       prefix-index-analyzer
     :query-analyzer query-analyzer}))
```

```
(d/add-doc engine 1 "search")
(d/search engine "sea")
;; => (1)
```

For fuzzy or substring-style matching, use ngrams. The example below indexes 3-character grams, so `"Datalevin"` contributes tokens such as `"dat"`, `"ata"`, `"tal"`, and `"ale"`.

> **Short n-grams grow quickly**
>
> Prefer ngrams of 3 characters or longer unless the domain is small and tightly controlled. One- and two-character grams generate many short, high-frequency tokens, which makes posting lists less selective and can grow the full-text index quickly.

```
(def trigram-analyzer
  (su/create-analyzer
    {:tokenizer (su/create-regexp-tokenizer #"\W+")
     :token-filters [su/lower-case-token-filter
                     (su/create-min-length-token-filter 3)
                     (su/create-ngram-token-filter 3)]}))

(def trigram-engine
  (d/new-search-engine
    (d/open-kv "/tmp/search-trigrams")
    {:analyzer       trigram-analyzer
     :query-analyzer trigram-analyzer}))

(d/add-doc trigram-engine 1 "Datalevin")
(d/add-doc trigram-engine 2 "Search")
(d/search trigram-engine "tale")
;; => (1)
```

3.3 Analyzer UDFs

For Datalog full-text domains, `:analyzer` and `:query-analyzer` may be normal functions, UDF descriptor maps, or registered UDF id keywords. Use descriptor maps when the analyzer implementation should be resolved through a runtime registry.

Analyzer UDFs run during full-text indexing, including normal transactions, re-indexing, and DB-open recovery for pending async jobs. Query-analyzer UDFs run during `fulltext` query evaluation. The registry is process-local runtime state: the descriptor can be stored in options, but the process that indexes or queries must register the callable implementation before opening or using the database.

The pattern is:

1. Create a UDF registry.
2. Create an analyzer descriptor with kind `:analyzer`.
3. Optionally create a query-analyzer descriptor with kind `:query-analyzer`.
4. Register functions that return `[term position offset]` triples.
5. Pass the registry in `:runtime-opts`.

6. Put the descriptors in the target search domain's `:analyzer` and `:query-analyzer` options.

Here is the descriptor-backed shape. The tokenizer implementation is intentionally simple: the index analyzer stores only words that appear with a leading #, with the # removed. For example, `alpha #needle` indexes the term `needle` but not `alpha`. The query analyzer splits ordinary query text on whitespace.

The same descriptor-backed shape works in each embedded language. Direct analyzer functions are still fine for embedded-only programs, but descriptors are useful when options need to name a process-local registered implementation.

```
(require '[clojure.string :as str]
         '[datalevin.core :as d]
         '[datalevin.udf :as udf])

(def schema
  {:text {:db/valueType :db.type/string
          :db/fulltext true
          :db.fulltext/autoDomain true}})

(def analyzer-desc
  {:udf/lang :clojure
   :udf/kind :analyzer
   :udf/id   :text/hashtags})

(def query-analyzer-desc
  {:udf/lang :clojure
   :udf/kind :query-analyzer
   :udf/id   :text/plain-query})

(defn hashtag-analyzer
  [^String text]
  (let [matcher (re-matcher #"#\w+" text)]
    (loop [position 0
           tokens []]
      (if (.find matcher)
        (recur (inc position)
               (conj tokens [(subs (.group matcher) 1)
                             position
                             (.start matcher)]))
        tokens))))

(defn plain-query-analyzer
  [^String text]
  (->> (str/split text #"\s+")
       (remove str/blank?)
       (map-indexed (fn [position token]
                      [token position position]))
       vec))

(def registry
```

```
  (doto (udf/create-registry)
    (udf/register! analyzer-desc hashtag-analyzer)
    (udf/register! query-analyzer-desc plain-query-analyzer)))

(def conn
  (d/create-conn
    "/tmp/fulltext-udf"
    schema
    {:runtime-opts {:udf-registry registry}
     :search-domains
     {"text" {:index-position? true
              :analyzer analyzer-desc
              :query-analyzer query-analyzer-desc}}}))
```

When the analyzer is used for Datalog full-text indexing, transaction code can stay ordinary. The custom domain analyzer controls what terms are indexed, and the custom query analyzer controls how the user's query is translated into the same term space. The following example searches for `"needle"` after indexing two strings; only the string containing `#needle` matches, because plain `needle` was not stored by the index analyzer.

```
(d/transact! conn
  [{:db/id 1 :text "alpha #needle"}
   {:db/id 2 :text "needle without hash"}])

(d/q '[:find [?e ...]
       :in $ ?q
       :where [(fulltext $ :text ?q) [[?e ?a ?v]]]]
     (d/db conn)
     "needle")
;; => [1]
```

4. Querying with `fulltext`

In a Datalog query, the built-in `fulltext` function returns matching datoms as `[e a v]` triples, ordered by relevance. These triples can then be destructured and bound into variables in the query.

In a query, `fulltext` behaves like a relation-producing function: it takes a database and a search expression, then yields rows that can be joined with ordinary Datalog clauses through shared variables such as `?e`.

```
(d/q '[:find ?e ?a ?v
       :in $ ?q
       :where [(fulltext $ ?q) [[?e ?a ?v]]]]
     db
     "red fox")
```

Use `fulltext-datoms` / `fulltextDatoms` / `fulltext_datoms` when you want matching datom triples directly instead of joining through a Datalog query. It accepts ranking and paging options such as `:top`, `:limit`, `:offset`, and `:domains`. Display modes that add scores, text, or offsets belong in the Datalog `fulltext` relation or standalone `search`.

```
(d/fulltext-datoms db "red fox" {:limit 10 :offset 20})
;=> ([123 :post/body "The red fox ..."] ...)
```

4.1 Attribute-Specific Search

To search within a specific attribute, that attribute must have `:db.fulltext/autoDomain true` in the schema. An auto domain is a separate per-attribute full-text index, so `:post/title` can be searched without also searching `:post/content`; Section 5 explains domains in detail.

```
(d/q '[:find ?e ?a ?v
       :in $ ?q
       :where [(fulltext $ :post/title ?q) [[?e ?a ?v]]]]
     db
     "clojure")
```

4.2 Search Options

Pass search options as the last argument. Python and JavaScript use `q.fulltext_options`/`q.fulltextOptions` so option names, display modes, and non-negative bounds are checked while the query is composed:

```
(d/q '[:find ?e ?a ?v
       :in $ ?q
       :where [(fulltext $ ?q {:top 5}) [[?e ?a ?v]]]]
     db
     "database")
```

Available options:

Option	Default	Meaning
`:top`	`10`	Number of results when `:limit` is not supplied.
`:limit`	none	Page size; when supplied, this overrides `:top` as the number of returned results.
`:offset`	`0`	Number of ranked results to skip before returning results.
`:paging-cache-pages`	`10`	Number of `:limit`-sized pages to score and cache as one top-k window for paged search.
`:display`	`:refs`	Result shape: `:refs`, `:refs+scores`, `:texts`, `:offsets`, or `:texts+offsets`.
`:domains`	all initialized domains	List of search domains to query.
`:doc-filter`	none	Predicate function to filter results.

Use `:top` for one bounded result set, such as the first page of a search UI. Use `:limit` and `:offset` when the interface needs stable page-sized slices of the ranked result list. For example, `{:limit 20 :offset 40}` returns the third page when each page contains 20 results. The offset is applied to the full-text ranked result stream before the rows are destructured and joined by the surrounding Datalog query. It is therefore different from a top-level Datalog `:limit` or `:offset`, which would page the final query result after all clauses have run.

When `:limit` is present, Datalevin scores and caches a top-k window of `max(:offset + :limit, :limit * :paging-cache-pages)` candidates. This lets ordinary page-by-page access reuse a ranked window instead of rescoring each page in isolation. When `:limit` is not present, Datalevin scores `:offset + :top` candidates and returns up to `:top` results after the offset.

`:doc-filter` receives each document reference as results are converted to the requested display shape. In Datalog, that reference is the indexed datom `[e a v]`. The filter runs after ranking and paging, so it is best for final visibility checks, not for changing the ranking itself; a restrictive filter may return fewer rows than the requested page size.

```
(d/q '[:find ?e ?a ?v
       :in $ ?q
       :where [(fulltext $ ?q {:limit 20 :offset 40})
               [[?e ?a ?v]]]]
     db
     "database")
```

Now consider `:display` options:

`:refs` returns document references. As mentioned, the document reference is the indexed datom itself, so it can be destructured into `?e`, `?a`, and `?v` component variables.

`:refs+scores` add relevance scores for the full-text query. Higher scores mean a better match within that query.

> **Search scores are local**
>
> Treat a relevance score as meaningful only within the query, domain, analyzer, and ranking settings that produced it. Do not compare raw scores across unrelated searches or use them as stable identifiers.

`:texts` are the original texts, available only when the search engine stores raw text with `:include-text? true`.

`:offsets` report where matched terms occur in the original document text. They are useful for highlighting snippets. This requires `:index-position? true` (default is `false`) because the index must store term positions and character offsets at indexing time.

By default, a full-text domain stores the inverted index and document references, not a second copy of the original document text. The source datom remains in the Datalog database, but `:display :texts` and `:display :texts+offsets` require the domain option `:include-text? true`. Section 5.2 shows how to configure it.

When `:display` is not `:refs`, destructure the extra values in the returned tuple. For example, relevance scores use `[e a v score]`:

```
(d/q '[:find ?e ?title ?score
       :in $ ?q
       :where [(fulltext $ ?q {:top 100 :display :refs+scores})
               [[?e _ _ ?score]]]
              [?e :post/title ?title]
       :order-by [?score :desc ?title :asc ?e :asc]
       :limit 5]
```

```
      db
      "database")
```

Source-local and query-level bounds

The structured `:post/title` clause keeps only hits whose entity has a title and supplies that title to the result. The `:top 100` above is the full-text function's candidate window; `:limit 5` is the final Datalog window after that join, projection, and deduplication. The optimizer never widens the source-local window, so choose `:top` large enough to preserve recall after surrounding clauses prune candidates.

Chapter 21, Section 1.4 explains when the optimizer can consume this source incrementally and preserve its score order.

4.3 Search Expressions and Boolean Logic

Datalevin uses EDN data structures for search expressions, enabling arbitrary boolean combinations of search terms. A string query is analyzed into search terms; a structured expression lets you say how those terms should combine. A **term** is the normalized token produced by the analyzer, not necessarily the exact word typed by the user. For example, a stemming analyzer may turn `"running"` into `"run"`.

The simplest expression is a plain string, which the analyzer splits into terms. To control how those terms combine, wrap them in one of three boolean operators, `:and`, `:or`, and `:not`:

```
[:and "clojure" "database"]                  ; both terms
[:or "fox" "red"]                            ; either term
[:not "java"]                                ; exclude term
[:and "clojure" [:not "java"]]               ; clojure but not java
```

Each operator takes one or more sub-expressions, where a sub-expression can be a bare term, a phrase map (Section 4.4), or another nested operator:

- `:and` matches documents that satisfy *every* sub-expression.
- `:or` matches documents that satisfy *at least one* sub-expression.
- `:not` excludes documents that satisfy its sub-expression. It is a filter, so combine it with at least one positive term; on its own it has nothing to retrieve or rank.

A search expression goes in the same argument position as a plain query string. For example, to find posts that mention Clojure but not Java:

```
(d/q '[:find ?e ?a ?v
       :where [(fulltext $ [:and "clojure" [:not "java"]])
               [[?e ?a ?v]]]]
     db)
```

When a query is a plain string with several words, the terms are combined with coverage-based ranking (Section 8.3): documents that match more of the terms rank higher, but no term is strictly required. Reach for the boolean operators when you need explicit control over how terms must combine rather than relying on ranking alone.

4.4 Phrase Search

Phrase search looks for analyzed terms that appear at consecutive token positions in the same order. A phrase query for `"little lamb"` should match text such as `"Mary had a little lamb"` but not `"the little boy has a lamb"`, where `"little"` and `"lamb"` occur apart. The phrase text is analyzed by the query analyzer before matching, so analyzer choices such as stemming and stop-word removal affect phrase behavior.

Phrases are encoded as maps inside the search expression:

```
[:and {:phrase "little lamb"} "fleece"]
```

Phrase search is enabled by an index option, not by a special query option. The search engine must store term positions, which means the relevant full-text domain or standalone search engine must be created with `:index-position? true`. For Datalog full-text search, set this in `:search-opts` for the default `"datalevin"` domain or in `:search-domains` for a named domain:

```
(d/create-conn
  "/tmp/search-phrases"
  schema
  {:search-domains {"public" {:index-position? true}}})
```

At query time, the usual search options still apply.

```
(d/q '[:find ?e ?a ?v
       :where [(fulltext $ {:phrase "little lamb"}
                         {:domains ["public"]
                          :top 10})
               [[?e ?a ?v]]]]
     db)
```

Using a phrase query against a search engine that was not created with `:index-position? true` raises an error.

4.5 Complex Expressions

Boolean operators can be arbitrarily nested to express complex search requirements:

```
[:or
 "fox"
 {:phrase "red fox"}
 [:and "black" "sheep" [:not "yellow"]]]
```

5. Search Domains

A search domain is a named full-text index. Each domain has its own search engine, term statistics, analyzer settings, position index setting, and indexing mode. Domains let one database support different search surfaces without running a separate service: public site search, private draft search, autocomplete, admin-only search, or an attribute-specific index can all live in the same Datalevin store.

Domains are strings. They are not Datalog namespaces, and they are not inferred from an attribute namespace unless you ask Datalevin to create an automatic attribute domain.

> **Domains are index names**
>
> A domain name selects a full-text index with its own analyzer, term statistics, position setting, and indexing mode. It may look like an attribute name, but it is still an index configuration name, not an EAV namespace.

There are two separate decisions:

- Schema properties assign attributes to domains.
- Store options configure the search engine used by each domain.

Figure 16.2 shows an example of assigning attributes to search domains.

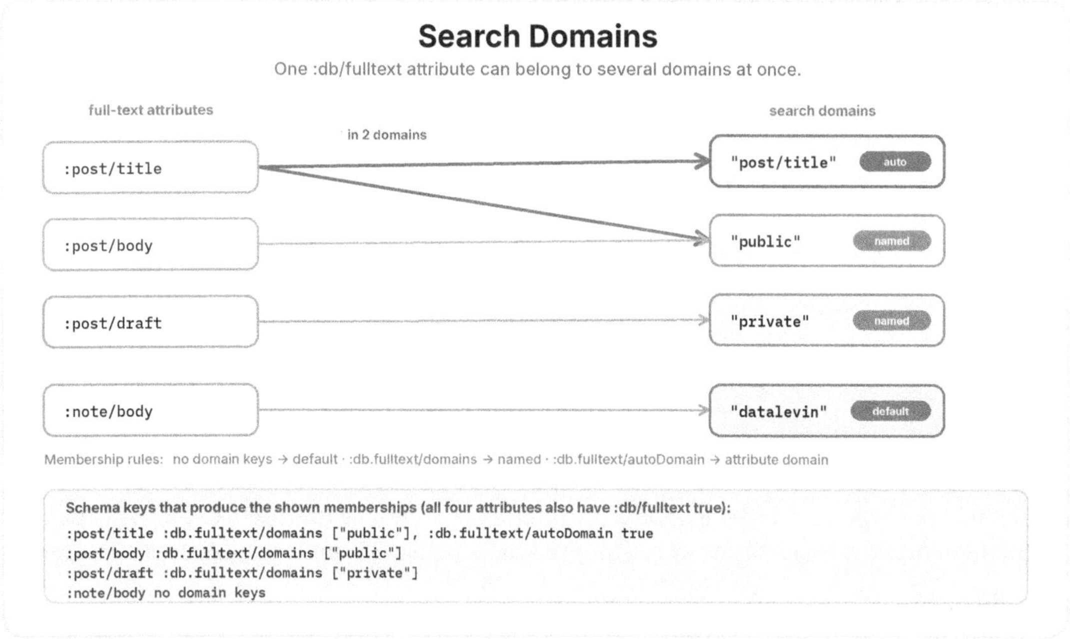

Figure 16.2: Search domain membership from schema keys: all shown attributes have :db/fulltext true; :post/title has public domains plus autoDomain true and joins public plus post/title; :post/body joins public; :post/draft joins private; :note/body has no domain keys and joins the default datalevin domain

5.1 Assigning Attributes to Domains

Full-text attributes enter domains according to these rules:

Schema on a `:db/fulltext` attribute	Domain membership
Neither `:db.fulltext/domains` nor `:db.fulltext/autoDomain`	The default `"datalevin"` domain
`:db.fulltext/autoDomain true`, with no `:db.fulltext/domains`	The default `"datalevin"` domain plus the attribute domain, such as `"post/title"`

Schema on a `:db/fulltext` attribute	Domain membership
`:db.fulltext/domains ["public"]`	Only the listed domains
`:db.fulltext/domains ["datalevin" "public"]`	The default domain plus `public`
`:db.fulltext/domains ["public"]` plus `:db.fulltext/autoDomain true`	`public` plus the attribute domain, such as `"post/title"`

`:db.fulltext/autoDomain true` does not replace the default domain by itself; it adds an attribute domain. When `:db.fulltext/domains` is present, the attribute participates in the listed domains. If `:db.fulltext/autoDomain true` is also present, it also participates in its attribute domain. Include `"datalevin"` in `:db.fulltext/domains` when an attribute with listed domains should also be searched through the default domain.

Full-text attribute domains use the attribute name without the leading colon: `:title` becomes `"title"` and `:post/title` becomes `"post/title"`. This rule is specific to full-text search. Vector and embedding domains use a different storage naming rule for namespaced attributes.

```
(def search-schema
  {:post/title {:db/valueType :db.type/string
                :db/fulltext  true
                :db.fulltext/domains ["public"]
                :db.fulltext/autoDomain true}
   :post/body  {:db/valueType :db.type/string
                :db/fulltext  true
                :db.fulltext/domains ["public"]}
   :post/draft {:db/valueType :db.type/string
                :db/fulltext  true
                :db.fulltext/domains ["private"]}
   :note/body  {:db/valueType :db.type/string
                :db/fulltext  true}})
```

In this schema, `:post/title` is indexed in `"public"` and `"post/title"`, `:post/body` is indexed in `"public"`, `:post/draft` is indexed in `"private"`, and `:note/body` is indexed in the default `"datalevin"` domain.

5.2 Configuring Domains

Use `:search-domains` to configure named domains. A domain mentioned in schema but omitted from `:search-domains` is still created with default search-engine settings. Use `:search-opts` for the default `"datalevin"` domain.

These are index-time settings. For example, `:index-position? true` stores term positions and offsets, `:include-text? true` stores raw text for `:display :texts`, `:display :texts+offsets`, and re-indexing, and `:indexing-mode :async` enables asynchronous indexing for a domain.

```
(def conn
  (d/create-conn
    "/tmp/search-domains"
    search-schema
    {:search-opts
```

```
{:index-position? true
 :include-text? true}

:search-domains
{"public"     {:index-position? true
               :include-text? true}
 "private"    {:indexing-mode :async}
 "post/title" {:index-position? true}}}))
```

Use per-domain configuration when search surfaces need different behavior. For example, a public documentation domain may enable phrase search with `:index-position? true`, while a high-ingestion private domain may use `:indexing-mode :async`. Autocomplete can be another domain over the same attribute with a prefix analyzer.

5.3 Querying Domains

An unqualified `fulltext` call searches all full-text domains in the database. Use `:domains` to limit the query to one or more named domains:

```
;; Search every full-text domain.
(d/q '[:find ?e ?a ?v
       :where [(fulltext $ "clojure") [[?e ?a ?v]]]]
     db)

;; Search only public text.
(d/q '[:find ?e ?a ?v
       :where [(fulltext $ "clojure" {:domains ["public"]}) [[?e ?a ?v]]]]
     db)

;; Search only private drafts.
(d/q '[:find ?e ?a ?v
       :where [(fulltext $ "roadmap" {:domains ["private"]}) [[?e ?a ?v]]]]
     db)

;; Search only :post/title through its auto-created attribute domain.
(d/q '[:find ?e ?v
       :where [(fulltext $ :post/title "clojure") [[?e _ ?v]]]]
     db)
```

The attribute-specific form is a convenience for querying the attribute domain. It is available only when the attribute has `:db.fulltext/autoDomain true`. Domain-specific search is more general: it can search any named domain, including domains that combine several attributes.

5.4 Application Pattern: Typed Search Records

For application search, do not assume that the thing you index must be the original domain entity. Often the better model is to create explicit search records: ordinary Datalevin entities whose job is to represent searchable units of the application.

Index the thing users can open

Search records work well when a hit should land on a section, rendered block, example, API entry, or other user-facing unit. The indexed entity can be a purpose-built search entity that points back to the larger domain object.

A documentation site is a good example. If you index an entire Markdown chapter as one string, a hit tells you that the chapter matched somewhere, and offsets are accurate only inside that Markdown source. They do not necessarily line up with rendered HTML, generated anchors, syntax-highlighted code, or extracted examples. A more precise design is to index the units users can actually land on: sections, examples, figures, API entries, and other rendered blocks.

The `:search/type` attribute is application metadata. It tells the result renderer what kind of thing matched. Full-text domains are index configuration. They decide which analyzer, visibility boundary, and search surface should be used.

The pattern is ordinary schema, transaction data, and query data:

```
(def search-schema
  {:search/key     {:db/valueType :db.type/string
                    :db/unique :db.unique/identity}
   :search/type    {:db/valueType :db.type/keyword}
   :search/doc     {:db/valueType :db.type/string}
   :search/anchor {:db/valueType :db.type/string}

   :search/title   {:db/valueType :db.type/string
                    :db/fulltext  true
                    :db.fulltext/domains ["site"]
                    :db.fulltext/autoDomain true}

   :search/text    {:db/valueType :db.type/string
                    :db/fulltext  true
                    :db.fulltext/domains ["site"]}

   :search/code    {:db/valueType :db.type/string
                    :db/fulltext  true
                    :db.fulltext/domains ["code"]}})
```

An indexing job can infer the type from the source structure. A heading becomes a section record. A code fence becomes an example record. A figure caption becomes a figure record. The user still types one search query; the application uses the type internally to render better results.

```
(d/transact! conn
  [{:search/key "08-query#rules"
    :search/type :section
    :search/doc "08-query"
    :search/anchor "rules"
    :search/title "Rules"
    :search/text "Rules let you name reusable query logic..."}

   {:search/key "08-query#rules:example-basic"
    :search/type :example
    :search/doc "08-query"
```

```
  :search/anchor "rules"
  :search/title "Basic rule query"
  :search/code "(d/q '[:find ?e :where ...] db)"}])
```

Now the result entity already knows where it should send the user: `/docs/08-query#rules`. The snippet can be built from the same normalized text that was indexed, while the page link jumps to a stable rendered anchor.

Queries can search one domain or merge several domains. The offset display below assumes the `"site"` and `"code"` domains were created with `:index-position? true`:

```
(d/q '[:find (pull ?hit [:search/type
                          :search/doc
                          :search/anchor
                          :search/title])
             ?offsets
       :in $ ?q
       :where [(fulltext $ ?q {:domains ["site" "code"]
                               :top 20
                               :display :offsets})
               [[?hit _ _ ?offsets]]]]
     (d/db conn)
     "rules")
```

Use this pattern when search results need precise destinations, type-aware rendering, or different analyzers for prose and code. Keep using the original domain entities directly when the searchable field naturally belongs to the thing being returned, such as a post title, product description, or customer message.

6. Standalone Search Engine

Datalevin can be used as a standalone embedded search engine outside of Datalog:

```
(def lmdb (d/open-kv "/tmp/search-db"))
(def engine (d/new-search-engine lmdb {:index-position? true}))

(d/add-doc engine 1 "The quick red fox jumped over the lazy red dogs.")
(d/add-doc engine 2 "Mary had a little lamb whose fleece was red as fire.")

(d/search engine "red")
;=> (1 2)

(d/search engine "red" {:limit 1 :offset 1})
;=> (2)

(d/search engine "red" {:display :offsets})
;=> ([1 (["red" [10 39]])] [2 (["red" [40]])])
```

Document References: Full-text search returns document references by default. In Datalog's `fulltext`, the document reference is the datom `[e a v]`. In standalone mode, the `doc-ref` can be any value, including numbers, strings, maps, or whatever uniquely identifies a document in your application. This flexibility lets you index external content without storing source text in the search index.

Standalone search engines follow the same rule: by default they store the `doc-ref` and the inverted index, assuming the application can recover the source document from `doc-ref`. Enable `:include-text? true` when the search engine should keep raw text for `:display :texts`, `:display :texts+offsets`, or `re-index`:

```
(def engine
  (d/new-search-engine
    lmdb
    {:include-text? true
     :index-position? true}))
```

6.1 Document Lifecycle and Bulk Loading

The standalone search engine exposes the normal lifecycle operations you expect from a search index:

```
(d/doc-indexed? engine 1)
;=> true

(d/doc-count engine)
;=> 2

(d/remove-doc engine 2)

(d/clear-docs engine)
```

`add-doc` checks for an existing `doc-ref` by default and updates the index when the document is already present. During a trusted initial import, pass `false` as the fourth argument to skip that existence check:

```
(d/add-doc engine doc-ref doc-text false)
```

For larger embedded imports, use `search-index-writer`, `write`, and `commit`. The writer batches index changes and flushes final term metadata on `commit`, so call `commit` before relying on search results:

Listing 16.1. Bulk loading a standalone full-text search index

```
(def lmdb (d/open-kv "/tmp/search-db"))
(def writer
  (d/search-index-writer
    lmdb
    {:domain "docs"
     :index-position? true
     :include-text? true}))

(doseq [[doc-ref doc-text] docs]
  (d/write writer doc-ref doc-text))

(d/commit writer)

(def engine
```

```
(d/new-search-engine
  lmdb
  {:domain "docs"
   :index-position? true
   :include-text? true}))
```

The bulk writer is an embedded local API. It is not available through the Datalevin client/server API; use Datalog full-text attributes and normal transactions there, or run the writer in a local indexing process.

Treat the bulk writer as a short-lived loading handle for one search domain. In all APIs, `commit` is the required final flush. After commit, treat the writer as finished and create a new writer for a later bulk load. Do not keep multiple active writers for the same domain; use one writer for the load, then query with a search engine or use normal `add-doc` / `remove-doc` operations for incremental changes.

7. Asynchronous Indexing

Full-text indexing is synchronous by default: a transaction updates source datoms and the full-text index before returning. This preserves read-your-writes behavior for `fulltext`.

For high-ingestion workloads, a search domain can opt into async indexing:

```
(def conn
  (d/create-conn
    "/tmp/search-db"
    {:post/content {:db/valueType :db.type/string
                    :db/fulltext  true}}
    {:search-opts {:indexing-mode :async}}))
```

or per domain, when the attribute is assigned to that domain:

```
(def conn
  (d/create-conn
    "/tmp/search-db"
    {:post/content {:db/valueType :db.type/string
                    :db/fulltext  true
                    :db.fulltext/domains ["content"]}}
    {:search-domains
     {"content" {:index-position? true
                 :indexing-mode   :async}}}))
```

> **Async search is eventually consistent**
>
> In async mode, newly committed text may not be searchable until the background worker has applied the search-index job, including after DB-open recovery. Use this mode only for ingestion paths that can tolerate that lag.

Chapter 22 covers testing and waiting for async secondary indexes, and Appendix E lists the helper APIs.

8. Implementation Details

Because Datalevin implements full-text search on its own KV substrate instead of embedding Lucene, the storage layout and ranking algorithm are part of the database design. This section explains the main pieces.

8.1 Storage Layout

The search indices are stored in LMDB sub-databases:

- `terms` - map of term → term-info (id, max-weight, doc-frequency list)
- `docs` - map of document id → doc reference and norm
- `positions` - inverted lists for proximity (optional)
- `rawtext` - original document text (optional, only when `:include-text? true`)

An **inverted index** maps from a term to the documents that contain it. A **document frequency list** is the posting list for one term: the set of documents where the term appears, plus enough statistics to rank them. A **norm** is a precomputed document-length normalization factor used during ranking. Datalevin loads norms into memory on initialization and loads term information as queries need it. Document frequency lists use compressed bitmaps (Roaring Bitmaps) for efficient intersection and union operations.

8.2 Scoring: TF-IDF and the Vector Space Model

Full-text ranking uses a different kind of vector from Chapter 17's embedding vectors. In the full-text **Vector Space Model**, a document is a sparse vector whose coordinates are terms. A query is another sparse vector over the same term space. Ranking asks: which document vectors are most similar to the query vector?

Datalevin uses TF-IDF weighting described by Manning, Raghavan, and Schütze in *Introduction to Information Retrieval* [2], with the `lnu.ltn` weighting scheme. The important ideas are:

- **Term frequency (TF)**: a term that appears more often in a document is more important to that document, with logarithmic damping so repetition does not dominate.
- **Inverse document frequency (IDF)**: a term that appears in fewer documents is more discriminating than a term that appears almost everywhere.
- **Normalization**: document length affects ranking, so Datalevin normalizes document vectors to avoid blindly favoring longer documents.

In `lnu.ltn` notation, the document side and query side use different weighting:

- **Document vectors**: log-weighted term frequency, no idf, pivoted unique normalization
- **Query vectors**: log-weighted term frequency, idf weighting, no normalization

This weighting scheme is designed to handle document length without blindly penalizing longer documents.

This subsection defines the score. The next subsection explains how Datalevin uses that score to retrieve the top results without scoring every matching document.

8.3 Top-k Retrieval: T-Wand Algorithm

Datalevin uses **Tiered WAND** (T-Wand) to find the highest-scoring documents efficiently. The algorithm is described in the T-Wand blog post [3]. To understand what T-Wand adds, first understand ordinary WAND [4].

WAND is a top-k retrieval algorithm. A query such as `"red fox database"` has one posting iterator per query term. Each iterator walks the document references that contain that term, and each term has an upper bound: the largest score that term could possibly contribute to any document. During search, the engine keeps a priority heap of the current best `k` documents. Once the heap is full, its lowest score becomes the threshold a new document must beat to enter the results.

Operationally, WAND repeatedly asks which document id is still worth scoring. It orders the posting iterators by their current document id, then accumulates each term's maximum possible contribution in that order. The first iterator whose accumulated bound can beat the current top-k threshold is the pivot, and the pivot's document id becomes the next candidate. Any smaller document id can appear only in iterators before the pivot, whose combined upper bound is too low to enter the heap, so those iterators can jump directly to the pivot document id. If every iterator up to the pivot is now positioned on that same id, the engine computes the real score and updates the heap if needed. Otherwise it reorders the iterators and repeats. No result quality is traded away: a document is skipped only after its maximum possible score has been proven too low.

T-Wand keeps that WAND score pruning and adds a relevance constraint based on term coverage. It divides matching documents into tiers:

1. First, documents containing all `n` query terms.
2. Then documents containing `n - 1` query terms.
3. Then documents containing `n - 2` query terms, and so on until enough results are found.

Within a tier, documents are still ranked by TF-IDF score. Across tiers, term coverage wins: a document that contains all query terms is considered before documents that contain only some of them. This addresses a common search frustration where a high-scoring partial match outranks a complete match.

The tiering also gives the engine more ways to prune. Suppose a query has `n` terms and the current tier requires at least `t` matching terms. Sort the query terms from rarest to most common. Any document with `t` matches must contain at least one of the rarest `n - t + 1` terms. If it contained none of those rare terms, it could match at most the remaining `t - 1` terms. Therefore the engine can use the union of those rare-term posting lists as the candidate set for the tier, instead of considering every document that matches any query term.

There is a second tier-specific pruning rule while checking a candidate. If the candidate has matched `h` terms so far and only `r` unchecked query terms remain, then its best possible final coverage is `h + r`. If `h + r < t`, the candidate can no longer reach the tier and can be discarded before exact scoring. Combined with WAND's score upper bounds, this lets Datalevin avoid both low-coverage candidates and high-coverage candidates that still cannot enter the current top-k result set. This coverage check is one of the key optimizations in the T-Wand algorithm.

8.4 Term Proximity Scoring

When `:index-position? true` is enabled, a two-stage ranking applies:

1. TF-IDF scoring produces top `m * k` candidates.
2. Proximity scoring re-ranks those candidates and returns the top `k` results.

This reflects the intuition that query terms appearing closer together indicate higher relevance. Here `k` is the requested result count, normally `:top`. `m` is controlled by `:proximity-expansion`; raising it considers more candidates before re-ranking, which can improve quality at higher cost. `:proximity-max-dist` controls how far apart terms may be while still contributing to a proximity span.

Benchmark context. Datalevin's self-published comparison with Apache Lucene 8.10.1 indexed 4.1 million English Wikipedia articles and ran 40,000 TREC queries [1]. On the reported six-core Linux machine, Datalevin's median and mean query latencies were 0.8 and 3.5 ms versus Lucene's 2.7 and 4.7 ms, and Datalevin had higher throughput at every tested concurrency level.

Lucene indexed faster and had better tail latency on broad queries. This is a single-machine snapshot, not a general performance result; see [1] for the full protocol.

Summary

Full-text search in Datalevin is a native database capability: integrated with Datalog, synchronous by default, and configurable by domain. Enable `:db/fulltext`, use the `fulltext` function with EDN boolean expressions, choose analyzers that match your data, and use search domains for organized indexing without operating a separate search cluster. Within a Datalog query, the full-text function's ranking and paging bounds remain separate from the final query bounds. Chapter 21, Section 1.4 explains the corresponding physical access paths.

References

[1] Datalevin project, "Datalevin Search Benchmark," benchmark implementation. URL: https://github.com/datalevin/datalevin/tree/master/benchmarks/search-bench.

[2] Christopher D. Manning, Prabhakar Raghavan, and Hinrich Schütze, *Introduction to Information Retrieval*, Cambridge University Press, 2008, Chapter 6, "Scoring, Term Weighting and the Vector Space Model." URL: https://nlp.stanford.edu/IR-book/html/htmledition/scoring-term-weighting-and-the-vector-space-model-1.html. Section 6.3, "The Vector Space Model for Scoring." URL: https://nlp.stanford.edu/IR-book/html/htmledition/the-vector-space-model-for-scoring-1.html.

[3] Huahai Yang, "T-Wand: Beat Lucene in Less Than 600 Lines of Code," yyhh.org, November 5, 2021. URL: https://yyhh.org/blog/2021/11/t-wand-beat-lucene-in-less-than-600-lines-of-code/.

[4] Andrei Z. Broder, David Carmel, Michael Herscovici, Aya Soffer, and Jason Zien, "Efficient Query Evaluation Using a Two-Level Retrieval Process," in *Proceedings of the Twelfth International Conference on Information and Knowledge Management* (CIKM '03), 2003. DOI: https://doi.org/10.1145/956863.956944.

CHAPTER 17

Vector Search

Modern applications often need to search by semantic relatedness, not only by exact values or graph structure. In Datalevin, vector search is another indexed access path beside EAV, AVE, full-text search, and idoc indexes. Datalevin provides two related similarity-search features:

- `:db.type/vec` stores user-supplied dense vectors and indexes them for nearest-neighbor search.
- `:db/embedding true` indexes string datoms by embedding similarity. Datalevin computes the vectors during transaction processing and returns the original source datoms in queries.

This chapter explains how to configure both paths, perform semantic searches, and understand the underlying Hierarchical Navigable Small World (HNSW) engine.

A **vector** is an ordered array of numbers. A **dense vector** usually has a fixed number of dimensions, such as 384, 768, or 1536, and every position has a numeric value. An **embedding** is a vector produced by a model from some input, often text. Texts with similar meanings should land near each other in the embedding space.

Vector search is **nearest-neighbor search**: given a query vector, find stored vectors that are closest under a chosen metric. The metric defines what "closest" means. For example, cosine distance compares direction, Euclidean distance compares geometric distance, and dot product is often used when vectors have already been normalized by the model or application.

1. Configuring Vector Indexes

To store vectors in Datalevin, define an attribute with type `:db.type/vec`. Vector dimensions and similarity metrics are configured at the domain level (see Section 1.2):

```
(def conn (d/create-conn
           "/tmp/mydb"
           {:id        {:db/valueType :db.type/string
                        :db/unique    :db.unique/identity}
            :embedding {:db/valueType :db.type/vec}}
           {:vector-opts {:dimensions  300
                          :metric-type :cosine}}))
```

With this schema, a transaction that adds or replaces `:embedding` stores the vector as the datom value and updates the corresponding HNSW vector index. The datom remains ordinary Datalog data, while the vector index is a secondary index used by `vec-neighbors`.

1.1 Vector Options

Configure these in `:vector-opts`:

- **`:dimensions`** - Number of dimensions. **Required.**
- **`:metric-type`** - Similarity metric:
  - `:euclidean` (default) - Euclidean distance
  - `:cosine` - Cosine-based distance, comparing vector direction
  - `:dot-product` - Dot product
  - `:haversine` - Great-circle distance for geo coordinates
  - `:divergence` - Jensen-Shannon divergence for probability distributions
  - `:pearson`, `:jaccard`, `:hamming`, `:tanimoto`, `:sorensen` - other specialized metrics
- **`:quantization`** - Scalar type: `:float` (default), `:double`, `:float16`, `:int8`, `:byte`
- **`:connectivity`** - Connections per node (default 16). Range 5-48.
- **`:expansion-add`** - Candidates during indexing (default 128)
- **`:expansion-search`** - Candidates during search (default 64)

Quantization controls how vector numbers are stored. `:float` and `:double` preserve more numeric precision. Smaller formats such as `:float16`, `:int8`, and `:byte` reduce memory and storage, but may slightly reduce search quality because each coordinate is represented less precisely.

1.2 Vector and Embedding Domains

Chapter 16 introduced domains as named secondary indexes. Vector search uses the same idea, but the backing index is an HNSW index rather than an inverted full-text index, so the operational rules are different.

The most important difference is that vector search is always domain-scoped. Unlike `fulltext`, `vec-neighbors` and `embedding-neighbors` do not search every domain when no domain is specified. Use either the attribute-specific form or an explicit `:domains` option.

> **Vector search always chooses a domain**
>
> Full-text search can search all initialized domains by default. Vector and embedding search do not: use an attribute-specific call for the attribute domain or pass `:domains` when searching named vector or embedding domains.

For `:db.type/vec` attributes:

- Every raw vector attribute has an attribute domain, and raw vector writes always participate in it.
- Namespaced attributes use underscores in domain names: `:embedding` becomes `"embedding"`, and `:user/profile` becomes `"user_profile"`.
- `:db.vec/domains` lists shared vector domains in addition to the attribute domain. Datalevin indexes writes into all listed domains and the attribute domain, so configure every participating domain with compatible vector options when using per-domain overrides.
- All vectors in a vector domain must have the same dimensions, metric, and quantization settings.

For `:db/embedding` attributes:

- Embedding domains are configured separately from raw vector domains.
- If no `:db.embedding/domains` is specified, the attribute participates in the default `"datalevin"` embedding domain.
- `:db.embedding/autoDomain true` adds the attribute domain and enables attribute-specific `embedding-neighbors`.
- Each embedding domain is tied to an embedding provider as well as vector search options such as metric and dimensions.

Configure defaults with `:vector-opts` and `:embedding-opts`; use `:vector-domains` and `:embedding-domains` for per-domain overrides. The same domain string can appear in full-text, vector, and embedding configuration, but those are separate indexes.

2. Text Embedding Indexes

Use `:db/embedding true` when the source data is text and you want Datalevin to maintain the embedding vector index for you. Unlike `:db.type/vec`, the stored datom remains the original string. The embedding vector is a secondary index detail.

An embedding index has two moving parts:

- An **embedding provider** turns text into vectors. It may be the built-in local llama.cpp provider or an OpenAI-compatible HTTP provider.
- A **vector index** stores those generated vectors and finds nearest neighbors.

Figure 17.1 shows the text embedding indexing pipeline.

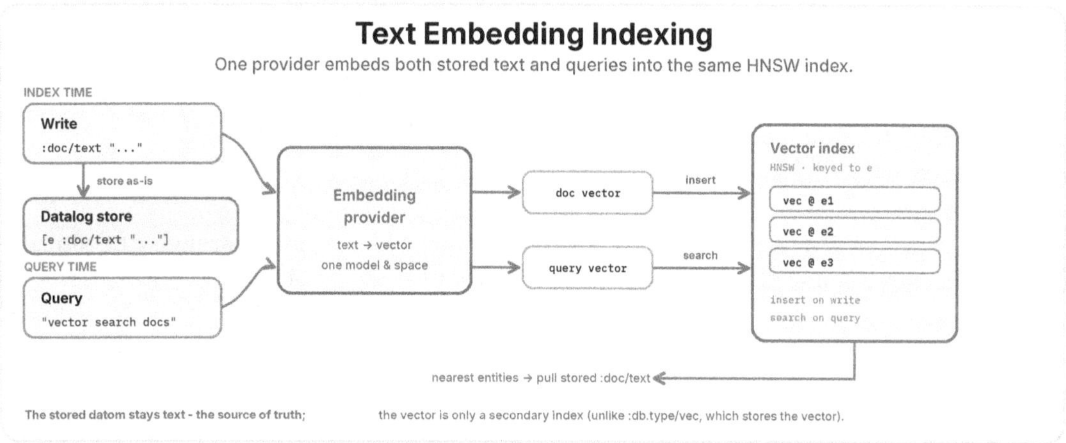

Figure 17.1: Text embedding indexing pipeline: on write, transacting :doc/text stores the datom as-is in the Datalog store (the text stays the source of truth) and also sends the text through the embedding provider to a vector that is inserted into the HNSW index keyed to the entity - the vector is only a secondary index, unlike :db.type/vec which stores the vector itself; on query, embedding-neighbors runs the query text through the same provider and embedding space to a query vector, the HNSW index does nearest-neighbor search, and the nearest entities' stored :doc/text is pulled

The provider, model, dimensions, metric, and quantization together define an embedding space.

> **Keep one embedding space per domain**
> Values from different providers, models, dimensions, metrics, or quantization settings should not be mixed in one domain. If any of those settings change for populated data, rebuild or re-index before trusting nearest neighbors.

```
(def embedding-schema
  {:doc/id    {:db/valueType :db.type/string
               :db/unique    :db.unique/identity}
   :doc/text {:db/valueType              :db.type/string
               :db/embedding             true
               :db.embedding/domains     ["docs"]
               :db.embedding/autoDomain true}})

(def conn
  (d/create-conn
    "/tmp/embedding-db"
    embedding-schema
    {:embedding-opts {:provider    :default
                      :metric-type :cosine}}))
```

With this schema, a transaction that adds or replaces `:doc/text` stores the string datom as-is, sends the text through the configured embedding provider, and inserts the resulting vector into the embedding domain's HNSW index. The generated vector is not the source of truth; it is a secondary index derived from the stored text.

> **Embeddings are derived index data**
> For `:db/embedding`, the durable fact is still the source string datom. The generated vector exists so the secondary HNSW index can answer similarity queries; application code should treat the text as the canonical value.

The built-in default provider uses the local llama.cpp embedder bundled through `dtlvnative`. If no model path is supplied, Datalevin uses `multilingual-e5-small-Q8_0.gguf`; the model is downloaded under the database root on first use when missing. GGUF is the local model file format used by llama.cpp. For larger models or hosted embedding APIs, use an OpenAI-compatible provider:

```
{:embedding-opts
 {:provider            :openai-compatible
  :model               "text-embedding-3-small"
  :base-url            "https://api.openai.com/v1"
  :api-key-env         "OPENAI_API_KEY"
  :request-dimensions 1536
  :metric-type         :cosine}}
```

Important rules:

- `:db/embedding` applies to string attributes.
- If no embedding domain is specified, the attribute participates in the default `"datalevin"` embedding domain.

- :db/embedding may coexist with :db/fulltext on the same attribute.
- Changing embedding-related schema on populated attributes requires an explicit rebuild workflow. In practice, changing the provider, model, dimensions, metric, or quantization means old indexed vectors were produced in a different embedding space; rebuild or re-index before relying on search results. Test the embedding configuration separately before adding it to the production database.

2.1 Direct Embedding Provider API

Most applications use :db/embedding and let Datalevin embed string datoms during transactions. The direct provider API is useful when application code needs embeddings before a transaction, wants to check token counts, or needs to truncate text to a model limit.

```
(def provider
  (d/new-embedding-provider
    {:provider :llama.cpp
     :model    "/models/embed.gguf"}))

(try
  {:metadata   (d/embedding-metadata provider)
   :dimensions (d/embedding-dimensions provider)
   :tokens     (d/token-count provider "Datalevin vector search")
   :vectors    (d/embed-texts
                 provider
                 [(d/truncate-text provider
                                   "Datalevin vector search"
                                   512)])}
  (finally
    (d/close-embedding-provider provider)))
```

embed-text returns one float array. embed-texts returns one float array per input string. token-count, token-counts, truncate-item, and truncate-text are the Clojure provider helpers for preparing inputs before indexing or prompt assembly. The Java, Python, and JavaScript local llama.cpp wrappers expose the same concepts as language-native methods such as embed, tokenCount / token_count, and truncateText / truncate_text.

3. Querying Vector and Embedding Indexes

Once the vector or embedding index exists, Datalog queries use neighbor functions to turn similarity search results into ordinary query rows. Raw vector attributes use vec-neighbors; text embedding attributes use embedding-neighbors.

3.1 Similarity Search with vec-neighbors

The Datalog built-in vec-neighbors function takes a query vector and returns matching datoms as [e a v] triples, ordered by similarity:

```
(d/q '[:find ?i ?v
       :in $ ?q
       :where
```

```
    [(vec-neighbors $ :embedding ?q {:top 4}) [[?e ?a ?v]]]
    [?e :id ?i]]
  (d/db conn) query-vector)
```

In Datalog vector search, the returned reference is the datom itself. Similar to `fulltext`, the returned triples are destructured and bound to variables, so Datalog joins can use them.

3.2 Search Options

Both neighbor functions accept these common options:

- `:top` - Number of results (default 10)
- `:domains` - List of domains to search
- `:display` - Result format:
  - `:refs` (default) - `[e a v]` triples
  - `:refs+dists` - `[e a v dist]` with the metric distance

For raw vector search, `vec-neighbors` also accepts `:vec-filter`, a predicate applied to each candidate vector reference before the chosen display shape is returned.

Python's `q.vector_search_options` and JavaScript's `q.vectorSearchOptions` build these maps for both neighbor functions. The dedicated `q.vec_neighbors`/`q.vecNeighbors` clause also checks that a static display mode matches the result binding width.

When `:display :refs+dists` is used, `dist` is a metric distance. Datalevin returns nearest neighbors first. Lower distances usually mean closer neighbors, but the numeric scale depends on the metric, model, quantization, and domain.

> **Do not mix distance scales**
>
> Vector distances are not comparable to full-text scores, and distances from one embedding space are not comparable to another. Fuse ranked lists by rank or evidence, or keep one index as the candidate generator and the other as a filter.

3.3 Domain-Specific Vector Search

The attribute-specific form from Section 3.1, such as `(vec-neighbors $ :embedding ?q ...)`, searches only the attribute domain for `:embedding`. Use `:domains` when the query should search named vector domains directly:

```
(d/q '[:find ?e ?id ?dist
       :in $ ?q
       :where [(vec-neighbors $ ?q
                              {:top 100
                               :domains ["embedding"]
                               :display :refs+dists})
                [[?e _ _ ?dist]]]
               [?e :id ?id]
       :order-by [?dist :asc ?id :asc ?e :asc]
       :limit 4]
     db
     query-vec)
```

Source-local and query-level bounds

The structured `:id` clause keeps only neighbors attached to an entity with an application identifier and supplies that identifier to the result. The vector function's `:top 100` is the source-local approximate result window; `:limit 4` is the final Datalog window after that join, projection, and deduplication. The optimizer never widens the source-local window, so `:top` remains a recall-versus-search-cost choice when surrounding clauses prune candidates.

Chapter 21, Section 1.4 explains when the optimizer can consume this source incrementally and preserve its distance order. That optimization can reduce downstream Datalog work, but it does not improve recall beyond `:top`.

3.4 Text Search with `embedding-neighbors`

Query an embedding index with `embedding-neighbors`. The query input is text, not a vector. Datalevin embeds the query text with the same provider and embedding space configured for the target domain, then searches the HNSW index:

```
(d/q '[:find ?id ?text
       :in $ ?q
       :where
       [(embedding-neighbors $ ?q {:domains ["docs"] :top 5})
        [[?e _ ?text]]]
       [?e :doc/id ?id]]
     (d/db conn)
     "vector search docs")
```

3.5 Attribute-Specific Embedding Search

Attribute-specific embedding search requires `:db.embedding/autoDomain true`:

```
(d/q '[:find ?id ?dist
       :in $ ?q
       :where
       [(embedding-neighbors $ :doc/text ?q {:top 5 :display :refs+dists})
        [[?e _ _ ?dist]]]
       [?e :doc/id ?id]]
     (d/db conn)
     "semantic database")
```

4. Standalone Vector Database

Datalevin can be used as a standalone vector database. Use the standalone vector-index API when you want a KV-backed vector index without a Datalog schema attribute.

```
(def lmdb (d/open-kv "/tmp/vector-db"))
(def index (d/new-vector-index lmdb {:dimensions 300}))

(d/add-vec index "cat" cat-vector)
(d/add-vec index "dog" dog-vector)
```

```
(d/search-vec index query-vector {:top 2})
;=> ("cat" "dog")
```

Vector References: In Datalog's `vec-neighbors`, results are datoms `[e a v]`. In standalone mode, `vec-ref` can be any bridge-supported value whose encoded key is at most 511 bytes: strings, numbers, maps, or any identifier meaningful to your application.

Multiple vectors can share the same `vec-ref`. For example, different image embeddings might all reference the same tag `"cat"`, or document chunks in a RAG system might all reference the same document ID.

4.1 Vector Index Lifecycle and Introspection

Standalone vector indexes also expose cleanup and inspection functions:

```
(d/vector-index-info index)
;=> {:size 2, :dimensions 300, :metric-type :euclidean, ...}

(d/remove-vec index "dog")

(d/force-vec-checkpoint! index)
(d/vector-checkpoint-state index)

(d/close-vector-index index)
```

`remove-vec` removes all vectors associated with a `vec-ref`. `clear-vector-index` closes the index and deletes all vectors. `close-vector-index` only releases index resources. `vector-index-info` reports size, memory, configuration, hardware, domain, and checkpoint metadata.

5. The Core Engine: HNSW

Datalevin's vector search uses **Hierarchical Navigable Small World (HNSW)** graphs, the graph-based approximate nearest-neighbor method introduced by Malkov and Yashunin [1], implemented via the usearch (https://github.com/unum-cloud/usearch) library with SIMD optimizations.[1]

5.1 Performance Trade-offs

HNSW is an **approximate nearest-neighbor** index. It trades exactness for faster search on large vector sets: depending on tuning and data shape, it may miss an exact nearest vector. The quality measure for this trade-off is **recall**: the fraction of true nearest neighbors that the approximate search returns. Higher recall usually costs more memory, indexing time, or query time.

> **Tune for recall, memory, and latency**
>
> HNSW settings are workload knobs, not universal constants. Validate recall against examples that represent your product's retrieval needs, then trade memory, indexing time, and query latency deliberately.

[1] SIMD means "single instruction, multiple data": CPU vector instructions that operate on several numeric coordinates at once. It is an implementation detail, but it is one reason vector distance calculations can be fast on modern CPUs.

- **`:connectivity`** (M) - Higher = better recall, more memory
- **`:expansion-add`** (efConstruction) - Higher = better index quality, slower writes
- **`:expansion-search`** (ef) - Higher = better recall, slower queries

`connectivity` controls how many graph neighbors each vector can keep. `expansion-add` controls how thoroughly the graph is searched while inserting a new vector. `expansion-search` controls how many candidates are explored during a query. If recall is too low, raise `:expansion-search` first; if newly inserted data has poor recall, tune `:connectivity` and `:expansion-add` before rebuilding.

5.2 How HNSW Works

HNSW builds a multi-layer graph where each node is a vector:

- **Higher layers** have long-range connections for fast "skipping"
- **Lower layers** have short-range connections for precision

Search starts at the top layer and "zooms in" through descending layers to find nearest neighbors. Figure 17.2 shows the search process.

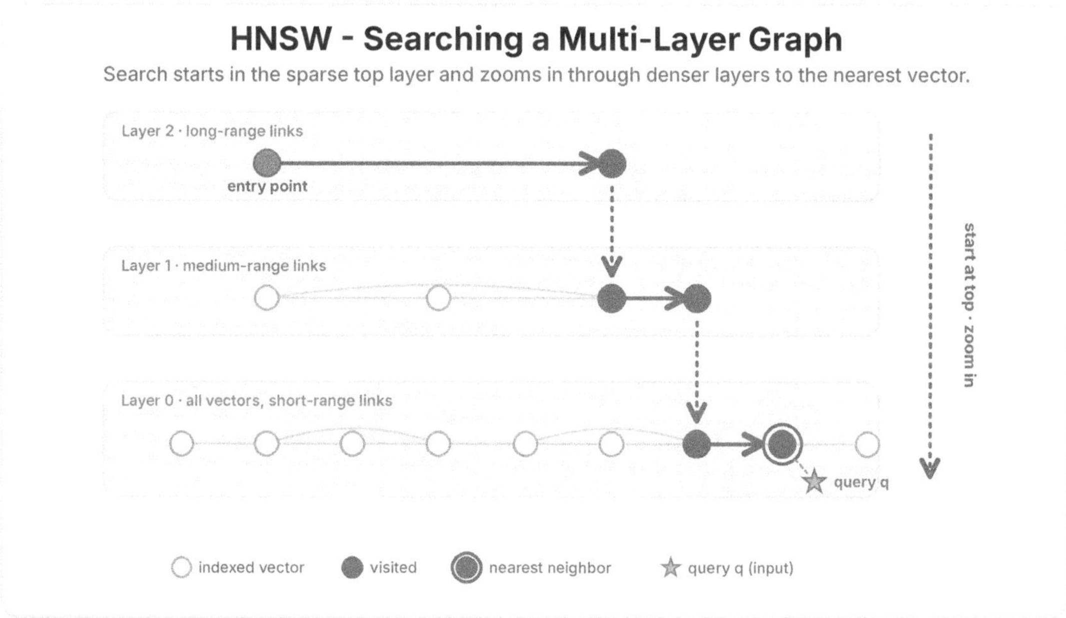

Figure 17.2: HNSW multi-layer graph search: the sparse top layer has long-range links for fast skipping and the dense bottom layer holds every vector with short-range links; search starts at the entry point and descends layer by layer, hopping to closer nodes until it reaches the nearest neighbor in Layer 0

5.3 Storage Layout

Datalevin does not store each HNSW graph edge as an ordinary datom or as a separate LMDB key. Each vector domain has a native usearch HNSW index, optimized for nearest-neighbor search. Datalevin keeps that native index open in memory while the database is open, and persists it into the same LMDB environment as checkpoint data.

The durable storage has three main pieces:

- A per-domain reference map, stored in an LMDB list database named like `<domain>/vec-refs`, maps each application-level `vec-ref` to one or more internal vector ids.
- The serialized native HNSW index is stored as chunked blobs in the `datalevin/vec-index` database. Chunking avoids requiring one large contiguous byte buffer for very large indexes.
- Checkpoint metadata is stored in `datalevin/vec-meta`. It records information such as chunk count, total serialized bytes, checkpoint LSNs, and vector replay state.

On open, Datalevin reconstructs the in-memory native index from the LMDB checkpoint and reloads the `vec-id -> vec-ref` map from the per-domain reference database. If the transaction log is enabled, the checkpoint metadata also records the replay floor so Datalevin can apply vector changes that happened after the last checkpoint.

This design keeps vector search responsive while keeping the index tied to the same database lifecycle as the rest of Datalevin. Adding or removing a vector first updates the durable LMDB reference mapping, then mutates the native HNSW index. Periodic or explicit vector checkpoints serialize the native index back into LMDB. For `:db.type/vec`, the vector value is also the datom value in the Datalog store. For `:db/embedding`, the source datom remains text, and the generated embedding vector lives only in the secondary vector index.

6. Hybrid Retrieval: Combining Logic and Similarity

Vector search integrates with Datalog, enabling hybrid queries that combine semantic similarity with structured filters:

```
(d/q '[:find ?title ?v
       :in $ ?target-vec
       :where
       [(vec-neighbors $ :product/embedding ?target-vec {:top 10}) [[?e _ ?v]]]
       [?e :product/title ?title]
       [?e :product/status :product.status/in-stock]
       [?e :product/price ?price]
       [(< ?price 100.0)]]
     db query-vec)
```

This is a common **Retrieval-Augmented Generation (RAG)** pattern: find semantically similar documents, then filter by metadata, access control, or recency. Chapter 18 develops these hybrid retrieval patterns in depth.

7. Asynchronous Vector and Embedding Indexing

Vector and embedding indexing are synchronous by default. A transaction updates the source datoms and the secondary index before returning.

For ingestion-heavy workloads, configure async indexing:

```
{:vector-opts {:dimensions    300
               :metric-type   :cosine
               :indexing-mode :async}}
```

or for embedding providers:

```
{:embedding-opts
 {:provider       :openai-compatible
  :model          "text-embedding-3-small"
  :api-key-env    "OPENAI_API_KEY"
  :metric-type    :cosine
  :indexing-mode :async}}
```

> **Async vector indexes can lag**
>
> In async mode, newly committed vectors or text may not affect neighbor search until the background worker has applied the index job. Failed jobs retry with bounded backoff and worker leases, so a later worker can reclaim stalled work.

Chapter 22 covers testing and waiting for async secondary indexes, and Appendix E lists the helper APIs.

Summary

Vector search adds semantic similarity as another Datalevin access path:

- **Raw vectors**: Use `:db.type/vec` and `vec-neighbors` when your application owns vectors directly.
- **Text embeddings**: Use `:db/embedding` and `embedding-neighbors` when Datalevin should embed string datoms and maintain the vector index for you.
- **Domains**: Vector and embedding queries are domain-scoped. Use attribute domains for focused searches and named domains when several attributes share a vector search surface.
- **Standalone indexes**: Use standalone vector indexes when vectors are not part of a Datalog schema or when an application needs direct lifecycle control.
- **HNSW tuning**: Dimensions, metric, quantization, connectivity, and expansion settings define the search space and the speed, memory, and recall tradeoffs.
- **Hybrid retrieval**: Combine nearest-neighbor results with Datalog filters, joins, and access control facts for application search and RAG.
- **Query bounds**: Vector or embedding `:top` defines the approximate source window, while a final Datalog limit applies after surrounding clauses. Chapter 21, Section 1.4 explains how the optimizer costs these access paths.
- **Async indexing**: Keep the default synchronous path for immediate consistency; use async indexing for ingestion-heavy workloads and monitor pending secondary-index jobs explicitly.

References

[1] Yu. A. Malkov and D. A. Yashunin, "Efficient and Robust Approximate Nearest Neighbor Search Using Hierarchical Navigable Small World Graphs," arXiv:1603.09320, 2016; *IEEE Transactions on Pattern Analysis and Machine Intelligence* 42(4):824-836, 2020. URL: https://arxiv.org/abs/1603.09320. DOI: https://doi.org/10.1109/TPAMI.2018.2889473.

CHAPTER 18

Hybrid Queries

Datalevin's indexes are most useful when they work together. Because Datalog, full-text search, vector similarity, text embedding search, document indexes, and the KV store can all live in the same storage system, you can combine them in a single query.

By default, full-text, vector, embedding, and idoc indexes are updated synchronously with source datoms. Full-text, vector, and embedding domains can opt into `:indexing-mode :async`; in that case source datoms commit with durable index jobs and index queries are eventually consistent until the worker catches up.

In Datalevin, **indexes are programmable capabilities.** This chapter demonstrates how to build hybrid queries that bridge structured logic and unstructured search.

Unless stated otherwise, snippets that call `fulltext` with an attribute argument, such as `:doc/content`, assume that attribute has `:db/fulltext true` and `:db.fulltext/autoDomain true`.

1. Unified Retrieval: The Single-Engine Advantage

In Datalevin, a hybrid query is a single Datalog expression. The engine manages joins between different index types automatically.

This single-engine model has consequences beyond convenience:

- **Consistency.** In the default synchronous mode, specialized indexes are updated in the same transaction as the source datoms, so a query over one database snapshot does not combine a full-text index from one committed state with a vector index from another. If a domain uses async indexing, source datoms may be visible before its secondary-index job has caught up.
- **No dual writes.** A Datalevin transaction records the source datoms and, in synchronous mode, the secondary-index updates. In async mode it records the source datoms and durable index jobs. You do not run change-data-capture pipelines or reconciliation jobs to keep separate search and vector stores in step with the system of record.
- **No fan-out.** Results are joined inside Datalevin over local indexes rather than serialized to several services and merged afterward. The query joins on entity ids instead of doing an application-level merge.
- **One operational surface.** One database environment to back up and restore, one access-control boundary, and one failure domain, instead of several systems that can fail or drift independently.

2. Index Return Formats

All specialized index functions return datoms for consistent destructuring. By default, they all return `[e a v]`. Different options return additional values.

Function	Score/Distance Option
`fulltext`	`:refs+scores` returns `[e a v score]`
`vec-neighbors`	`:refs+dists` returns `[e a v dist]`
`embedding-neighbors`	`:refs+dists` returns `[e a v dist]`
`idoc-match`	-

The shared entity id (`?e`) enables seamless joins across all index types, as depicted in Figure 18.1.

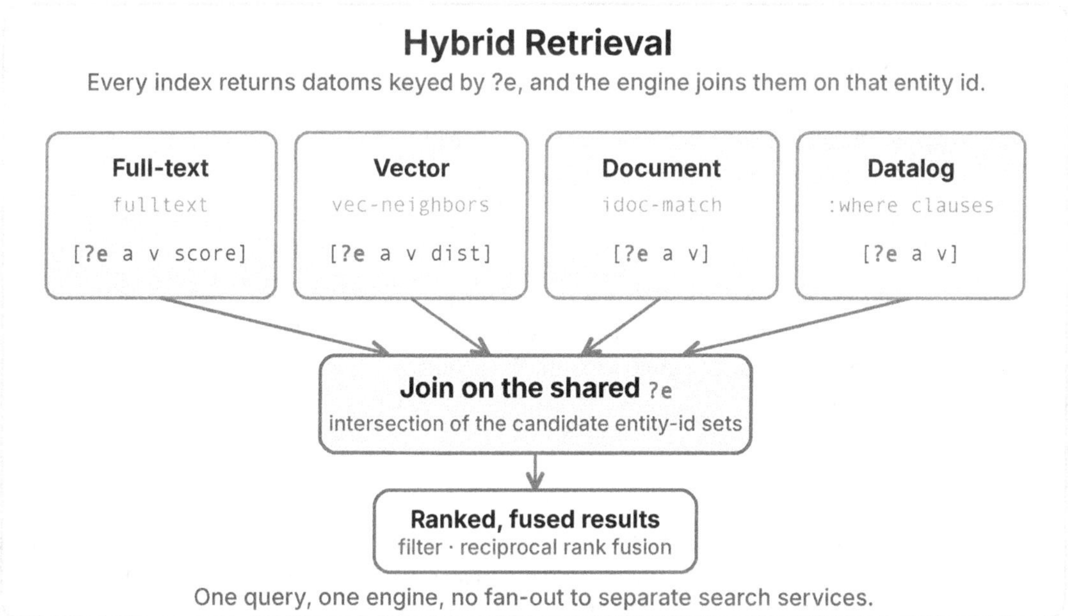

Figure 18.1: Hybrid retrieval: full-text, vector, document, and Datalog lenses each return datoms keyed by ?e, which the engine joins through shared entity ids

The implication is that the entity id is a *universal join key*. In a multi-store architecture, each system returns identifiers in its own space, such as Lucene document ids, vector-store ids, or SQL primary keys, and the application maintains translation tables to line them up. In Datalevin, every index returns datoms over the same entity-id space, so joining a full-text hit to a vector hit to a graph fact is just unification on `?e`. A practical consequence is that adding a new index type to an existing model, such as embeddings over text you already store, can compose with existing facts without building an id-mapping layer first.

Full-text scores and vector distances are different quantities. A full-text score is a relevance score where higher is better. A vector distance is a metric distance where lower is usually closer. They are not on the same scale, so production code should not combine the raw numbers directly.

> **Fuse ranks, not raw scores**
>
> Prefer a rank-based method such as reciprocal rank fusion (RRF), discussed next, or keep one index as the candidate generator and use the other as a filter. Raw scores and distances only make sense inside the scoring system that produced them.

3. Evidence Fusion

When a query combines full-text search and vector similarity, there are two different problems you might be solving.

The first problem is **intersection**: find entities that satisfy every retrieval condition. A single Datalog query with shared `?e` variables does exactly that. If a full-text clause and a vector clause both bind `?e`, the result contains entities that appear in both candidate sets.

The second problem is **fusion**: accept candidates from several ranked lists and combine their ranks into one final ranking. This is useful when either keyword evidence or semantic evidence can be enough to make an item relevant. Reciprocal rank fusion (RRF) is a simple, robust way to do this without pretending that full-text scores and vector distances share a numeric scale [1].

> **Intersection and fusion answer different questions**
>
> Intersection asks for entities that satisfy every retrieval condition. Fusion asks how to rank entities that may have evidence from any of several retrieval lenses. Choose the question first; the Datalog shape follows from that choice.

3.1 Reciprocal Rank Fusion

RRF fuses ranked lists by rank position. A common formula is `1.0 / (60 + rank)`, where `rank` is 1 for the first item, 2 for the second item, and so on. The constant `60` dampens the effect of any one list. The same entity can appear in the full-text list, vector list, and graph list; add its reciprocal-rank contributions from each list, then sort by the fused score.

The practical pattern is:

1. Run each retrieval lens with enough `:top` headroom: full-text, vector, embedding, idoc, or graph expansion.
2. Convert each result list to `(entity-id, rank, source)` records.
3. Sum `1 / (60 + rank)` per entity id.
4. Sort by the fused score.
5. Hydrate or filter the fused entity ids with ordinary Datalog facts: permissions, recency, status, ownership, and business rules.

Use the same entity id as the fusion key. This is where Datalevin's return format matters: every index returns datoms in the same entity-id space, so there is no external id-mapping table to maintain.

```
(defn rrf-score [rank]
  (/ 1.0 (+ 60 rank)))
```

```
(defn ranked [source hits]
  (map-indexed
    (fn [idx hit]
      (assoc hit
             :source source
             :rank   (inc idx)
             :rrf    (rrf-score (inc idx))))
    hits))

(defn rrf-fuse [& source->hits]
  (->> source->hits
       (mapcat (fn [[source hits]] (ranked source hits)))
       (group-by :e)
       (map (fn [[e hits]]
              {:e         e
               :rrf-score (reduce + (map :rrf hits))
               :sources   (set (map :source hits))}))
       (sort-by :rrf-score >)))

(rrf-fuse
  [:text   [{:e 10} {:e 20} {:e 30}]]
  [:vector [{:e 20} {:e 40} {:e 10}]])
```

3.2 Datalog-Native Evidence Fusion

The application-side RRF example above is useful when you want literal rank-position fusion over lists returned to application code. There is also a fully Datalog-native alternative: model each retrieval lens as evidence, assign each kind of evidence a weight, and aggregate those weights inside the query.

This is not exactly RRF unless rank positions are already represented as data. Datalog relations are unordered, so a query does not derive "the third result in the vector list" as a rank fact by itself. But for many retrieval pipelines, weighted evidence fusion is the right in-query solution: keyword evidence, semantic evidence, tag evidence, permissions, and business rules all remain in one Datalog query.

```
(def fusion-rules
  '[[(evidence ?e ?weight ?source ?query ?query-vec ?query-tags)
     [(fulltext $ :doc/content ?query {:top 50}) [[?e _ _]]]
     [(ground 1.0) ?weight]
     [(ground :text) ?source]]

    [(evidence ?e ?weight ?source ?query ?query-vec ?query-tags)
     [(vec-neighbors $ :doc/embedding ?query-vec {:top 50}) [[?e _ _]]]
     [(ground 1.0) ?weight]
     [(ground :vector) ?source]]

    [(evidence ?e ?weight ?source ?query ?query-vec ?query-tags)
     [?e :doc/tag ?tag]
     [(contains? ?query-tags ?tag)]
     [(ground 0.25) ?weight]
```

```
    [(ground :tag) ?source]]])

(d/q '[:find ?e ?title (sum ?weight)
       :with ?source
       :in $ % ?query ?query-vec ?query-tags ?allowed-owner-set
       :where
       (evidence ?e ?weight ?source ?query ?query-vec ?query-tags)
       [?e :doc/title ?title]
       [?e :doc/owner ?owner]
       [(contains? ?allowed-owner-set ?owner)]
       :order-by [2 :desc 1 :asc]]
     db
     fusion-rules
     user-query
     query-embedding
     query-tags
     allowed-owner-set)
```

The `:with ?source` clause is important. It keeps one row per evidence source before aggregation, so text evidence and vector evidence for the same entity are both counted even when they have the same numeric weight. The `:order-by` clause uses result column 2, the aggregate score, to rank entities by total evidence.

The `ground` clauses in the rules introduce literal values such as the evidence weight and source label. Each branch mentions only the inputs it actually uses. The outer query binds `?query`, `?query-vec`, and `?query-tags` before calling `evidence`, so those bound rule-call arguments are carried through even when a branch does not inspect them. The rule head itself gives all `evidence` definitions the same arity.

Literal rank-position RRF is usually an application-side fusion step unless ranks have already been represented as data. Datalog-native weighted fusion is the better fit when the fusion policy itself should be queryable data: rules produce evidence, aggregation sums it, and strict filters remain ordinary Datalog clauses. If a condition must be mandatory, keep it as a Datalog clause rather than treating it as just another evidence source.

> **Keep requirements as clauses**
>
> Evidence weights are for ranking candidates that may be relevant. Access control, tenancy, status, and other mandatory requirements should remain normal Datalog clauses so noncompliant entities never enter the result.

4. Order of Execution: How the Optimizer Thinks

When you mix different index types, the query optimizer uses cost-based analysis for both ordinary Datalog clauses and eligible physical access paths over AVE, idoc, full-text, and vector indexes. A specialized query function remains a logical relation in the query, but the optimizer can choose how that relation is produced and joined.

A **candidate set** is the intermediate set of entities produced by one clause or index function. Specialized functions such as `fulltext`, `vec-neighbors`, `embedding-neighbors`, and `idoc-`

`match` create candidate sets from their own indexes. Ordinary Datalog clauses then join, filter, or extend those candidates.

The optimizer compares two physical alternatives, and both use indexes. A **conventional plan** uses the normal complete clause-and-join execution: it consumes the candidates made available by each logical source, then applies final query bounds. An **access-path plan** makes an eligible indexed source the physical driver and can avoid consuming an unused suffix once the final result demand is satisfied. "Conventional" means complete source execution, not unindexed execution; source-local bounds such as `:top` still apply to both alternatives.

- **Choose Source Bounds Deliberately**: `:top`, domains, and index-specific filters determine which candidates enter the Datalog query. Set them large enough for recall but no larger than the application needs.
- **Final Query Bounds**: A top-level `:limit` applies after structured filters and joins. It never widens a source `:top`.
- **Physical Access Choice**: The cost model compares the conventional plan with access-path plans driven by AVE, idoc, full-text, or vector sources, including the joins and filters that follow the candidate source.
- **Join Semantics**: Shared variables such as `?e` make search hits, document matches, vector neighbors, and ordinary facts normal Datalog joins — e.g. `[?e :product/title ?title]` naturally restricts vector or full-text candidates for `?e`.

You usually do not need to reorder clauses manually. State constraints clearly, then tune the candidate-producing options when a search/vector function returns too much or too little. See Chapter 21, Section 1.4 for the optimizer's detailed access-path eligibility, ordering, and execution rules.

> **Top-k trades recall for cost**
>
> A small `:top` keeps downstream joins cheap, but it can discard a real match before structured filters see it. A larger `:top` improves recall and may increase source search work. For an eligible finite query, the optimizer may avoid joining, filtering, and materializing the whole source window, but it cannot recover candidates that `:top` excluded.

5. Mixing KV and Datalog

KV and Datalog storage do not have to live in separate files or separate services. A Datalevin Datalog store is built on the same DLMDB substrate as the key-value APIs, so application-owned KV data can live beside Datalog data in the same database environment. Use separate DBIs for those KV datasets when they need their own key/value encoding, range access pattern, or lifecycle, while Datalog keeps managing its own internal DBIs.

> **Keep application KV data in separate DBIs**
>
> Use application-named DBIs for raw key-value structures that live beside Datalog. Treat Datalevin's internal DBIs as owned by the Datalog, full-text, vector, embedding, and idoc machinery.

For performance-sensitive lookups, combine raw key-value layer access with Datalog using function bindings or descriptor-backed runtime UDFs. The example below registers `:kv/check-blacklist`, a small deterministic function that reads the relevant same-store KV DBI.

This pattern is useful when a check is naturally key-addressed and operational: deny lists, rollout flags, short-lived suppression tables, or availability markers that are maintained outside the main entity model. Keep durable domain facts in Datalog. Use the KV DBI for the compact lookup table, then let the query apply that table as one more filter.

The underlying KV layer is a first-class capability, allowing custom logic and specialized indexes. At application startup, register the UDF against the same registry used to open the connection. The registered function can close over the backing KV handle. That borrowed handle is safe to retain and reuse for as long as its owning Datalog connection or raw DB remains open. Unlike the Datalog DB object returned by `d/db`, the handle returned by `d/datalog-kv` does not need to be obtained again after writes. Close the owner, not the borrowed KV handle. The complete setup looks like this:

```
(require '[datalevin.core :as d]
         '[datalevin.udf :as udf])

(def blacklist-fn
  {:udf/lang :app
   :udf/kind :query-fn
   :udf/id   :kv/check-blacklist})

(def registry (udf/create-registry))

(def conn
  (d/get-conn "/tmp/app" schema
    {:runtime-opts {:udf-registry registry}}))

(let [kv (d/datalog-kv conn)]
  (d/open-dbi kv "blacklisted-users")
  (udf/register! registry blacklist-fn
    (fn [user-id]
      (boolean
        (d/get-value kv "blacklisted-users" user-id :string :data)))))
```

Once registered, the descriptor can be passed to queries and called through the portable `udf` form:

```
(d/q '[:find ?e
       :in $ ?blacklist-fn
       :where [?e :user/id ?id]
              [(udf ?blacklist-fn ?id) ?blocked?]
              [(false? ?blocked?)]]
     db blacklist-fn)
```

The Datalog query uses a function binding through the portable `udf` form: the function receives a value from the query and returns another value that the query can bind. The callable receives only the arguments that appear after the descriptor in the `udf` clause. In the query above, it receives `?id`; it does not receive `db` as an implicit first argument. Descriptor-backed UDFs are the portable shape for server deployments and code that should be configured through an application registry.

Keep query UDFs deterministic and cheap. If a function performs network calls, writes state, or scans a large KV range, it can make query behavior surprising and hard to optimize.

> **Query UDFs are read-time hooks**
>
> Use query functions for deterministic lookup and filtering. If a write must update Datalog datoms and application-owned KV DBIs atomically, perform that write in an explicit transaction rather than during query evaluation.

When a write must update Datalog datoms and application-owned KV DBIs atomically, use the `with-transaction` pattern from Chapter 6.

6. Practical Pattern: Hybrid Retrieval with Structured Filters

The database pattern behind Retrieval-Augmented Generation (RAG) is the same pattern this chapter has been building toward: use full-text and vector indexes to produce candidates, then join those candidates with ordinary Datalog facts for permission, recency, ownership, status, and business rules.

The important database point is that the filter should run in the same query as retrieval. That avoids a separate step where the system fetches candidates and then has to remember to filter them before handing context to another component. In a fan-out architecture, filters often run after results are merged from several services, which is where access-control leaks and stale data can slip in. In a single Datalevin query, an entity that fails a permission or policy clause is not part of the result, no matter which index surfaced it.

Chapter 25 returns to this shape in the agent-memory setting. There, the problem is no longer just candidate retrieval; it is recall and context assembly: choosing the right lenses, ranking evidence, preserving provenance, and formatting a small packet for an LLM. This chapter stays at the database level: how to make heterogeneous indexes participate in one planned query.

7. End-to-End Example: The "Smart" Product Search

After the retrieval pieces are clear, a larger query can combine all of them. This example finds products that:

1. Match the term "clojure" in the description (full-text)
2. Are semantically similar to a user's interest vector or query text
3. Are in stock and in a specific category (Datalog)
4. Have specific nested metadata (idoc)
5. Pass a fast operational policy check stored in a KV DBI (KV)

This is the intersection pattern from Section 3: every retrieval and structured condition must bind the same `?e`. Use fusion when either textual or semantic evidence should be enough to admit a candidate.

This example assumes `:product/desc` has `:db/fulltext true` and `:db.fulltext/autoDomain true`, `:product/embedding` is a vector attribute, `:product/metadata` is an idoc attribute, and the KV-backed policy check is registered as the descriptor-backed query UDF `:kv/sellable-sku?`, using the pattern from Section 5.

Chapter 14 introduced `idoc-match`: it searches inside a path-indexed document value. In this example it filters products whose nested metadata says `{:discounted true}` without requiring that nested field to become a separate Datalog attribute.

Listing 18.1. Hybrid product search with text, vector, logic, KV, and idoc

```
(def sellable-fn
  {:udf/lang :app
   :udf/kind :query-fn
   :udf/id   :kv/sellable-sku?})

(d/q '[:find ?title ?dist
       :in $ ?search-term ?target-vec ?category ?sellable-fn
       :where
       ;; full-text search
       [(fulltext $ :product/desc ?search-term) [[?e _ _]]]
       ;; vector search with distance
       [(vec-neighbors $ :product/embedding ?target-vec
                       {:top 20 :display :refs+dists}) [[?e _ _ ?dist]]]
       ;; structured filters
       [?e :product/status :product.status/in-stock]
       [?e :product/category ?category]
       [?e :product/sku ?sku]
       [?e :product/title ?title]
       ;; KV-backed policy check
       [(udf ?sellable-fn ?sku) ?sellable?]
       [(true? ?sellable?)]
       ;; idoc metadata filter
       [(idoc-match $ :product/metadata {:discounted true}) [[?e _ _]]]
       :order-by [?dist :asc ?title :asc]
       :limit 10]
     db "clojure" user-embedding :books sellable-fn)
```

The shared `?e` variable joins all results together. The _ placeholders discard returned attribute and value positions that this query does not need. The final `:order-by` and `:limit` ask for the ten nearest qualifying products. The vector function's `:top 20` remains the approximate candidate universe; the full-text, idoc, structured, and UDF clauses determine which of those candidates qualify for the final limit. The `?sellable-fn` input names ordinary application code that can read a small KV DBI, for example a same-file operational table of suppressed SKUs, real-time availability flags, or rollout policy. Register it with the UDF registry used to open the connection, as in Section 5.

Summary

Datalevin's index model is about **composition**. Full-text, vector, embedding, idoc, graph, relational, and KV-backed checks can participate in one Datalog query because they all return or constrain the same entity-id space.

- **Universal join key**: Specialized indexes return datoms keyed by entity id, so a full-text hit, vector neighbor, idoc match, and ordinary fact can meet on `?e` without an application-side id mapping table.
- **Intersection vs. fusion**: Use a single query with shared variables when all conditions must hold. Use reciprocal-rank fusion or weighted Datalog evidence when several retrieval lenses should contribute evidence and any one lens may be enough to surface a candidate.
- **Planner-friendly composition**: AVE, idoc, full-text, and vector operations can compete as costed physical access paths, while source bounds remain separate from final query bounds. Chapter 21 explains the execution details.
- **Operational composition**: Descriptor-backed UDFs let query-time logic read small same-store KV DBIs without giving up the portable server-safe UDF model.
- **Consistency**: Indexes are synchronous by default, with explicit async modes for expensive secondary indexing when ingestion throughput matters more than immediate index visibility.

Treating every index as a programmable capability lets Datalevin answer queries that would otherwise require several services and an application-level merge step.

References

[1] Gordon V. Cormack, Charles L. A. Clarke, and Stefan Buettcher, "Reciprocal Rank Fusion Outperforms Condorcet and Individual Rank Learning Methods," *Proceedings of the 32nd International ACM SIGIR Conference on Research and Development in Information Retrieval* (SIGIR '09), 2009. DOI: https://doi.org/10.1145/1571941.1572114.

PART V

Performance and Operations

CHAPTER 19

Storage Tuning and Durability

Performance and durability are the two sides of the same operational problem: the database must be fast, bounded, recoverable, and predictable under failure. This chapter starts with durability, backup, and maintenance practices, then covers memory and storage tuning.

1. Durability Model

Performance only matters if the data remains recoverable. Datalevin's storage engine (LMDB/DLMDB) is designed around copy-on-write commits and fast recovery after system crashes.

This section covers default LMDB durability, sync behavior, and flags that deliberately weaken the crash boundary.

1.1 Durability: The Power of Copy-on-Write (CoW)

Traditional databases such as PostgreSQL use a Write-Ahead Log (WAL) with an update-in-place storage strategy. If the system crashes before dirty pages are checkpointed, the database recovers by replaying the log before normal service resumes.

Datalevin's storage engine (LMDB) uses a **Copy-on-Write (CoW) B+Tree.**

1. **Never Overwrite**: When you write a new datom, LMDB doesn't modify existing pages. Instead, it creates a *new version* of the affected pages.
2. **Atomic Visibility**: The new pages do not become the current database state just because they have been written. They become visible only when LMDB publishes metadata that points to the new B+Tree root. Until that metadata update succeeds, readers and restart recovery still use the previous committed root.
3. **Durability Boundary**: In the default durable mode, commit confirmation follows `pwrite` page writes with the required `fsync` calls so the new pages and the metadata that publishes the new root are safely on disk. The "atomic" part is the switch from the old committed root to the new committed root; the durability part is making the pages and metadata reach stable storage before the commit is reported as durable.
4. **Instant Recovery**: If new pages reach disk but power fails before the root metadata is updated, those pages are not part of the committed database. On restart, LMDB chooses the last valid committed root. The interrupted transaction is lost from the application's point of view, just as if it had never committed. The bytes for the unpublished pages may still be present in the file, but no committed root reaches them, so they are not queryable database state and may be overwritten or reused later. If the root metadata was updated and flushed successfully, the new version is the committed state. There is no long log-replay process for the direct LMDB file because recovery chooses between valid committed roots rather than repairing an in-place update.

1.2 Syncing to Disk

In standard mode (the default for local stores), the speed of your writes is primarily limited by the speed of synchronous storage flushes.

- **Durable by default**: On every transaction commit, direct LMDB writes changed pages with `pwrite` and calls `fsync` before reporting a durable commit.
- **Hardware Impact**: On high-speed NVMe drives, this flush is very fast. On older magnetic drives or cloud-based block storage (like AWS EBS), it can be a major bottleneck.

1.3 Non-Durable Environment Flags

Chapter 20 shows the concrete LMDB environment flags for repeatable imports and cache rebuilds. From an operations perspective, the rule is simple: every flag that reduces commit-time flushing changes the crash boundary. `:nometasync` is the least aggressive option because database integrity is retained, though the last transaction may be lost. `:nosync` and `:writemap`/`:mapasync` trade much more safety for speed and can leave the database corrupted after an untimely system crash.

Use these modes only when the data can be rebuilt from a durable source or when the application has an explicit recovery plan. For raw KV stores, call the KV `sync` operation at explicit checkpoints if you need to force a flush. For the full flag table, setup examples, and ingestion trade-offs, see Chapter 20.

> **Non-durable flags change the crash boundary**
> Use reduced-sync LMDB modes only for rebuildable data, repeatable imports, temporary stores, or jobs with a separate source of truth. They are performance tools, not general production defaults for irreplaceable writes.

2. WAL Mode

While standard LMDB commits are durable by default, they can be limited by disk I/O. Datalevin's **Write-Ahead Log (WAL) mode** is an explicit opt-in for local embedded Datalog and KV stores, is required on the primary for non-HA async read replicas, and is forced on for consensus-lease HA. Use WAL when you need higher write throughput from concurrent callers, WAL replay, replication, or HA behavior.

The operational loop is: choose a durability profile, monitor LSN watermarks, optionally consume committed records with `open-tx-log`, create snapshots or checkpoints, and garbage-collect WAL segments that are no longer retained.

2.1 Durability Profiles

WAL mode durability is specified with a database open-time option called durability profile.

- **`:strict`**: The database waits for durable WAL acknowledgement for every transaction using the configured WAL sync mode. This is the default for new stores whenever WAL is enabled, including local embedded stores and consensus-lease HA.

- `:relaxed`: Transactions are batched before syncing. This is significantly faster but risks losing the last few milliseconds of work in a crash. This is Datalevin's group commit profile: several transactions' WAL records can be collected and forced to storage with one sync call, so those transactions become durable together. It does not make the database corrupt after a crash: recovery opens a consistent database and replays durable WAL records, but the latest committed records that had not yet crossed the durability boundary may be absent. This is an explicit opt-in for local WAL; consensus-lease HA does not accept it.
- `:extra`: Uses even stricter durability guarantees, such as `fcntl(F_FULLFSYNC)` on macOS, to protect against hardware write-cache failures.

Specify the profile in the same open-time options map that enables WAL. The following opens a local Datalog database with strict WAL durability:

```
(def conn
  (d/get-conn "/data/app-db"
              schema
              {:wal? true
               :wal-durability-profile :strict}))
```

Use the same option keys with `open-kv` / `openKV` / `open_kv` / `openKv` for raw KV stores. For a new store, WAL uses `:strict` when no profile is specified. Select `:relaxed` explicitly only when the workload accepts a short crash-loss window for higher throughput, or select `:extra` when the platform-specific stronger flush is required.

An existing WAL database retains its persisted durability profile when it is reopened. To move a database created under the earlier `:relaxed` default to `:strict`, reopen it with `:wal-durability-profile :strict` explicitly.

2.2 The LSN Lifecycle

In WAL mode, every transaction is assigned a **Log Sequence Number (LSN)**. LSNs are positive, strictly increasing commit positions in the WAL stream. They are the unit Datalevin uses for durability tracking, recovery replay, snapshots, replication, and WAL garbage collection.

Figure 19.1 should be read from left to right. A transaction first becomes **committed** when Datalevin accepts it into the WAL commit stream and assigns an LSN. It becomes **durable** when the WAL record containing that LSN has reached the durability boundary selected by the current profile. It becomes **applied** when Datalevin knows the corresponding state is present in the LMDB file or covered by recovery-marker state. Later, `create-snapshot!` records a checkpoint that covers applied state, and `gc-txlog-segments!` can reclaim WAL segments that are older than all retention floors.

Each watermark answers a different operational question. Has Datalevin accepted the transaction into the WAL stream? Has the WAL record reached storage under the selected durability policy? Has the LMDB file or recovery-marker state caught up to that record? Under `:relaxed` group commit, a transaction can be committed before it is durable. A durable transaction can also be ahead of the applied marker until checkpoint work catches up. This is normal operational lag, not logical inconsistency: committed data is visible according to the transaction semantics of the open database, durable data survives the crash boundary, and unapplied durable WAL can be replayed during recovery.

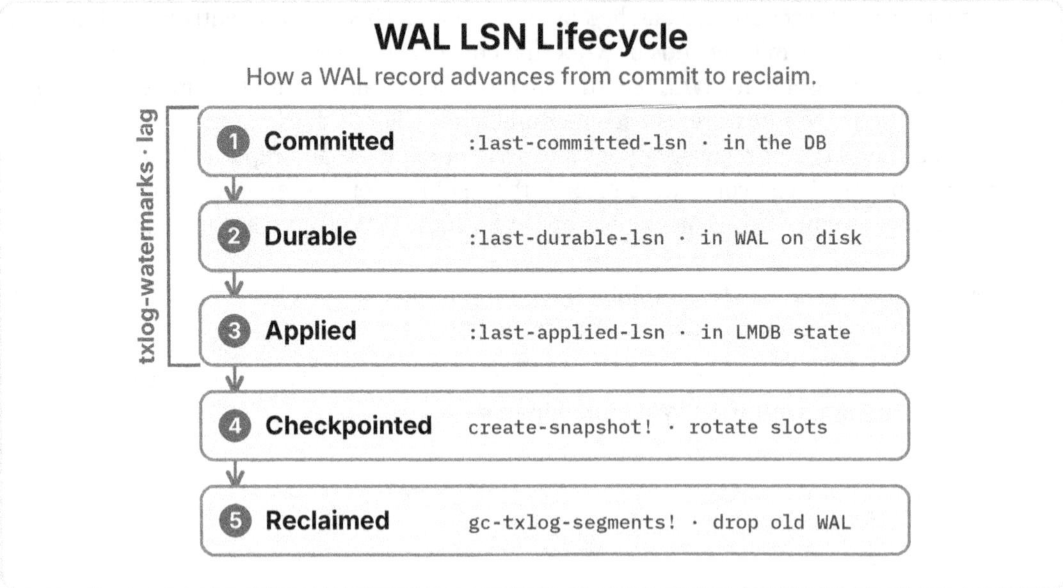

Figure 19.1: The WAL LSN lifecycle: a transaction advances from committed to durable to applied (the three watermarks read with txlog-watermarks), then is checkpointed by create-snapshot! and finally reclaimed by gc-txlog-segments!

> **Committed, durable, and applied are different**
> WAL watermarks let operators distinguish visible committed work from work that has crossed the selected durability boundary and work already reflected in LMDB checkpoint state. Lag between those positions can be normal, especially under relaxed group commit.

- **`:last-committed-lsn`**: The latest transaction that has been successfully committed in the database.
- **`:last-durable-lsn`**: The latest transaction that has reached the WAL durability boundary selected by the current profile and sync mode.
- **`:last-applied-lsn`**: The latest transaction known to be applied to the LMDB state or covered by recovery marker state.

By monitoring these watermarks with `txlog-watermarks`, you can precisely track the "lag" between application commits and physical disk durability.

In Clojure, call `txlog-watermarks` on the local KV handle. For a Datalog connection, use `datalog-kv` to get the backing KV handle. The Java, Python, and JavaScript Datalog bindings expose this as a connection-level operation, and call through to the connection's backing store internally:

```
(def conn
  (d/get-conn "/data/app-db" schema {:wal? true}))

(d/transact! conn [{:db/id -1 :event/name "started"}])
```

```
(-> (d/txlog-watermarks (d/datalog-kv conn))
    (select-keys [:wal?
                  :durability-profile
                  :last-committed-lsn
                  :last-durable-lsn
                  :last-applied-lsn]))
;; => {:wal? true
;;     :durability-profile :strict
;;     :last-committed-lsn 1
;;     :last-durable-lsn 1
;;     :last-applied-lsn 1}
```

If `:last-committed-lsn` is ahead of `:last-durable-lsn`, the WAL has accepted transactions that have not yet reached the selected durability boundary. If `:last-durable-lsn` is ahead of `:last-applied-lsn`, durable WAL records exist that are not yet covered by the LMDB applied marker used for recovery. Under `:strict`, the committed-to-durable gap is usually small or zero in steady state. Under `:relaxed`, short committed-to-durable lag is expected during group commit. The durable-to-applied gap is reported as `:durable-applied-lag-lsn` and `:checkpoint-stale?` in `txlog-watermarks`; it is controlled by LMDB checkpoint cadence (`:lmdb-sync-interval-ms` / `:lmdb-sync-interval` and `:wal-checkpoint-stale-threshold-ms`), not by the durability profile alone. These checkpoint options are explained in Section 2.4 and summarized in Appendix F.

On startup, a WAL-enabled store compares the LSN applied to the LMDB file with the durable LSN in the log. If the log is ahead, Datalevin validates newer log records and replays them into the LMDB environment before opening the database for normal reads and writes.

2.3 Reading Committed Transactions

Use **`open-tx-log`** when you need to consume committed WAL records, for example to build an audit export, a change-capture process, or an external replication bridge. The function reads records by LSN. The lower bound is inclusive, and the optional upper bound makes catch-up batches explicit:

```
;; Read committed records from LSN 5000 onward.
(d/open-tx-log kv 5000)

;; Read a bounded inclusive range.
(d/open-tx-log kv 5000 5999)
```

Each returned record includes its `:lsn` and an `:ops` collection containing the KV operations recorded for that committed transaction. A consumer should store the last processed LSN in its own state and resume from the next LSN on restart. If crash-surviving audit output matters, compare progress with `txlog-watermarks` and treat records at or below `:last-durable-lsn` as the durable boundary.

Retention matters for long-running consumers. `gc-txlog-segments!` may remove old segments after snapshots, replicas, vector checkpoints, backup pins, and operator retention floors no longer require them. If a consumer may lag, run GC with an explicit retain floor, such as `(d/gc-txlog-segments! kv next-lsn)`, or use deployment-level retention policy so the consumer's next unread LSN remains available.

2.4 Snapshots and Checkpoints

WAL mode introduces checkpoint, snapshot, and retention maintenance. Checkpoints are LMDB environment syncs. Use `force-lmdb-sync!` to force one immediately, and use `:lmdb-sync-interval-ms` or `:lmdb-sync-interval` in seconds to set the regular cadence. `:wal-checkpoint-stale-threshold-ms` sets the age at which `txlog-watermarks` reports `:checkpoint-stale?`; by default, Datalevin uses twice the sync interval.

Datalevin can run snapshot creation from its built-in scheduler when the WAL-enabled store is opened with `:snapshot-scheduler? true`; otherwise, `create-snapshot!` remains available for explicit operator-controlled snapshots.

When the snapshot scheduler is enabled, Datalevin creates snapshots automatically according to `:snapshot-interval-ms`, `:snapshot-max-lsn-delta`, `:snapshot-max-log-bytes-delta`, and `:snapshot-max-age-ms`. Calling **`create-snapshot!`** manually is for an on-demand snapshot or for deployments that choose to run their own scheduler.

Snapshot creation forces the WAL and LMDB state to disk, copies the LMDB environment into the `current` snapshot slot, rotates the previous `current` snapshot into the `previous` slot, and advances snapshot-floor bookkeeping used by WAL retention. WAL maintenance helpers are available directly on local KV handles and Datalog connections:

```
(def kv
  (d/open-kv "/data/app-kv"
             {:wal? true
              :snapshot-scheduler? true}))

(d/txlog-watermarks kv)
(d/list-snapshots kv)

;; Optional: force an on-demand snapshot now.
(d/create-snapshot! kv)
```

2.5 WAL Segment Garbage Collection

Use **`gc-txlog-segments!`** to reclaim disk space by deleting old WAL segments that are no longer needed by snapshots, replicas, vector checkpoints, or operator retention pins. Datalevin respects the `:wal-retention-bytes` and `:wal-retention-ms` policies, but deletion happens when this cleanup operation runs.

```
;; Reclaim disk space.
(d/gc-txlog-segments! kv)

;; Keep WAL records from LSN 5000 onward.
(d/gc-txlog-segments! kv 5000)
```

WAL snapshots bound recovery and retention inside a store.

> **Snapshots are not backup copies**
>
> WAL snapshots help one store recover and decide which WAL segments can be retained or collected. They do not replace independent backup copies, off-site retention, or restore drills.

3. Backup and Maintenance

This section covers online copies, compaction, command-line maintenance, logical dump/load, and re-indexing.

3.1 Online Backup Copies: d/copy

In other databases, backing up a live database can be tricky. If you simply copy the file while a write is happening, the copy might be corrupted.

Datalevin's **d/copy** function creates a transactionally consistent backup copy from a Datalog connection or DB object, or from a KV handle. This is the ordinary online backup path: use it when you want a recoverable point-in-time copy of the current database without changing the source. It can run while the database is active. Readers can continue using the source, and writers do not have to stop for the copy to be consistent.

This section is about the default, non-compacting copy. It preserves the database content and layout closely enough for fast backup creation. If the goal is to reclaim free pages after large deletes or to rewrite the destination into a smaller file, use the compacting form in Section 3.2.

```
;; Copy a live Datalog database from a connection
(d/copy (d/db conn) "/path/to/backup-db")

;; Copy a raw KV store
(d/copy kv "/path/to/backup-kv")
```

The same operation is available from the command line:

```
$ dtlv -d /data/companydb copy /backup/companydb-2026-01-15
```

- **Backup purpose**: Use ordinary `copy` for routine online backup when the destination does not need compaction.
- **Consistency**: The copy represents a single transactionally consistent point in time.
- **Operational impact**: LMDB's MVCC architecture lets the copy run while the source remains available. A long-running copy still consumes I/O and can keep older pages live until it finishes, so schedule large copies deliberately.

3.2 Compacting with d/copy

The compacting form of `copy` uses the same consistency guarantees as Section 3.1, but it serves a different maintenance purpose. LMDB reuses deleted pages inside the source file; it does not shrink the file in place. A compacting copy rewrites live pages into a new destination, leaving free pages behind in the source. Use it after large deletes, after major data reshaping, or before archiving a smaller backup artifact.

```
;; Copy and compact a Datalog database
(d/copy (d/db conn) "/path/to/compact-backup-db" true)

;; Copy and compact a raw KV store
(d/copy kv "/path/to/compact-backup-kv" true)
```

In Java and Python, pass `true` / `True` as the compact argument to `copy`. In JavaScript, pass `{ compact: true }`.

The command-line form uses `-c`:

```
$ dtlv -d /data/companydb -c copy /backup/companydb-compact
```

Compaction is not an in-place operation. It creates a new compacted destination. If you want the compacted copy to replace the production database, treat that as a controlled maintenance step: stop writers, close the old connection, move or restore the compacted directory into place, and reopen.

3.3 The `dtlv` Command-Line Tool

The `dtlv` CLI tool is the operator entry point for maintenance tasks. Use it when a backup, compaction, dump/load, or inspection task belongs in a shell script or runbook rather than in application code. The command-specific help is usually the best reference while operating a system:

```
$ dtlv help
$ dtlv help copy
$ dtlv help dump
$ dtlv help load

# View database statistics
$ dtlv -d /data/companydb stat
```

The examples above this section showed `dtlv copy` for backup and `dtlv -c copy` for compaction. The next section covers `dtlv dump` and `dtlv load`.

3.4 Dump and Load Formats

Dump/load is a logical export/import path. Use it when you need a portable data file, a version-independent migration path, or a way to move data into a new database directory. For routine same-version backups, `copy` is usually simpler and faster.

Datalevin supports two dump/load encodings:

- **EDN (default)**: Human-readable, version-independent text.
- **Nippy**: Binary serialization for faster dump/load of large databases.

For a Datalog database, pass `-g` / `--datalog`:

```
# EDN Datalog dump/load
$ dtlv -d /data/companydb -g -f ~/company.edn dump
$ dtlv -d /data/newdb -g -f ~/company.edn load

# Nippy Datalog dump/load
$ dtlv -d /data/companydb -g -n -f ~/company.nippy dump
$ dtlv -d /data/newdb -g -n -f ~/company.nippy load
```

For a named KV sub-database, omit `-g` and pass the DBI name to `dump` or `load`:

```
# EDN dump/load for a named KV DBI
$ dtlv -d /data/companydb -f ~/sales.edn dump sales
$ dtlv -d /data/newdb -f ~/sales.edn load sales
```

Use `-n` / `--nippy` when speed and file size matter more than human inspectability. Keep EDN when the dump needs to be reviewed, transformed, or stored as a long-lived interchange artifact.

3.5 Database Upgrades

When upgrading Datalevin to a new minor version, you may need to migrate your database. Datalevin provides automatic migration for databases newer than version 0.9.27.

In general, Datalevin only supports newer versions opening databases created or previously opened by older versions. The reverse is not supported. For example, if a database has been opened by Datalevin 0.10.18, it may not be possible to open that database again with Datalevin 0.10.5.

Automatic migration: For databases created with Datalevin 0.9.27 or later, opening with a newer version triggers automatic migration. This process downloads the old version's uberjar to dump the data, then loads it with the new version. Automatic migration detects Datalog stores by their `datalevin/eav` DBI; if the same file also contains extra KV DBIs, dump and load those manually.

Manual migration: For older databases, use the command-line tool:

```
# 1. Backup and compact with old version
$ dtlv-0.4 -d /src/dir -c copy /backup/dir

# 2. Dump with old version
$ dtlv-0.4 -d /src/dir -g -f dump.edn dump

# 3. Load with new version
$ dtlv -d /dest/dir -f dump.edn -g load
```

3.6 Re-Indexing

Datalevin provides a `re-index` function that dumps and reloads data to rebuild indexes. This can be useful in recovery scenarios, when changing index options, or when index structures need to be rebuilt. It is only safe when no other thread or process is accessing the same database.

> **Re-indexing is exclusive maintenance**
>
> Schedule re-indexing as an operator task on the host that owns the files. Close other connections and stop server access first; rebuilding indexes while another process uses the same database is not a live online operation.

```
;; Re-index a Datalog connection with existing schema/options
(def reindexed-conn (d/re-index conn {}))

;; Re-index with an updated schema and options
(def reindexed-conn
  (d/re-index conn new-schema {:backup? true}))
```

KV stores and standalone search engines expose corresponding re-index operations. Client/server users should schedule re-indexing as an operator task on the host that owns the local database files.

4. Durability and Maintenance Checklist

For recoverable operations, follow this checklist:

1. **Choose the right write mode**: Use default LMDB commits for simple local durability; enable WAL when you need WAL throughput, replay, replication, or HA.
2. **Use WAL snapshots**: Enable `:snapshot-scheduler? true`, or run your own scheduled `create-snapshot!`, so WAL/LMDB state is checkpointed and snapshot slots are rotated.
3. **Track `open-tx-log` consumers**: If you use `open-tx-log`, persist each consumer's next LSN and keep retention floors ahead of garbage collection.
4. **Run `gc-txlog-segments!`**: Reclaim disk space regularly.
5. **Monitor disk space**: LMDB and WAL need enough free space to perform their operations.
6. **Automate `d/copy` or `dtlv copy`**: Run a daily or hourly backup copy and move the result to an off-site location (e.g., AWS S3).
7. **Test your restores**: Regularly practice restoring from a backup copy; if using WAL, verify recovery and replay behavior.
8. **Use NVMe for speed**: The durability of your database is directly tied to the IOPS and latency of your disk.
9. **Compact periodically**: Use `d/copy` with `true` as the third argument, or `dtlv -c copy`, to reclaim disk space in a destination copy after large deletions.
10. **Plan upgrades**: When upgrading Datalevin versions, use dump/load for databases older than 0.9.27.

The next operational concern is sizing the local store so these durability and maintenance choices remain predictable under production load.

5. Memory Layout and Storage Tuning

Datalevin is a "zero-copy" database, which means its memory management is very different from an ordinary application that keeps its working data as managed runtime objects. As discussed in the preface and Chapter 4, Datalevin is layered: native LMDB storage, a JVM bridge, database semantics, and host APIs. When this section mentions runtime garbage collection, it is talking about Datalevin's runtime heap, not a requirement on the application implementation language. The important point is that the database pages themselves are file-backed pages managed by the operating system, not objects managed by Datalevin's runtime heap.

This section covers practical defaults for tuning Datalevin's memory and storage parameters for stable production performance.

5.1 The LMDB Map: `:mapsize`

The most critical parameter for Datalevin is **`:mapsize`**. This is the OS-allocated size for the memory-mapped database file, which defines the maximum size the file can grow to in the address space.

5.2 Virtual Memory vs. Resident Memory

Unlike a traditional managed application heap, the `:mapsize` is a **virtual memory** reservation, as shown in Figure 19.2. Setting a 1TB mapsize does *not* mean the database will consume 1TB of RAM. It simply means the OS will reserve a 1TB "address space" for the database file.

The practical ceiling is the process's usable virtual address space, which is CPU- and OS-dependent. For example, many modern 64-bit systems implement 48-bit virtual addresses and give a user process about 128 TiB of address space. Keep `:mapsize` comfortably below that ceiling and leave room for other memory mappings in the process.

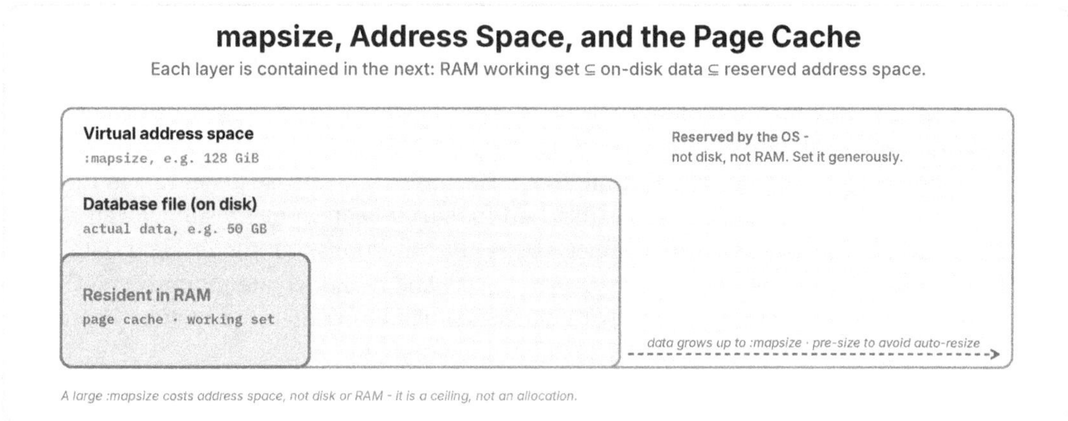

Figure 19.2: mapsize, address space, and the page cache as three nested scales: :mapsize reserves a large virtual address space (not disk, not RAM, safe to set generously); the database file grows on disk up to that mapsize (pre-size to avoid a costly auto-resize); and only the hot working set is resident in RAM via the OS page cache (zero-copy reads, outside Datalevin's runtime heap). Each layer is a subset of the one above

- **Recommendation:** Set `:mapsize` to be significantly larger than your expected data size. For example, if you expect 50 GiB of data, reserving 256 GiB gives room for growth and large transactions.
- **Default:** The default initial mapsize is 1000 MiB and grows automatically. For production, set a larger value upfront when growth is predictable.

```
;; Set mapsize to 128 GiB (in MiB)
(d/get-conn path schema {:kv-opts {:mapsize 131072}})
```

> **Set mapsize before production growth**
>
> Datalevin can automatically grow the mapsize if it runs out of space, but that resize is expensive and causes a performance spike. Set an appropriately large mapsize up front when production growth is predictable.

Chapter 22 explains how this internal resize path differs from application-level transaction failures such as validation, lookup-ref, unique, and CAS errors.

5.3 Leveraging the OS Page Cache

Because Datalevin uses **memory-mapped files (`mmap`)**, it does not manage its own buffer pool. Instead, it relies on the **Operating System Page Cache**. Here "page cache" means the operating system's cache of file-backed pages in main memory, not CPU L1/L2/L3 hardware caches. Operationally, a Datalevin database path names a directory, but the durable LMDB pages normally live in one memory-mapped data file, `data.mdb`. The directory may also contain LMDB's lock/readers file (`lock.mdb`) and Datalevin-managed WAL or snapshot files when those features are enabled. Named DBIs are logical key-value spaces inside the same LMDB data file; they are not separate data files.

- **Zero-copy reads**: When you perform a read, the OS handles fetching the required pages from disk into the page cache. LMDB maps those pages into the Datalevin process. The native layer hands the address of the LMDB value memory to Java as a `DirectByteBuffer`, and Datalevin's Clojure runtime decodes from that buffer into typed values without first copying the bytes into a JVM heap byte array.
- **Database pages outside the runtime heap**: Because the database pages live in the page cache, they are outside Datalevin's JVM-managed heap. A query still allocates normal runtime objects for decoded values and result collections, but the resident database pages themselves do not become managed objects and do not create proportional GC pressure. This lets Datalevin work with datasets much larger than the runtime heap when the active working set fits the available memory.

> **RAM should fit the active working set**
>
> Datalevin's effective buffer pool is the operating system page cache. For steady read performance, leave enough RAM for the pages your workload touches repeatedly, not for the full virtual mapsize.

5.4 Monitoring Page Cache Usage

There is no Datalevin buffer-pool hit ratio to watch, because Datalevin does not own the buffer pool. The useful signals come from the operating system: whether the LMDB data file is resident, whether queries cause major page faults, whether the machine is swapping, and whether reads are blocked on storage.

On Linux, start with the host-level view:

```
# Overall memory. Watch "available" and "buff/cache".
$ free -h

# Swap and I/O pressure. Sustained si/so or high wa is a warning sign.
$ vmstat 1
```

Then look at the Datalevin process:

```
# VSZ can be large because of :mapsize. RSS is resident memory.
$ ps -o pid,vsz,rss,maj_flt,comm -p "$DATALEVIN_PID"

# Per-process page faults. Sustained majflt/s under normal queries means the
```

```
# active working set is not staying resident in RAM.
$ pidstat -r -p "$DATALEVIN_PID" 1
```

If the Linux `fincore` tool is available, inspect the LMDB data file directly:

```
# Show how much of the mapped data file is currently in the page cache.
$ fincore /data/companydb/data.mdb
```

The interpretation is more important than any single number. A large virtual size is expected when `:mapsize` is large, and by itself is not a problem. A healthy steady-state read workload should show enough `available` memory, little or no swap activity, low sustained `majflt/s`, and low disk read pressure after the working set has warmed. If major faults and disk reads remain high during ordinary repeated queries, the hot working set is larger than available RAM or other processes are evicting Datalevin's file-backed pages.

On macOS, the built-in tools are less direct, but the same logic applies:

```
$ vm_stat 1
$ memory_pressure
```

Watch page-ins, swap activity, and memory pressure while running representative queries. For an exact per-file residency check on either Linux or macOS, use a tool such as `vmtouch` when it is available:

```
$ vmtouch /data/companydb/data.mdb
```

5.5 Reader Threads and Locking: `:max-readers`

LMDB uses a "lock-free" reader model, but it still needs to track active readers to prevent writers from overwriting pages that are still being read.

- **`:max-readers`**: This parameter (default: 1024) defines how many concurrent reader slots can access the database.
- **When to Increase**: If your application uses highly concurrent web servers or bounded worker pools above the default reader count, increase this before deployment.

Datalevin also tracks cached reader transactions and the owning thread so reader slots can be released when threads disappear. This reduces the common LMDB failure mode where abandoned thread pools exhaust reader slots. Virtual-thread handling is hardened by disabling thread-local read reuse for short-lived virtual threads, the lightweight JVM threads.

```
(d/get-conn path schema {:kv-opts {:max-readers 512}})
```

5.6 Storage Efficiency: Prefix Compression

Datalevin's storage engine (DLMDB) uses **prefix compression** to minimize the on-disk footprint of your indexes.

Prefix compression is most effective when a large portion of neighboring encoded keys is identical. Some KV workloads with long structured keys can therefore see large savings. Datalevin's Datalog indexes also benefit, but the effect is more modest than the largest KV cases because the shared part is usually a portion of the encoded 8-byte entity id in EAV and a portion of the encoded 4-byte attribute id in AVE. The `DUPSORT` nesting described in Chapter 15 is the main reason Datalog indexes avoid repeating full entity ids or full `(A, V)` prefixes.

- **Reduces storage size**: Workload dependent; high reductions are possible when key prefixes are substantial, while Datalog index savings are usually smaller because their common prefixes are compact encoded ids.
- **Improves cache locality**: More keys fit into memory pages and processor caches, which can make index scans faster.

6. Tuning Checklist

When deploying Datalevin to production, follow this checklist:

1. **Set a large `:mapsize`**: Reserve enough virtual address space for your future growth.
2. **Monitor Page Cache usage**: Ensure your server has enough RAM to keep your working set in memory.
3. **Adjust `:max-readers`**: If you have a high-concurrency application, increase the reader limit.
4. **Account for disk growth**: Leave room for LMDB free pages, compacted copies, snapshots, and WAL retention if WAL mode is enabled.

By tuning these parameters, you give Datalevin's zero-copy architecture the address space, page-cache headroom, reader slots, and disk space it needs to remain predictable as datasets grow.

Together with the durability and maintenance checklist in Section 4, this closes the chapter's two operational concerns: recoverable writes and predictable local storage.

CHAPTER 20

High-Throughput Ingestion

High write throughput is a different problem from low-latency reads or interactive transactions. This chapter collects the tools Datalevin offers for loading large volumes of data quickly: WAL mode and asynchronous transactions for durable online ingestion, non-durable LMDB flags for workloads that accept weaker crash behavior, pre-sorting to keep B+Tree inserts cheap, and the `init-db`/`fill-db` bulk-load path that bypasses normal transaction processing entirely. It closes with a decision flow for choosing among them. For the durability trade-offs behind these choices, see Chapter 19.

The short version is:

1. If this is an offline load and you can prepare final datoms, use `init-db` or `fill-db`.
2. If weaker crash behavior is acceptable, use non-durable LMDB flags.
3. If the workload is interactive or modest, use ordinary `transact!` batches.
4. If durable online ingestion needs more throughput, use manual batches with `transact-async`, often with WAL mode.

1. Online Ingestion: WAL and Async Transactions

For ingestion while the database remains queryable, the usual recipe is manual batches submitted through `transact-async`, often with WAL mode enabled. Chapter 19 explains WAL durability profiles and LSN maintenance; here the point is how to submit work without making every caller wait for its own commit. The crash-durability boundary still comes from the selected LMDB or WAL durability settings.

The async transaction API submits ordinary transaction data and returns a future or promise. Datalevin can then batch concurrent submissions internally. In the example below, each `batch-N` is a manually formed transaction chunk; sizing is covered next.

```
;; Fire multiple async transactions
(def f1 (d/transact-async conn batch-1))
(def f2 (d/transact-async conn batch-2))
(def f3 (d/transact-async conn batch-3))

;; Block on the last one as a queue barrier.
@f3
```

Block on the last submitted future when later work depends on the whole burst. Since async transactions for the same database are queued in order, the last realized future is a barrier for prior submissions in that burst. Keep the individual futures or promises when you need to observe each transaction's success or failure separately.

1.1 Manual Batching

Manual batching means grouping many transaction values into each `transact!` or `transact-async` call instead of writing one row or entity at a time. The goal is to amortize transaction setup and commit overhead while keeping memory use, validation cost, retry cost, and transaction-report size bounded.

There is no universal batch size. Start with batches large enough to amortize per-transaction overhead, such as thousands of simple entity maps, then measure with representative data. Each batch is still one transaction: if it fails, the whole batch fails, and the retry unit is that batch. `transact-async` can batch concurrent submissions internally, and that adaptive batching compounds with manual batching; it does not replace shaping the input into sensible chunks. The sorting and cache-tuning sections below use this same batch shape after the input has been ordered for cheaper B+Tree writes.

1.2 Commit Barriers After Async Work

For embedded applications that need both good throughput and deterministic handoff points, use a sequence of async transactions followed by an empty blocking transaction:

```
;; Async writes for throughput
(d/transact-async conn batch-1)
(d/transact-async conn batch-2)
(d/transact-async conn batch-3)

;; Empty d/transact queues behind the prior async work and blocks
(d/transact conn [])
```

The empty transaction is a barrier because `transact` is the blocking form of `transact-async`: it is submitted through the same async transaction queue after the earlier async transactions and blocks until its own future is realized. A successful return means the barrier itself has committed and all earlier submissions have finished; if those earlier transactions succeeded, they have committed too. Keep the futures or promises when you need to observe each result separately; waiting on the last submitted async result is also a queue barrier. Crash durability still follows the current durability mode. Under explicitly selected WAL `:relaxed`, committed LSNs can temporarily be ahead of durable LSNs; use the watermarks from Chapter 19 when an operator needs to confirm the durable boundary.

> **Async barriers do not change durability**
>
> Use the empty blocking transaction as a handoff point for commit ordering, not as a stronger crash-durability guarantee. Under explicitly selected relaxed WAL, use watermarks when the operator needs to confirm what is durable on storage.
> In the default direct LMDB path, writes still coordinate through the single-writer B+Tree transaction model. WAL mode can improve online ingestion through sequential append and group commit, but most jobs should combine manual batches with `transact-async` and measure before adding writer threads.

2. Non-Durable LMDB Flags

WAL mode and async transactions are the first tools to try when data must remain durable. This section is for workloads where weaker crash behavior is acceptable: one-time imports, cache rebuilds, temporary KV stores, derived indexes, test fixtures, disposable data, jobs with an external source of truth, or cases where losing recent writes after a crash simply does not matter. For those cases, LMDB exposes environment flags that reduce disk-sync work at commit time. These flags can apply to both Datalog connections and raw KV stores because both ultimately write through LMDB. When you supply a full `:flags` value, include Datalevin's default flags unless you intentionally want to replace them.

Flags	What changes	Use when	Risk
`:nometasync`	Sync data pages but not meta pages on commit	You want a modest speedup while preserving database integrity after a crash	The last transaction may be lost
`:nosync`	Keep normal page writes but skip commit-time `fsync`, leaving flushing to the OS	Lost or corrupted recent work is acceptable	A system crash may corrupt the database
`:writemap` + `:mapasync`	Use a writable memory map and defer dirty-page writeback to the operating system	High raw write speed, when the weaker crash boundary is acceptable	A crash may corrupt the database; buggy native code can overwrite mapped DB memory; some OSes preallocate the full map size

Non-durable flags have workload-specific effects. Measure them with your own data and storage before using them, and do not treat environment flags as a substitute for sensible batches.

```
(require '[datalevin.core :as d])
(require '[datalevin.constants :as c])

;; Raw KV store using non-durable flags
(def kv
  (d/open-kv "/tmp/import-kv"
             {:flags (-> c/default-env-flags
                         (conj :writemap)
                         (conj :mapasync))}))

;; Datalog connection using the same LMDB flags.
(def conn
  (d/get-conn "/tmp/import-dl" schema
              {:kv-opts {:flags (conj c/default-env-flags :nometasync)}}))
```

If you use `:nosync` or `:writemap`/`:mapasync`, make the crash boundary an explicit application choice. The data might be disposable, reproducible from a durable source, protected by a higher-level log, or simply not worth preserving after a system crash. For raw KV jobs that need a handoff point, call the KV `sync` operation at explicit checkpoints or before handing the database to normal durable workloads. For online systems with irreplaceable writes, prefer WAL mode with async transactions and tune durability there instead.

> **Use weak sync only for rebuildable data**
>
> Non-durable LMDB flags can make imports much faster, but they move failure handling out of the database and into your recovery plan. Use them for repeatable loads, caches, and derived data, not for primary OLTP writes that must survive a crash.

3. Preparing Transaction Batches

After you decide to use transaction-based ingestion, prepare each batch so it is cheap to apply. Two practical levers are ordering the input before transaction submission and disabling cache work that does not help a one-pass bulk job.

3.1 Pre-Sorting Your Data

A B+Tree performs best when keys are inserted in **sorted order**. If you insert random entity ids and random values, the engine must "jump around" the B+Tree, leading to frequent cache misses and page splits. A **page split** happens when a B+Tree page becomes full and must be divided into two pages, with parent pages updated to point at the new key ranges. Random insert order causes more of this rebalancing work; sorted input lets the tree append and fill pages more predictably.

If you are performing a bulk import (e.g., from a CSV or a SQL dump), sort the transaction data *before* it reaches the database.

> **Sort before the write path**
>
> Pre-sorting is an import-shaping step, not something the database should discover after work arrives. Group entity-local facts for EAV locality, and sort prepared datoms by attribute and value when AVE locality matters.

1. **Group by entity id**: This ensures attributes for a single entity are written to the EAV index contiguously.
2. **For prepared datoms, also sort by attribute and value where practical**: this makes updates to the AVE index more localized. For entity maps, entity grouping is usually the easiest useful first step.

```
;; Sort data by entity id
(let [sorted-datoms (sort-by :db/id raw-data)]
  (doseq [batch (partition-all 5000 sorted-datoms)]
    (d/transact! conn batch)))
```

3.2 Disable the Datalog Index Cache During Bulk Transactions

Datalevin keeps a small Datalog index cache for normal interactive workloads. During a large bulk transaction job, that cache can add memory pressure and cache maintenance work without helping much, especially when the import touches a wide range of entities and attributes only once. In every binding, set the cache limit to `0` for the import phase, then restore the previous limit:

```
(let [db             (d/db conn)
      previous-limit (d/datalog-index-cache-limit db)]
  (try
    (d/datalog-index-cache-limit db 0)
    (doseq [batch (partition-all 5000 sorted-datoms)]
      (d/transact! conn batch))
    (finally
      (d/datalog-index-cache-limit (d/db conn) previous-limit))))
```

This is a bulk-ingestion tuning knob, not a general query-performance setting. Leave the cache enabled for mixed read/write application traffic unless a measured workload shows that disabling it helps.

> **Restore cache settings after imports**
>
> Disabling the Datalog index cache is useful for one-pass bulk transaction jobs. Save the previous limit and restore it afterward so normal interactive traffic keeps the read-side cache behavior it expects.

4. Bulk Load: `d/init-db` and `d/fill-db`

For the highest-throughput bulk load into a database, bypass normal transaction processing using `d/init-db` and `d/fill-db`. Note that these functions **bypass the WAL** and will not appear in the transaction log.

> **Bulk load bypasses transaction semantics**
>
> `init-db` and `fill-db` are for prepared datoms with final numeric ids. Use normal transactions when you need tempids, lookup refs, upserts, transaction functions, transaction-log append, or transaction-level integrity checks.

- **`d/init-db`**: Load a collection of prepared datoms into a fresh, empty database
- **`d/fill-db`**: Bulk load additional datoms into an existing database

4.1 Assigning Entity Ids for Prepared Datoms

`init-db` and `fill-db` do not accept transaction maps, tempids, lookup refs, or upserts. They accept prepared `Datom` values. That means the subject entity id and every `:db.type/ref` value must already be the final numeric entity id.

A simple technique works well for relational imports whose source tables already have integer primary keys: reserve a non-overlapping entity-id range for each source table, then convert every primary key and foreign key with the same function [1,2]. For example:

Listing 20.1. Preparing stable entity ids for bulk datom ingestion

```
(def schema
  {:user/id        {:db/valueType :db.type/long
                    :db/unique    :db.unique/identity}
   :user/name      {:db/valueType :db.type/string}
   :order/id       {:db/valueType :db.type/long
                    :db/unique    :db.unique/identity}
   :order/customer {:db/valueType :db.type/ref}
   :order/total    {:db/valueType :db.type/double}})

;; Reserve disjoint numeric ranges. This is the same idea used by the
;; JOB benchmark, where each imported table has its own eid base.
(def user-base  1000000)
(def order-base 2000000)

(defn user-eid [source-id]
  (+ user-base (parse-long source-id)))

(defn order-eid [source-id]
  (+ order-base (parse-long source-id)))

(def users
  [{:id "1" :name "Alice"}
   {:id "2" :name "Bob"}])

(def orders
  [{:id "10" :customer-id "1" :total 42.5}
   {:id "11" :customer-id "2" :total 19.0}])

(defn user-datoms [{:keys [id name]}]
  (let [eid (user-eid id)]
    [(d/datom eid :user/id (parse-long id))
     (d/datom eid :user/name name)]))

(defn order-datoms [{:keys [id customer-id total]}]
  (let [eid (order-eid id)]
    [(d/datom eid :order/id (parse-long id))
     (d/datom eid :order/customer (user-eid customer-id))
     (d/datom eid :order/total total)]))

(def prepared-datoms
  (vec (concat (mapcat user-datoms users)
               (mapcat order-datoms orders))))

;; Load into empty database
(def db (d/init-db prepared-datoms "/tmp/import-db" schema))
```

The same id functions are used on both sides of the relationship: `:order/customer` stores `(user-eid customer-id)`, not the external customer id and not a lookup ref. The source ids are also stored as unique identity attributes (`:user/id`, `:order/id`) so later normal transactions can use lookup refs such as `[:user/id 1]`. Treat Datalevin's numeric entity ids as internal implementation ids even when you assign them during a bulk load. They are not a durable interchange format: another database loaded from the same source data may use different eids, while the unique source-id attributes stay stable.

If the source does not have dense integer ids, build an `eid-by-key` map first: collect stable keys such as `[:user external-id]` and `[:order external-id]`, sort or otherwise deterministically order them, assign fresh positive integers, and use that map when emitting subject ids and reference values.

4.2 Streaming with `fill-db`

For a large import, it is common to stream one table at a time with `fill-db`. Use the same id functions for every chunk so references remain consistent:

```
(def db
  (-> (d/init-db (mapcat user-datoms users)
                 "/tmp/import-db"
                 schema
                 {:closed-schema? true})
      (d/fill-db (mapcat order-datoms orders))))
```

4.3 Bulk-Load Safety Checks

With `fill-db`, new subject ids must not collide with existing entities. The simplest plan is to use ids greater than the database's current maximum eid, or to reserve non-overlapping ranges up front. In Clojure, inspect this with `(d/max-eid db)`; in Java, use `conn.maxEid()`. In Python and JavaScript, prefer deterministic reserved ranges in the import plan.

> **Prepared datoms must already be valid**
>
> Bulk-load helpers still expect values that match the declared schema. Ensure there are no duplicate entity-id ranges, no unresolved references, and every item uses the prepared datom shape accepted by the binding, such as `d/datom`, `Datalevin.datom`, `datom(...)`, or compact `[eid attr value]` tuples/arrays.

5. Choosing an Ingestion Strategy

After the individual tools are clear, Figure 20.1 expands the decision preview from the chapter introduction into a complete flow.

Summary: The Ingestion Checklist

When you need to ingest data at scale, follow this checklist:

1. **Use `init-db`/`fill-db` when prepared datoms are available:** This is the highest-throughput path for static data sets, but it bypasses normal transaction processing and WAL.

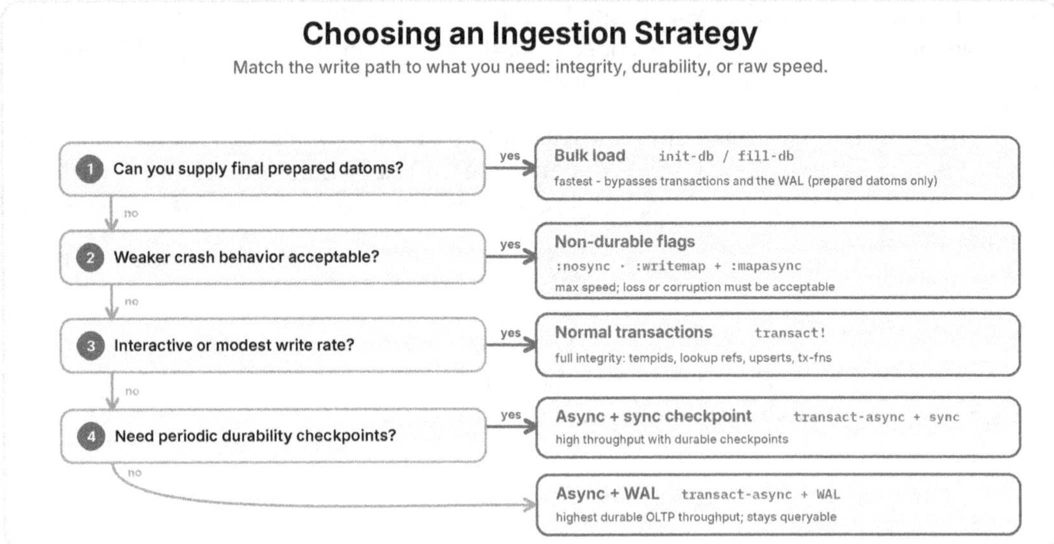

Figure 20.1: Choosing an ingestion strategy: if you can supply final prepared datoms for an offline load, use bulk load (init-db / fill-db); otherwise, if weaker crash behavior is acceptable, use non-durable flags (:nosync / :writemap+:mapasync); otherwise, for interactive or modest write rates use normal transactions (transact!); otherwise use async transactions, with an empty blocking transact barrier when later work must wait, or async plus WAL for higher durable online throughput

2. **Use non-durable LMDB flags only when weaker crash behavior is acceptable**: `:nometasync`, `:nosync`, and `:writemap`/`:mapasync` can improve write speed, but they change durability (i.e., crash behavior).
3. **Use ordinary transaction batches for interactive or modest write rates**: Keep tempids, lookup refs, upserts, transaction functions, and integrity checks in the normal transaction path.
4. **Use WAL and async for higher durable online throughput**: Choose the durability profile using Chapter 19's guidance. `d/transact-async` provides adaptive batching; wait on its future or promise, or use an empty blocking transaction as a queue barrier, before dependent work.
5. **Batch and prepare before submitting**: Manual batches compound async auto-batching. Sort by entity id, and disable the Datalog index cache only when measured bulk workloads justify it.

See Chapter 19 for durability and maintenance trade-offs.

References

[1] Huahai Yang, "Competing for the JOB with a Triplestore," yyhh.org, 2024. URL: https://yyhh.org/blog/2024/09/competing-for-the-job-with-a-triplestore/.

[2] Viktor Leis, Andrey Gubichev, Atanas Mirchev, Peter Boncz, Alfons Kemper, and Thomas Neumann, "How Good Are Query Optimizers, Really?" *Proceedings of the VLDB Endowment* 9(3):204-215, 2015. URL: https://www.vldb.org/pvldb/vol9/p204-leis.pdf.

CHAPTER 21

Query Planning and Diagnostics

Chapter 20 focused on getting data into Datalevin quickly. Once the data is loaded, performance work shifts from the write path to the read path: which query plan will run, how many facts each clause is expected to match, whether the optimizer's cardinality estimates match the current data, and what `explain` shows when they do not.

Datalevin's query optimizer leverages the strengths of sorted triple indexes for cardinality estimation, one of the hardest problems in database design. With that foundation, Datalevin can choose good plans for complex joins without requiring users to hand-order clauses.

A fast query depends on both the plan Datalevin chooses and the evidence you can collect when that plan surprises you. This chapter combines optimizer mechanics, recursive rule evaluation, and the `explain` workflow used to diagnose real query behavior.

If you are debugging a slow or surprising query right now, start with Section 3 and run `explain` first. Then return to Sections 1 and 2 to interpret why the planner chose a particular join order, access path, or recursive-rule strategy. Benchmark results are kept with their workload context: Chapter 1 covers JOB, Chapter 9 covers OpenRuleBench, and Chapter 13 covers LDBC SNB.

> **Measure the plan before rewriting the query**
>
> When a query surprises you, start with plan-only `explain`. It is usually faster to inspect join order, access paths, and estimates than to guess from clause order or rewrite a correct query blindly.

1. Query Planning and Optimization

As discussed in Chapter 8, Datalog is **declarative**. You describe *what* you want, and Datalevin's **Query Optimizer** decides *how* to find it. This decoupling is critical for performance because even a simple change in the **order of joins** can make a query orders of magnitude faster or slower, even though the results are the same.

This section explores how Datalevin's **Cost-Based Optimizer (CBO)** uses the counted indexes and ordered storage in DLMDB to create efficient execution plans.

Before a query executes its result-producing plan, Datalevin rewrites it into a simpler execution shape. Predicate pushdown, inequality conversion, constant parameter plugging, and complex-clause dependency analysis all happen before join planning. The optimizer then chooses access paths and join methods against that rewritten query.

Query optimizers evolve much like compilers and Boolean satisfiability (SAT) solvers: a stable core supports many specialized transformations and execution strategies. Datalevin follows the same pattern. New optimizations usually do not replace the storage or planning model; each recognizes another safe query shape and exposes it to the same sorted EAV and AVE indexes, counted ranges, query-specific samples, graph planner, and cost model. When a rewrite does not apply or its complete estimated cost is not lower, Datalevin retains the existing complete execution strategy. The examples below are selected to show the range and depth of Datalevin's optimizer—logical rewrites, costed access paths, physical execution, and recursive evaluation—not to inventory every optimization added in each release.

1.1 The Selinger-Style Optimizer

Datalevin uses a **Selinger-style cost-based query optimizer**, following the System R tradition [1]. It uses dynamic programming over join alternatives, in the same broad family as enterprise relational optimizers.

1.1.1 How it Works

Figure 21.1 describes the query planner pipeline.

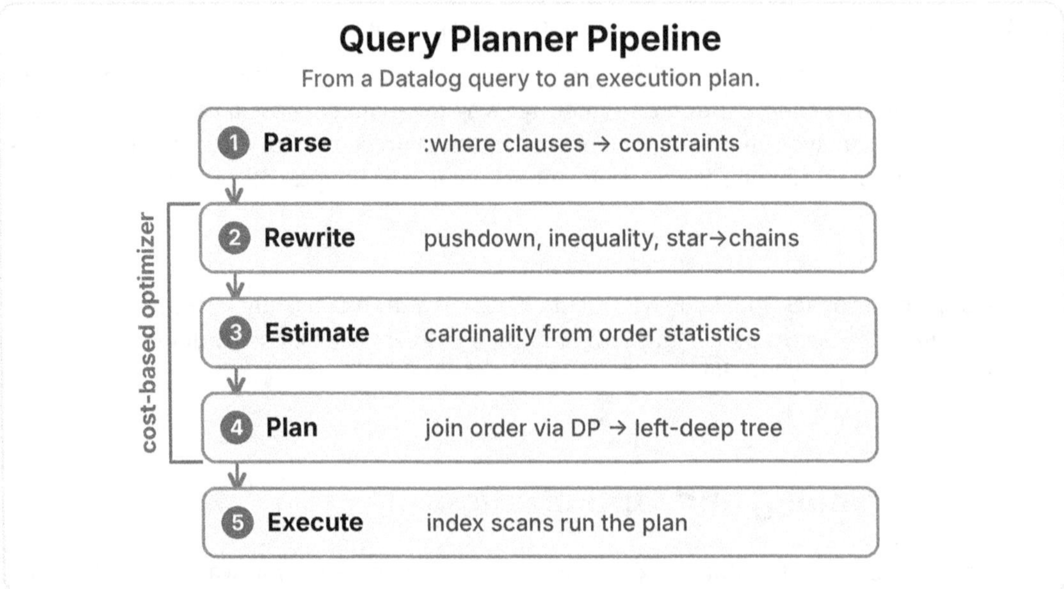

Figure 21.1: The query planner pipeline: parse, rewrite, estimate, plan, and execute, with rewrite through plan forming the cost-based optimizer

When a query is submitted:

1. **Parsing**: The engine breaks down the `:where` clauses into individual constraints.
2. **Rewrite and Specialization**: Safe inputs become constants, indexable predicates become scan bounds, and selective bindings may be resolved before graph planning.
3. **Query Graph Simplification**: Star-like attributes (multiple attributes on the same entity) are handled via merge scan, defined in Section 1.5, reducing the graph to chains between stars.
4. **Cardinality Estimation**: The engine uses DLMDB's order statistics to estimate how many results each clause will produce.

5. **Join Planning**: It explores possible join orders using dynamic programming, generating **left-deep join trees.**
6. **Bounded Search Policy**: The planner keeps dynamic-programming plan tables, but shrinks the search space when the number of alternatives exceeds a cap based on the number of ordered pairs, `P(n, 2) = n * (n - 1)`. This keeps planning bounded while preserving careful comparison of the early joins that determine later intermediate sizes.

A **left-deep join tree** is a plan shape where each step joins one new base relation into the intermediate result built so far. The alternative is a bushy tree, where two independently built subplans are joined together. Datalevin uses left-deep plans because its join methods are index-scan oriented: keeping one side as a base relation preserves accurate counts and keeps planning cost bounded [2,3].

1.1.2 A Concrete Join-Order Example

Join order matters because every join produces an intermediate relation. A bad plan can create millions of temporary candidates and then throw almost all of them away. A good plan starts with the most selective facts and keeps the intermediate relation small.

Suppose an e-commerce database has:

- 10,000,000 orders.
- 80,000,000 line items.
- 500,000 products priced above `$100`, stored as `10000` cents.
- A unique order id, so `:order/id` lookup returns exactly one order.
- About 12 line items per order.

The logical query is simple: "For one order, find the expensive products in its line items."

```
(d/q '[:find ?sku ?qty
       :in $ ?order-id
       :where
       [?order :order/id ?order-id]
       [?line :line-item/order ?order]
       [?line :line-item/product ?product]
       [?line :line-item/quantity ?qty]
       [?product :product/sku ?sku]
       [?product :product/price ?price]
       [(> ?price 10000)]]
     db
     "o-2026-000001")
```

These clauses have one logical meaning, but the physical join order can be wildly different.

Plan step	Good plan: start from order id	Approx. candidates
1	AVE lookup `[:order/id "o-2026-000001"]`	1 order
2	Reverse AVE lookup for `:line-item/order`	12 line items
3	Join each line item to its product	12 products
4	Check each product's price	a few rows

The good plan does work proportional to one order.

Plan step	Bad plan: start from product price	Approx. candidates
1	AVE range scan for `:product/price > 10000`	500,000 products
2	Find line items for those products	millions of line items
3	Join those line items to orders	millions of order links
4	Keep only order `"o-2026-000001"`	same final rows

The result is identical, but the bad plan may process millions of candidates to answer a question about one order. This is why Datalevin invests in cardinality estimation and join planning. The textual order of `:where` clauses does not matter; the optimizer is choosing this physical order for you.

The good plan is a left-deep tree: the optimizer starts from the single order, then folds in one base relation at a time, keeping the intermediate result small at every step. Figure 21.2 shows the tree for the example above.

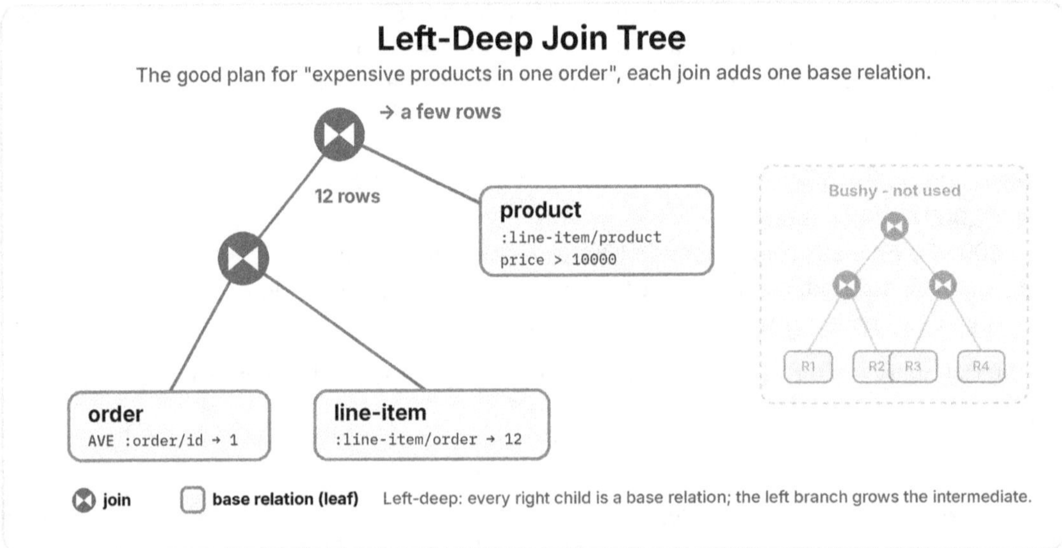

Figure 21.2: The good plan as a left-deep join tree: the order leaf (AVE :order/id, one order) joins the line-item leaf (:line-item/order, 12 rows), and that 12-row intermediate joins the product leaf (:line-item/product, price > 10000) to yield a few rows; every right child is a base relation, unlike a bushy tree

1.2 Accurate Cardinality Estimation

The quality of a query plan depends on accurate cardinality estimates. Datalevin excels here because counting is cheap in its nested triple storage. The resulting join order is often similar to a hand-written join plan.

1.2.1 Direct Counting

Some counts can be obtained in **O(1) time** directly from the index without scanning. For example, `[?e :user/city "London"]` returns an exact count from the AVE index. For range queries, DLMDB's order statistics provide **O(log n)** counting.

1.2.2 Query-Specific Sampling

Direct counts are most useful for individual indexed clauses and range bounds. After several clauses have been joined, the next estimate depends on the size and distribution of an intermediate relation. Small errors can compound: if the planner underestimates users in one city, then underestimates how many of those users have active orders, the combined estimate can become much too small. For these plan-dependent estimates, Datalevin uses **online reservoir sampling** under actual query conditions. This is query-time work, not a background histogram. It:

1. Collects sample entity ids
2. Performs merge scans to get selectivity ratios
3. Applies a conservative envelope to sampled fan-out estimates

The envelope is the statistical guardrail. For a sampled link, the planner uses the largest of the query-local sample mean, the storage-derived default fan-out for that attribute, and a minimum fan-out of one. This prevents a small or unrepresentative sample from producing an overly optimistic join estimate.

1.2.3 Directional Estimation

Many relational optimizers fall back to independence assumptions when they lack multi-column or join-specific statistics. Datalevin's estimation is **directional** because different join directions produce different estimates. This matters for `:ref` and `:_ref` joins, where starting from the source entity or the referenced value can have very different selectivity.

> **Planner statistics are counts and samples**
>
> Datalevin uses exact index counts, query-time sampling, per-attribute eid samples, per-attribute cardinality counters, and default fan-out ratios. These are not background histogram statistics, so `analyze` refreshes the planner evidence Datalevin actually uses.

1.3 Query Rewrites and Predicate Push-Down

Datalevin rewrites queries to expose selective work to index scans before large intermediate relations are produced. Some rewrites become part of the query graph; others seed the graph with a small relation or reconsider work at the boundary between the main plan and late clauses.

1.3.1 Indexable Predicates and Bound Inputs

Comparison operators involving one attribute-value variable and constants are converted to AVE range boundaries. For example:

```
[(> ?age 21)]  ;; Becomes a range scan with an open lower bound at 21
```

Equality becomes an exact range, while `in` and `not-in` can become exact or complementary ranges. A suitable `like` or `not-like` pattern contributes a prefix range and keeps a residual predicate when the range alone is not an exact test. Multiple bounds on the same variable are intersected, so the scan uses the narrowest interval and a contradictory interval can produce an empty result without reading attribute values.

Not every predicate is safe to encode as an index boundary. A single-variable predicate that cannot be converted exactly can still be attached to the value scan, while multi-variable predicates and dynamic-source expressions follow the normal dependency and late-clause paths.

Bound query inputs participate in the same rewrite:

```
[:find ?e ?age
 :in $ ?min-age
 :where
 [?e :user/age ?age]
 [(>= ?age ?min-age)]]
```

For a particular invocation, `?min-age` is already known. Datalevin plugs that value into the query before planning, so the comparison can become a concrete lower bound on the AVE scan instead of remaining a separate predicate over a bound input.

1.3.2 Selective Anchors Before Graph Planning

Sometimes a selective fact is more useful as a small relation *before* the optimizer builds the query graph. If an input relation already binds one side of a pattern, Datalevin can perform the corresponding EAV or AVE probes, materialize the small result, and propagate the new bindings through other indexed patterns. This avoids representing a logically constrained lookup as a broad scan.

Multiple unique constants provide a related opportunity. In this path query, suppose both `:node/id` and `:node/code` are declared unique:

```
[:find ?start ?middle ?end
 :where
 [?start :node/id 1]
 [?end :node/code "three"]
 [?edge1 :edge/from ?start]
 [?edge1 :edge/to ?middle]
 [?edge2 :edge/from ?middle]
 [?edge2 :edge/to ?end]]
```

The two endpoint lookups can seed the same connected component, constraining the path from both ends instead of expanding from one endpoint and applying the other as a late filter. A missing unique value produces an empty relation and short-circuits the query. A single unique literal remains with the ordinary planner because it is already an ideal selective root.

Pre-materialization is not limited to unique attributes. A constant AVE lookup on a non-unique attribute can also be considered when its entity variable joins the rest of the query. The optimizer uses the exact AVE fan-out, projects the cost of propagating the resulting bindings, and compares the complete candidate with the unchanged plan. It materializes the lookup only when the whole path is estimated to be cheaper; a selective first lookup is not enough if its downstream expansion would be expensive.

1.3.3 Costed Reordering at the Late-Clause Boundary

Rules, disjunctions, and other binding dependencies sometimes keep useful work outside the main query graph. Datalevin can still compare a currently executable bound pattern with a compatible indexed producer among those late clauses. Two representative producers are an indexed union from an `or-join` and a bounded AVE range defined by lower and upper inequalities.

For example, suppose an earlier rule or disjunction has produced the current bindings for `?person`, while `?from` and `?to` are bound query inputs. These clauses remain to be resolved:

```
[?message :message/hasCreator ?person]
[?message :message/creationDate ?date]
[(<= ?from ?date)]
[(< ?date ?to)]
```

The creator pattern can probe messages for the bound people. Alternatively, the engine can fuse the date pattern and its two inequalities into one AVE interval scan. A few bound people may make the creator probes cheaper; many bound people paired with a narrow date window may make the range cheaper. An indexed `or-join` presents the same kind of choice: Datalevin can form a small union with AVE point probes before expanding a broader bound pattern. These alternatives are costed with the values available at runtime, so the logical query need not encode either physical order.

These alternatives preserve a conservative fallback. If the clause shape is not eligible or the full switch cost is not lower, Datalevin keeps the dependency-ordered late plan. With measured `explain`, the selected runtime strategy is reported in `:late-clause-decisions`.

1.4 Cost-Based Top-k Access Paths

Datalevin's optimizer treats an ordered AVE scan and the `idoc-match`, `fulltext`, `vec-neighbors`, and `embedding-neighbors` query functions as physical alternatives for their logical clauses. It costs those alternatives with their residual joins and filters, propagates ordering and resumability as physical properties, and compares two complete physical alternatives:

Plan alternative	How it consumes a candidate source	Effect of the final `:limit`
Conventional plan	Uses normal indexed clause and join execution to consume the complete logical source relation, including any source-local bound such as `:top`.	Trims the final result and may reduce sort memory, but does not stop source work through the bounded access-path mechanism.
Access-path plan	Makes an eligible AVE, idoc, full-text, or vector source the physical driver, then applies residual joins and filters as candidates are consumed.	May stop source work once the requested final result is complete; otherwise it can return to the conventional plan.

"Conventional" does not mean unindexed: both alternatives use Datalevin's indexes and join machinery. The distinction is normal complete source execution versus an incremental source plan whose work can be bounded by the final result demand. Merely having a specialized index does not force it to drive the query.

Glossary. *Finite* = the query has a positive `:limit` and a relation-shaped `:find` (plain tuples, no aggregates, pull, `:with`, `:having`, or result maps). *Correlated* = the function's arguments depend on a variable bound by an earlier clause. *Resumable* = the source can be consumed batch-wise and continued from an opaque position. A *certified frontier* additionally provides a bound proving that unseen candidates cannot improve the final window. Resumability alone does not imply that proof.

Two bounded execution modes are important:

- **Adaptive top-k** — the access path supplies the leading `:order-by` property and a certified frontier. Datalevin reads batches until it has `:offset + :limit` distinct final tuples and the frontier proves unseen candidates cannot change the window, consuming through the tie at the cutoff before applying secondary order terms.
- **Adaptive limit** — the access path is complete and resumable but unordered. Datalevin reads batches until enough distinct final tuples survive the rest of the query. The selected window is intentionally unordered.

The current access methods expose these properties:

Access method	Ordered top-k property	Unordered bounded use	Important condition
AVE	Attribute value ascending or descending	Complete AVE execution remains available conventionally	The leading order variable is the value in a simple default-source `[?e :attr ?value]` pattern, and an inequality supplies a scan boundary in that direction.
`idoc-match`	None	Complete, resumable adaptive limit	The match arguments are uncorrelated and the result uses the normal `[e a v]` relation shape.
`fulltext`	Score descending	Complete, resumable adaptive limit	Ranked order requires exactly one domain, `:display :refs+scores`, and score as the leading order term.
`vec-neighbors`, `embedding-neighbors`	Distance ascending	Complete, resumable adaptive limit	Ranked order requires exactly one domain, `:display :refs+dists`, and distance as the leading order term.

Ordered example — AVE supplies the leading order. The scan starts at the inequality boundary and proceeds backward, testing structured clauses per batch:

```
[:find ?title ?score
 :in $ ?max-score
 :where [?e :item/score ?score]
        [(<= ?score ?max-score)]
        [?e :item/active? true]
        [?e :item/title ?title]
 :order-by [?score :desc ?title :asc]
 :offset 20
 :limit 10]
```

The inequality is eligibility, not a hint about clause order. Only the leading `?score :desc` comes from the AVE order; a large tie at the cutoff still requires scanning through the tie before `?title`

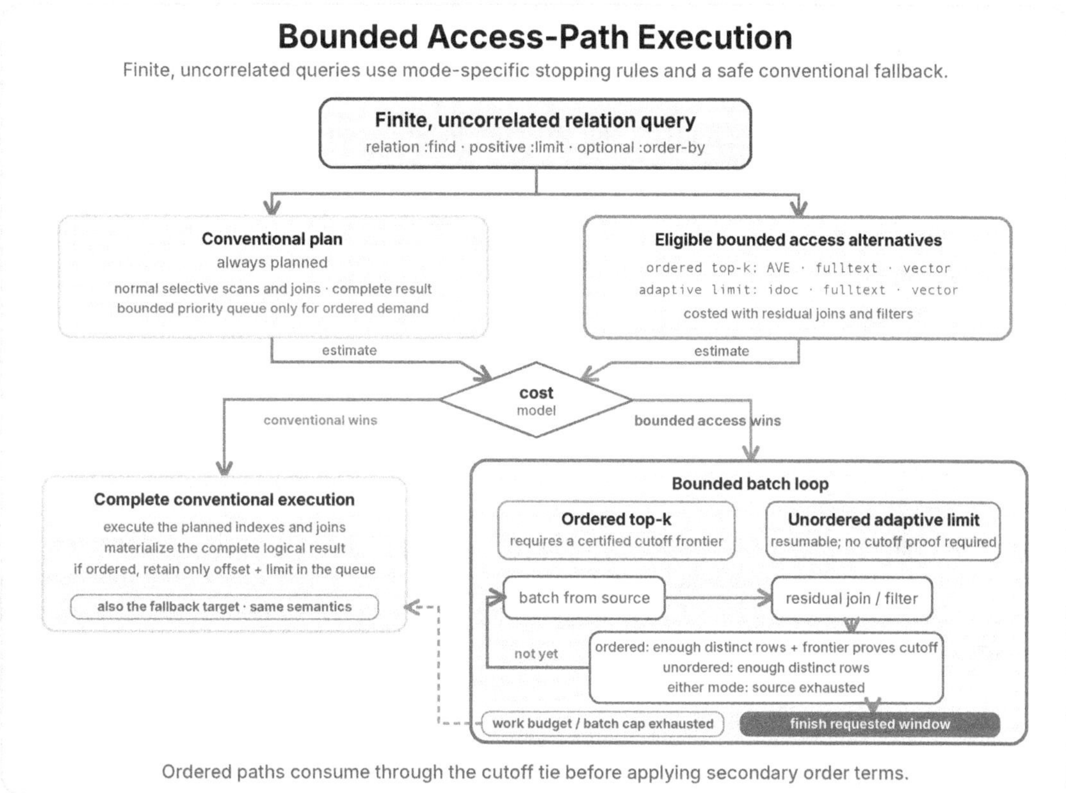

Figure 21.3: Bounded access-path execution for a finite, uncorrelated relation query: the optimizer costs an always-available conventional plan against eligible bounded access alternatives, including ordered top-k from AVE, full-text, or vector search and unordered adaptive limit from indexed-document, full-text, or vector search. When the access alternative wins, it fetches a source batch and applies residual joins and filters. Ordered top-k stops after enough distinct final tuples survive and a certified frontier proves that unseen candidates cannot change the cutoff; unordered adaptive limit needs only enough distinct surviving tuples. Either mode also finishes when the source is exhausted. Otherwise execution reads another batch, or falls back to the already-planned conventional plan with unchanged semantics when its work budget or batch cap is exhausted.

`:asc` is applied.

Unordered example — idoc has no ranked order but can still stop early:

```
[:find ?e ?name
 :where [(idoc-match $ :user/metadata {:theme "dark"}) [[?e _ _]]]
        [?e :user/name ?name]
        [?e :user/active? true]
 :limit 20]
```

Batches of document matches are joined and filtered until 20 distinct final tuples survive. Without `:order-by` the window has no ordering guarantee and an offset does not define stable pagination.

Source-local and query-level bounds remain separate. Full-text functions use `:top`, `:limit`, and `:offset` to bound their candidate relations; vector functions use `:top`. Query-level `:offset` and `:limit` apply after residual joins, filters, projection, and duplicate elimination. The optimizer can avoid consuming an unused suffix, but it never widens the source window or changes vector-search approximation and recall.

Early stopping requires a finite relation-shaped query as defined above. Correlated arguments (full-text, idoc, vector, embedding) use normal complete query-function execution. A physical join must preserve the source order for adaptive top-k; if a cheaper hash join would destroy that property, the optimizer keeps the unordered access-path plan or the conventional plan.

Adaptive execution runs under a bounded candidate budget and a hard candidate-batch safety cap, retaining the already planned conventional plan as a fallback. If either limit is exhausted before a valid stopping proof, Datalevin runs the conventional plan with unchanged semantics — for an ordered query the same window, while for an unordered limited query no stable window identity exists. Non-access-path ordered queries still use the bounded priority queue described in Chapter 8.

1.4.1 Post-Top-k Projection Enrichment

An ordered, limited query can sometimes select its result window before loading optional properties that are used only in the projection. `get-some-else` makes that safe because it always produces exactly one tuple: the first present `[attribute value]`, or `[nil fallback]` when none of the listed attributes is present.

```
[:find ?id ?content ?score
 :where [?item :item/id ?id]
        [?item :item/score ?score]
        [(get-some-else $ ?item nil :item/content :item/image)
         [_ ?content]]
 :order-by [?score :desc ?id :asc]
 :limit 5]
```

Here `:item/id` is unique and remains in the projection, while `?content` is not used by ordering, filtering, or another clause. The planner can therefore rank and select the five distinct rows first, then look up content only for those rows. `get-some` is not interchangeable in this optimization: when neither attribute is present it removes the row, which can change the selected window.

The rewrite is applied only when the planner proves that the enrichment is cardinality-preserving and projection-only, and that the retained projection contains the entity itself or a unique value identifying it. `explain` reports the plan and proof under `:post-top-k-enrichment`; measured execution also adds the candidate and selected row counts. If a proof condition is absent, Datalevin evaluates the clause in its normal position.

1.5 Merge Scan

For star-like queries (multiple attributes on the same entity), Datalevin uses **merge scan**, a technique similar to pivot scan [4].

Instead of joining each attribute separately, an EAV-oriented merge scan can retrieve matching attributes for the same entity in one ordered pass. The optimizer can then plan over a simplified query graph whose nodes are star-shaped groups and whose edges are links between those groups, reducing the join search space [7,8,9]. For star-heavy queries, this removes a large amount of otherwise repetitive entity-local work.

1.6 Join Methods

Datalevin considers six join methods and picks the best based on cost estimation:

Method	Use Case
Forward ref `:ref`	`[?e :user/friend ?f]` - merge scans ref values
Reverse ref `:_ref`	`[?f :user/_friend ?e]` - scans AVE then retrieves entities
Value equality `:val-eq`	When variables unify via attribute values
Hash join `:hash-join`	Large non-selective joins where the cost model favors materializing a target relation
Or-join `:or-join`	Handles `or-join` clauses with sideways information passing (SIP)
Not-join `:not-join`	Optimizes conservative anti-join shapes instead of always deferring negative filters

Sideways information passing (SIP) means passing already-known variable bindings from one part of the plan into another part that is not simply the next linear data pattern. In ordinary joins this happens naturally through shared variables: once `?user` is bound, a later clause such as `[?order :order/customer ?user]` can use that value to do a narrow lookup. An `or-join` is more complicated because it contains several alternative branches. SIP lets the planner evaluate each branch with the outer bindings it is allowed to see, rather than evaluating every branch as an independent broad subquery and joining the results afterward.

For example, an `or-join` whose declared join variables include `?user` can receive the current `?user` bindings from earlier indexed clauses. Each branch then asks a bounded question such as "does this user match by email?" or "does this user match by phone?" instead of scanning all users that match either branch. The declared join-variable list matters: it is the contract that tells the optimizer which outer variables may be passed sideways into the alternatives and which branch-local variables must stay internal. If a complex clause cannot use the available bindings, it is less selective and may be planned later.

An **anti-join** is the negative counterpart of a join: it keeps a candidate row only when no matching row exists on the excluded side. In SQL this is the shape behind `NOT EXISTS`; in Datalevin it is usually expressed with `not` or `not-join`. A planned `not-join` is useful when the engine has already bound the declared join variables and can check the negative pattern with indexed lookups. For example, after `?start` and `?person` are bound, a `not-join [?start ?person]` can ask "is there a direct friendship edge between these two?" and discard the candidate only if that edge exists. The result is still declarative negation, but the physical plan can avoid treating the negative pattern as a late row-by-row filter.

In hash-join terminology, the **build side** is the input used to construct an in-memory hash table, and the **probe side** is the input scanned to look up matches in that table. Building on the smaller side usually saves memory and work. Datalevin considers a hash join when estimated intermediate sizes make repeated index probes expensive; for reverse-reference joins it can also combine hash joining with SIP when the target side is much larger than the current input.

When several links from the partial plan can reach the same next graph node, Datalevin first prefers the link from the **most recently resolved** source. Here, "recently" refers to its position in the partial plan, not the age of the stored data. That source is closest to the current intermediate relation, so its link estimate uses the freshest binding context; cost and estimated output size break ties among equally recent links.

1.7 Parallel Processing

Datalevin's query engine uses parallelism at several levels:

- **Planning**: Independent counts, samples, and connected components can be processed in parallel.
- **Storage scans**: Large batches for selected EAV and AVE operations can be divided among workers when their predicates are safe to run concurrently.
- **Plan execution**: Long eligible runs of indexed steps can use morsel-driven partition execution. Other plan segments use a bounded tuple pipeline, so downstream work can begin before the entire upstream relation is available.

1.7.1 Morsel-Driven Execution

A **morsel** is an independent chunk of an intermediate tuple relation. **Morsel-driven** means that these chunks, rather than individual plan steps, are the units scheduled across workers. For a sufficiently large input followed by a long linear run of `:link` and `:merge` steps, Datalevin divides the input into morsels and processes them concurrently [11]. Each worker carries its morsel through the remaining steps in that run. This keeps index operations batched within a partition instead of assigning a separate task to every step.

The payoff appears when an early step produces many tuples and several indexed joins remain: available cores can advance independent parts of that relation through the indexed suffix. With the current defaults, partitioning is considered at 8,000 input tuples when at least three eligible index steps remain. A target of roughly 2,000 tuples per participant bounds useful parallelism; the processor count and available common-pool capacity can reduce the participant count further. These thresholds choose an execution strategy and do not change query semantics.

The executor finds each maximal eligible run independently. A hash join, semi-join, branch, access operator, or another coordinating step ends one run, but a safe run before or after it can still be partitioned. Merge predicates must be forkable, and a merge that depends on a precomputed relation is not eligible. Writes, non-forkable predicates, small inputs, short runs, and single-core execution retain the pipelined or serial path.

Datalevin uses only one layer of parallelism inside a partitioned run. It disables storage-level chunking while a worker advances a morsel, preventing nested tasks from oversubscribing the shared pool. The caller processes one morsel while other morsels use available workers. Results are concatenated in partition order, and measured per-step row counts are summed across morsels.

Reading morsel execution in measured explain

Measured `explain` records each runtime split in the top-level `:partitioned-execution` vector. Each entry identifies the segment with `:segment-start-step-index`, `:split-step-index`, and `:segment-end-step-index`; reports `:input-rows` and `:partitions`; and describes the work with `:segment-step-count`, `:segment-step-types`, `:partitioned-step-count`, and `:partitioned-step-types`. The optional `:previous-step-type` and `:next-step-type` fields show the nonlinear steps at the segment boundaries. Plan-only explain omits this vector because the runtime split depends on the intermediate relation and available execution capacity.

1.8 Complex Clauses and Rules

The optimizer handles complex clauses in stages:

1. **Index access clauses** produce intermediate results first
2. **Heuristics and variable dependencies** reorder remaining complex clauses such as `and`, `or`, `not`, complex predicates, and function bindings that are not pushed down.
3. **Rules** are handled by the specialized bottom-up evaluation machinery described later in this chapter.

`or-join` participates in link planning, and common `not-join` shapes can become anti-join steps when their join variables are bound in a single plan component. More complex negative clauses still fall back to late resolution. Optimizer coverage at this boundary continues to grow: eligible indexed unions and bounded AVE ranges can now compete with an earlier bound pattern, while other clause shapes retain dependency-ordered late execution.

1.9 Optimizer Principles

In summary, Datalevin has a sophisticated query optimizer that compiles declarative query clauses into execution steps that use index scans where they help, take advantage of cheap counts from triple storage, and often produce plans similar to a hand-written one. Users can usually write the logical query first, then use `explain` when a specific workload needs inspection or tuning.

- **Clause order does not matter for the most part**: The optimizer chooses the physical order for most clauses. Deferred `:late-clauses` are also re-ordered according to their binding dependencies. If two late clauses are both fully bound at the same time, their original order does matter.

- **Core indexes are automatic; composite access paths are explicit**: Datalevin maintains EAV and AVE for every attribute, but it does not invent a maintained multi-attribute value from a query shape. For a measured hot query that repeatedly consumes the same cardinality-one attributes, define a `:db/tupleAttrs` attribute and make the query read it explicitly. This can replace a large intermediate relation produced from component EAV datoms with the usually smaller relation of qualifying joint combinations. The cost is another indexed datom and additional work on every component update; Chapter 11, Section 1.2 compares the two IC5 query variants.
- **Counts are live and samples are refreshable**: DLMDB gives the planner cheap current index counts, while background sampling and explicit `d/analyze` calls refresh per-attribute eid samples, per-attribute cardinality counters, and default fan-out ratios. These are not histogram-style distribution statistics.
- **Nested storage helps planning**: The nested triple storage makes counting cheap, giving the optimizer evidence that many systems have to approximate.
- **Properties are part of the choice**: Ordering, resumability, completeness, frontier guarantees, and result quality determine whether an AVE, idoc, full-text, or vector alternative can satisfy the root query safely.

2. Recursive Rules as Specialized Query Evaluation

Non-recursive rules are pulled out and planned as part of the query optimization described above. Recursive rules require specialized planning because the engine may have to derive new tuples over several rounds before the rule result is complete. This section explains the separate machinery: semi-naive evaluation, deltas, and rule-specific rewrites.

```
(def ancestor-rules
  '[[(ancestor ?x ?y)
     [?x :person/parent ?y]]

    [(ancestor ?x ?y)
     [?x :person/parent ?z]
     (ancestor ?z ?y)]])
```

For the ancestor rule set above, a naive evaluator would repeatedly join all known ancestors against all parents. That repeats work. In the third round, for example, it may rediscover grandparent facts already found in the second round. On large graphs, this redundant computation can dominate the query.

Datalevin therefore uses **semi-naive evaluation**, the standard bottom-up strategy for recursive Datalog [8]. The key idea is delta tracking: each round only uses the *new* tuples discovered in the previous round. Evaluation continues until a fixpoint, meaning a round produces no new tuples. The engine also builds rule-dependency components and evaluates them in dependency order. A later rule component can use facts computed by an earlier component, but not the other way around. Strongly connected recursive components run to a fixpoint before dependent components consume their results.

This bottom-up model is set-oriented. Joins operate over candidate sets and iterators, so the storage layer can use sequential page scans, merge-style operations, and bulk filtering rather than tuple-at-a-time random lookup.

Bottom-up evaluation has one obvious risk: it can compute more of the world than the user asked for. If a query asks only for Alice's ancestors, the engine should not materialize every ancestor relation in the database. Datalevin uses **magic-set rewriting**, a classic Datalog transformation for making bottom-up evaluation goal-directed [8,9]. Magic-set rules push bound variables from the outer query into the recursive rule, pruning intermediate results to the part of the graph relevant to the question.

2.1 Semi-Naive Evaluation

Consider a small directed graph, using letters as stand-ins for entity ids. The example EDB is the following set of base edge facts:

```
[a :link/to b]
[b :link/to c]
[b :link/to d]
[d :link/to e]
[p :link/to q]
[q :link/to r]
```

In a real Datalevin database, each row above corresponds to a stored `:link/to` datom. These base edge facts are the **EDB** (extensional database): facts read from Datalevin rather than derived by the recursive rule. The rule predicate `reachable` is the **IDB** (intensional database): derived tuples produced by the rules. The EDB stays fixed during the evaluation; each round adds a delta of new IDB tuples.

The recursive rule says that `?end` is reachable from `?start` if there is a direct edge, or if there is an edge to `?mid` and `?end` is reachable from `?mid`:

```
(def reachability-rules
  '[[(reachable ?start ?end)
     [?start :link/to ?end]]

    [(reachable ?start ?end)
     [?start :link/to ?mid]
     (reachable ?mid ?end)]])
```

Semi-naive evaluation tracks a **delta**, the new IDB tuples from the previous round, and joins only against that delta in the recursive step. The base rule derives round 1 directly from the EDB; later rounds join the previous IDB delta against the same EDB edge facts. For this graph:

Round	New IDB `reachable` tuples
1	`(reachable a b)`, `(reachable b c)`, `(reachable b d)`, `(reachable d e)`, `(reachable p q)`, `(reachable q r)`
2	`(reachable a c)`, `(reachable a d)`, `(reachable b e)`, `(reachable p r)`
3	`(reachable a e)`
4	No new tuples; fixpoint reached.

Figure 21.3 shows the delta flow for this example. Each round consumes only the previous round's delta; the fixpoint is reached when the next delta is empty.

A naive bottom-up evaluator would keep rejoining all known `reachable` tuples in every round, rediscovering many results it already had. Semi-naive evaluation uses only the previous round's

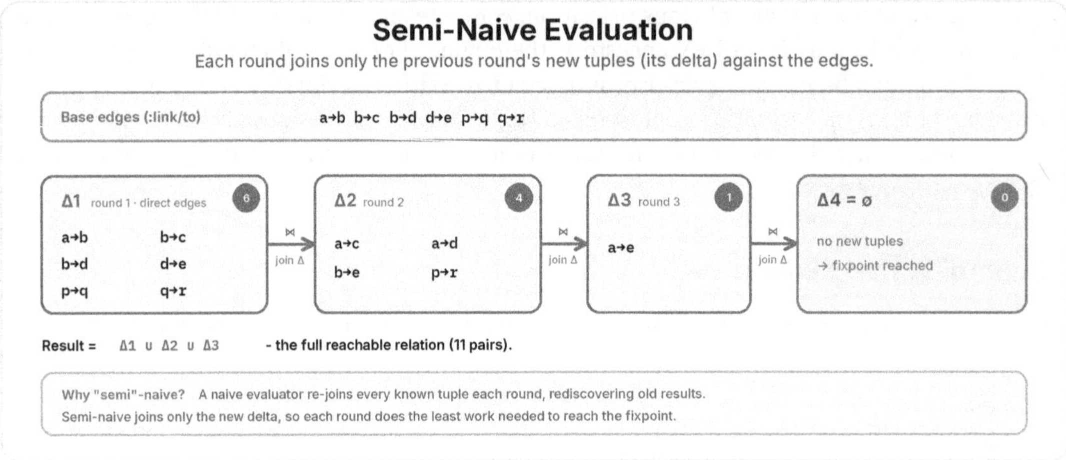

Figure 21.4: Semi-naive recursive evaluation with delta tracking: base edges a→b, b→c, b→d, d→e, p→q, and q→r produce delta round 1 with direct edges, delta round 2 with a→c, a→d, b→e, and p→r, delta round 3 with a→e, and round 4 with no new tuples; the result is the union of deltas, while a naive evaluator would keep rejoining all known tuples and rediscover old facts

delta to produce the next delta, then unions the deltas into the final IDB relation. Chapter 9 shows the same idea with a longer `reports-to` example.

2.2 Magic-Set Rewrite

Now suppose the query asks only for nodes reachable from one bound start node, represented as `a` in the graph trace:

```
(def reachable-from-query
  '[:find [?end ...]
    :in $ % ?start
    :where (reachable ?start ?end)])

(d/q reachable-from-query db reachability-rules start-id)
```

Plain bottom-up evaluation of `reachable` would also derive the disconnected `p -> q -> r` portion of the graph, even though it cannot affect the answer. A magic-set rewrite adds a temporary predicate that carries the bound arguments from the query into recursive evaluation. In this example, `magic-reachable` means "this start node can still contribute to the answer for the original query." Conceptually, the engine seeds that relation with the bound start node and threads it through the recursive rule:

```
;; Seed the temporary relation from the query binding:
;; only a is initially relevant.
(magic-reachable a)

;; If a relevant start has an outgoing edge, its neighbor can
;; become the start of a relevant recursive subcall.
[(magic-reachable ?mid)
 (magic-reachable ?start)
 [?start :link/to ?mid]]
```

```
;; Guard the original base rule so it only derives rows for relevant starts.
[(reachable ?start ?end)
 (magic-reachable ?start)
 [?start :link/to ?end]]

;; Guard the original recursive rule the same way.
[(reachable ?start ?end)
 (magic-reachable ?start)
 [?start :link/to ?mid]
 (reachable ?mid ?end)]
```

This is a conceptual rewrite, not Datalevin's literal internal representation, and `magic-reachable` is not a stored attribute, datom, or user-visible rule. It is an internal, temporary relation that marks which `?start` values are worth expanding. Starting from `a`, the propagation rule discovers `b`, `c`, `d`, and `e` as relevant starts. The disconnected `p`, `q`, and `r` component is never marked, so the guarded `reachable` rules never derive rows for that component. The answers for the original query are still `b`, `c`, `d`, and `e`; the rewrite only avoids recursive work that cannot feed those answers.

Datalevin applies magic-set rewriting when a recursive rule call has bound arguments that can actually restrict recursive branches. If those bindings would not filter the recursive work, or if the magic relations grow beyond the optimizer's guardrail, the engine falls back to ordinary dependency-ordered recursive evaluation.

Figure 21.4 visually compares plain bottom-up evaluation with magic-set rewriting.

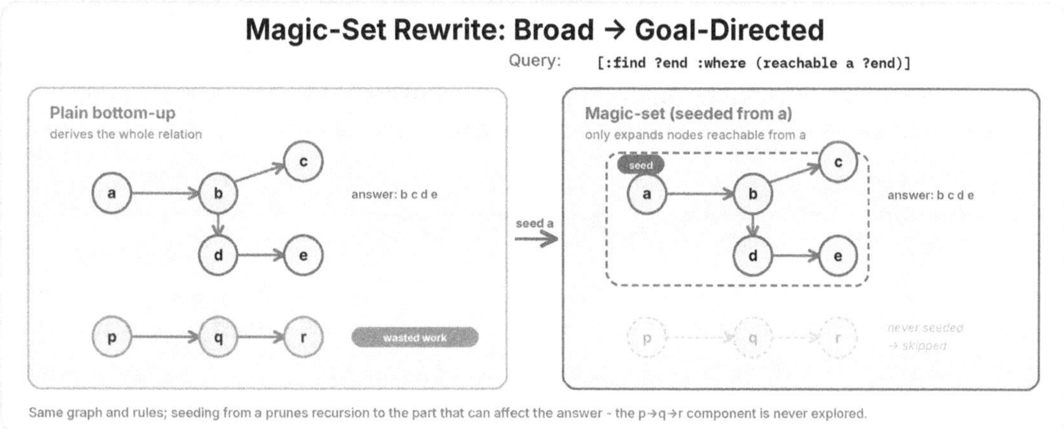

Figure 21.5: Magic-set rewrite turning broad recursion into goal-directed recursion: for the query reachable from a, plain bottom-up recursion derives the whole relation including the disconnected p→q→r component (wasted work), while the magic-set rewrite seeds the recursion at a and expands only the nodes reachable from a (a, b, c, d, e), leaving p, q, r never seeded and skipped; both return the same answer b c d e

2.3 Rule Evaluation Refinements

Datalevin also connects rule evaluation back to the cost-based optimizer:

1. **Seeding tuples:** Rule evaluation can receive bindings from earlier indexed query clauses. These seeds prevent unnecessary tuple generation and give recursion a warm start.
2. **Inlining non-recursive clauses:** Clauses not involved in recursion can be pulled back into the ordinary query plan, where the cost-based optimizer can use indexes and join estimates.
3. **Early set boundaries and fused distinct operators:** Each retained rule branch is projected to its head variables and deduplicated before branch results are combined. Where the shape permits, Datalevin fuses the join, projection, and distinct sink so duplicate proof paths are discarded as rows are produced instead of becoming a larger intermediate relation [12].
4. **Shape-specific EAV recursion:** Eligible linear EAV branches can stream new candidates into their seen sets and reuse cached adjacency. Canonical transitive closure with one singleton-bound endpoint can traverse through EA or AV index probes and adapt to an attribute scan when observed fanout makes that cheaper. Synchronized two-link closures—recursive shapes that follow two EAV links in lockstep—have guarded demand-driven and dense bitset paths; other shapes retain general semi-naive evaluation [12].
5. **Proof-driven stopping and result collection:** On a committed database, an eligible forward traversal can stop when its emitted answers saturate the exact AVE output domain. When a terminal result is also projected to the query's find variables and proven unique, Datalevin fully materializes it without repeating duplicate detection. Paths without those proofs keep the general evaluator and spillable set collector [12].
6. **Temporal elimination:** Recursive rules that meet T-stratification criteria can keep only the last iteration's results, reducing memory use for long chains [10].
7. **Conservative complex clauses:** `not`, `not-join`, `or`, and function clauses obey the same binding discipline as ordinary queries. Recursive rules with negative or complex clauses use conservative planning paths rather than the positive-recursion magic-set path.

The practical takeaway is that rules are not interpreted as repeated application callbacks. They are compiled into set-oriented recursive evaluation, constrained by outer query bindings, and integrated with the same storage and planning principles used by ordinary Datalog queries.

3. Inspecting Query Plans with `explain`

Datalevin's query optimizer has good evidence, but it is still making estimates. Sometimes a query that should be fast takes longer than expected. To debug those cases, inspect the execution plan instead of guessing from the query text.

Datalevin's public diagnostic API for Datalog queries is **`explain`**. By default, `explain` plans the query without running it. When called with `{:run? true}`, it executes the same optimizer-selected physical alternative that an ordinary query would use and adds runtime measurements to the explain output.

Entry point	Runs the query?	Physical-plan behavior
Ordinary `d/q`	Yes	Executes the selected conventional or access-path root and returns the query result.

Entry point	Runs the query?	Physical-plan behavior
Plan-only `d/explain`	No	Costs and selects a root, but does not execute it; `:executed-plan-alternative` has `:kind :not-run`.
Measured `d/explain` with `{:run? true}`	Yes	Executes the selected root, returns the result in the explain map, and reports the executed root and any runtime fallback.

3.1 Plan-Only Explain

Use plan-only `explain` when you want to inspect join order, access paths, and estimated cardinalities without running the query itself.

```
(def query
  '[:find ?name
    :where [?e :user/name ?name]
           [?e :user/age ?age]
           [(> ?age 30)]])

(d/explain {} query (d/db conn))
```

Plan-only output includes:

- **`:query-graph`**: The optimizer's graph of clauses and estimated counts
- **`:plan`**: The fully planned conventional clause-and-join root, including its steps, join methods, estimated `:cost`, and estimated `:size`. It is also the ready fallback when an access root wins. The order of the `:steps` is the chosen join order, and each step describes its join or index operation.
- **Access-path alternatives**: When an AVE, idoc, full-text, or vector access expression is applicable, `:access-plans` and `:preferred-access-plan` describe its method, strategy, ordering, capabilities, candidate budget, and estimate. `:physical-plan-alternatives` shows the conventional and access roots compared by the cost model; `:selected-plan-alternative` and `:access-path-selected?` identify the winner.
- **Selection versus execution**: `:selected-plan-alternative` and `:recommended-plan-alternative` report the optimizer's choice. `:executed-plan-alternative` has `:kind :not-run` for plan-only explain.
- **`:opt-clauses` / `:late-clauses`**: Clauses handled by the optimizer versus clauses processed after the optimized plan
- **Pre-planning rewrite decisions**: When a constant AVE lookup is considered for early materialization, `:pre-materialization-decisions` compares its fan-out and complete candidate cost with leaving the lookup in the graph.
- **Planning times**: `:parsing-time`, `:building-time`, `:planning-time`, and `:prepare-time`

The actual value returned by `explain` is ordinary structured data and is much larger than what is useful to print inline. It includes the query graph, plan steps, rewritten clauses, timing fields, and many implementation details that are useful when debugging a specific query. The following is an abbreviated, human-friendly excerpt from the small query above:

```
{:planning-time "3.217",
```

```
:opt-clauses
[[?e :user/name ?name]
 [?e :user/age ?age]
 [(> ?age 30)]],
:query-graph
{$
 {?e
  {:free
   [{:attr :user/name, :var ?name}
    {:attr :user/age,
     :var ?age,
     :range [[[:open 30] [:closed :db.value/sysMax]]]}]}}},
:plan
{$
 [[{:steps
    ["Initialize [?e ?age] by range ... on :user/age."
     "Merge [?name] by scanning [:user/name]."],
    :cost nil,
    :size nil}]]},
:late-clauses ()}
```

Read this in three parts. `:late-clauses` is empty, so the optimizer handled all clauses in the main plan. The predicate `[(> ?age 30)]` has been converted into a range bound on `:user/age`; it is not a separate row-by-row filter. The plan starts from that bounded age scan and then merges `:user/name` for the same entity. The `:cost` and `:size` fields can be `nil` for a single star-shaped component because there is no join-order competition to cost. In multi-component joins, `:cost` and `:size` are usually the main plan fields to inspect; in every case, the `:query-graph` count fields show the optimizer's estimated clause cardinalities.

Use plan-only explain to verify that a query is using the expected indexes and join strategies before running it on large data.

> **Use plan-only explain for large data**
>
> Plan-only diagnostics show the optimizer's intended access paths without executing the query. Use measured execution mode only when you are ready for the query to actually run.

3.2 Explain with Execution Measurements

Use `{:run? true}` when you also need measured execution information. This runs the query and augments the explain map with fields such as `:execution-time`, `:actual-result-size`, `:result`, and per-plan `:actual-size` where available.

```
(d/explain {:run? true} query (d/db conn))
```

`{:run? true}` is the runtime diagnostics mode of `explain`. It does not produce a separate clause-by-clause trace, but it lets you compare estimated sizes against actual sizes and distinguish planning cost from runtime cost.

For an ordered, limited AVE query like the one in Section 1.4, an abbreviated measured result can look like this. Unrelated fields are omitted:

```
{:selected-plan-alternative
 {:kind :access,
  :mode :adaptive-top-k,
  :size 7},
 :executed-plan-alternative
 {:kind :access,
  :mode :adaptive-top-k,
  :size 7},
 :plan
 {:kind :access,
  :mode :adaptive-top-k,
  :access-plan {:method :ave},
  :size 7,
  :actual-size 6,
  :candidate-count 360,
  :fragment-result-count 215,
  :residual-result-count 215},
 :actual-result-size 6}
```

The selected summary records the optimizer's choice; the executed summary confirms which root actually ran. Inside `:plan`, `:size 7` is the estimated output of that access root and `:actual-size 6` is its measured output. The source admitted 360 candidate tuples, the access fragment produced 215, and the residual query also produced 215. Applying final ordering and `:limit` returned six rows. Candidate, fragment, and residual counts describe successive stages of one pipeline; they are not alternative estimates of the same relation.

If the conventional root wins, `:plan` retains the grouped component plans and measured execution adds `:actual-size` where available. If an access root wins, `:plan` has `:kind :access` and identifies its `:mode` and `:access-plan`. Additional fields such as `:candidate-work`, `:batches`, and `:subqueries` provide progressively more detail when needed.

An adaptive access execution can still exhaust its candidate budget or safety cap and continue with the already planned conventional root. In that case, `:plan` preserves the attempted access work and adds `:fallback`; the same fallback is attached to `:executed-plan-alternative`, and the conventional fallback plan appears among `:subqueries`. Thus the measured plan describes the path actually taken rather than attributing conventional counts to the selected access root.

When eligible late indexed producers compete with a bound pattern, `:late-clause-decisions` records whether Datalevin kept the bound pattern first or selected an indexed union or range first, along with the runtime cost inputs. The decision can depend on values and relation sizes available only at that boundary. Morsel-specific `:partitioned-execution` fields are documented with morsel-driven execution in Section 1.7.1.

Compare estimated and actual fields at the same nesting level. Use `:actual-result-size` for the final measured cardinality and `:prepare-time` versus `:execution-time` to separate planning overhead from execution time. When corresponding estimated and actual sizes are close, the planner's model matched the current data. Datalevin is normally very accurate in the first few plan steps because it can combine exact index counts with query-specific sampling, and those early steps matter most for keeping intermediate results small.

If `:size` and `:actual-size` differ by orders of magnitude, cardinality estimation was wrong for that part of the plan. The cause may be a predicate or a join whose selectivity is hard to estimate, but it can also be stale or sparse attribute metadata: Datalevin refreshes per-attribute eid samples, attribute cardinality counters, and default fan-out ratios in the background, and a very fast load or major reshaping of data may outrun that process. Background sampling is enabled by default, but explicit `analyze` is useful when you need those statistics refreshed before judging a plan. After a bulk load or major reshaping of data, refresh statistics for the affected attribute, or for the database as a whole:

```
(d/analyze (d/db conn) :user/age)
(d/analyze (d/db conn))
```

3.3 Understanding Join Orders

The `:plan` section shows the **join order** chosen by the optimizer.

1. **Leading Clause**: This is the first clause the engine used to find the initial set of entities.
2. **Sequential Joins**: The order in which subsequent clauses were applied.

If you see that the engine is starting with a non-selective clause (e.g., searching for `:user/active? true`), it might be because the selective filter you expect is hidden inside a complex predicate or rule.

A common join-order bottleneck is a **large intermediate result set**. If the first join returns 1,000,000 entities, but the second join filters them down to 10, the engine still had to process 1,000,000 records.

The fix is to ask whether the query can express a more selective indexed constraint earlier in the plan. The optimizer tries to do this automatically, but a complex predicate or rule may hide the selective condition from estimation.

3.4 Diagnosing Slow Predicates

Some custom predicates are visible to the optimizer. A predicate that can be pushed down, typically one involving a single free variable and no source expression, is executed during query-specific sampling, so Datalevin can estimate its selectivity from the current data. For example, a predicate such as `[(my-ns/is-complex? ?x)]` can be planned differently from a late filter when `?x` is attached to a sampled value path.

The harder cases are predicates that cannot be pushed down, such as predicates over several variables, predicates that depend on an explicit source, or functions whose own runtime cost dominates even when selectivity is estimated well. `explain` can show whether the predicate became part of the optimized plan or remained in `:late-clauses`, but it does not time each predicate call separately. If a query becomes slow only after adding a predicate, isolate that predicate by comparing `explain` output and query runtime before and after adding it.

> **Filter before expensive predicates**
>
> Prefer indexed constraints that reduce the candidate set before expensive non-pushdownable predicates run. If a predicate makes a query slow, compare `explain` output and runtime before and after adding that predicate.

4. Related Benchmark Coverage

For performance evidence, see Chapter 1 for the Join Order Benchmark, Chapter 9 for OpenRuleBench, and Chapter 13 for LDBC SNB. This chapter stays focused on planner mechanics and the `explain` workflow.

Summary

The chapter has three layers: optimizer mechanics, recursive-rule evaluation, and the practical diagnostics workflow. When a query is slow, follow this workflow:

1. **Use plan-only `explain`**: Inspect join methods, index usage, and estimated cardinalities. For bounded queries, also inspect `:access-path-selected?`, the preferred method and strategy, and the selected adaptive mode. When present, inspect `:pre-materialization-decisions` for selective AVE anchors.
2. **Use `explain` with `{:run? true}`**: Confirm the executed root in `:executed-plan-alternative`, then compare estimates with actual result sizes and total execution time. For an access root, inspect its method, mode, candidate counts, batches, residual subqueries, and any fallback. Check `:late-clause-decisions` when the query has a late indexed union or bounded range. For a large read-only component plan, `:partitioned-execution` confirms whether an eligible linear segment was divided into morsels.
3. **Check cardinality gaps**: Large gaps between estimated `:size` and measured `:actual-size` are a strong hint that a predicate, rule, stale per-attribute sample, or rapidly changed data is hiding selectivity from the optimizer. If the data changed quickly, run `d/analyze` before drawing conclusions from the plan.
4. **Simplify and isolate**: Remove clauses one by one to find the specific part of the query that is slow.
5. **Refine the logic**: Prefer more selective indexed attributes to expensive custom predicates.

By using `explain` well, you gain the transparency needed to understand how Datalevin plans a query and why a query may be slower than expected.

References

[1] Patricia G. Selinger, Morton M. Astrahan, Donald D. Chamberlin, Raymond A. Lorie, and Thomas G. Price, "Access Path Selection in a Relational Database Management System," SIGMOD 1979, pp. 23-34. URL: https://research.ibm.com/publications/access-path-selection-in-a-relational-database-management-system. DOI: https://doi.org/10.1145/582095.582099.

[2] Guido Moerkotte and Thomas Neumann, "Dynamic Programming Strikes Back," SIGMOD 2008, pp. 539-552.

[3] Hai Lan, Zhifeng Bao, and Yuwei Peng, "A Survey on Advancing the DBMS Query Optimizer: Cardinality Estimation, Cost Model, and Plan Enumeration," *Data Science and Engineering*, 2021.

[4] Andreas Brodt, Oliver Schiller, and Bernhard Mitschang, "Efficient Resource Attribute Retrieval in RDF Triple Stores," CIKM 2011, pp. 1445-1454.

[5] Andrey Gubichev and Thomas Neumann, "Exploiting the Query Structure for Efficient Join Ordering in SPARQL Queries," EDBT 2014.

[6] Marios Meimaris et al., "Extended Characteristic Sets: Graph Indexing for SPARQL Query Optimization," ICDE 2017.

[7] Thomas Neumann and Guido Moerkotte, "Characteristic Sets: Accurate Cardinality Estimation for RDF Queries with Multiple Joins," ICDE 2011.

[8] Todd J. Green, Shan Shan Huang, Boon Thau Loo, and Wenchao Zhou, "Datalog and Recursive Query Processing," *Foundations and Trends in Databases* 5(2):105-195, 2013. URL: https://www.nowpublishers.com/article/Details/DBS-017. DOI: https://doi.org/10.1561/1900000017.

[9] Francois Bancilhon, David Maier, Yehoshua Sagiv, and Jeffrey D. Ullman, "Magic Sets and Other Strange Ways to Implement Logic Programs," PODS 1986, pp. 1-15. DOI: https://doi.org/10.1145/6012.15399.

[10] Amir Shaikhha et al., "Optimizing Nested Recursive Queries," *Proceedings of the ACM on Management of Data* 2(1), SIGMOD 2024, pp. 1-27.

[11] Datalevin project, "Query Processing," parallel-processing design and execution diagnostics, commit `ee208cec`. URL: https://github.com/datalevin/datalevin/blob/ee208cecb3378da5c843e4ebbcb9bcd372516ef4/doc/query.md#parallel-processing.

[12] Datalevin project, "Datalevin Rule Engine," implementation notes. URL: https://github.com/datalevin/datalevin/blob/master/doc/rules.md.

CHAPTER 22

Production Operations

After tuning storage durability, write ingestion, and query plans, production operations are about choosing the right boundary: where the database lives, who can reach it, how failures are isolated, how replicas are maintained, and what signals tell you the system is healthy. This chapter closes Part V by covering that operational lifecycle: how to test an application against Datalevin, choose a deployment mode, choose a tenancy shape, run a server when needed, secure remote access, operate replicas or HA, and monitor a production system.

It begins with testing, because the cheapest place to catch a modeling, schema, or query mistake is in a test, long before deployment. It also pulls together Datalevin's concurrency and failure model, because production code needs a clear policy for which transaction failures are data bugs, which are logical conflicts, and which are transient operational conditions.

1. Testing Datalevin Applications

Datalevin is easy to test because the same library that runs in production can run entirely inside a test process. There is no separate test server to manage and no external service to mock. Most application logic can be exercised against a real Datalevin database that is created and discarded inside a single test.

Three levels of test are useful, from fastest to most realistic:

1. **Pure logic tests** over in-memory tuples, with no database at all.
2. **In-memory database tests** that exercise schema, transactions, indexes, search, and queries against a real but ephemeral store.
3. **Temporary on-disk tests** for behavior that depends on files, persistence, or WAL.

Start at the cheapest level that can fail for the reason you care about, and move to a more realistic level only when the test needs it.

> **Choose the cheapest faithful test**
>
> Use tuple tests for pure query logic, in-memory databases for schema and transaction behavior, and temporary on-disk databases when persistence, WAL, file paths, or recovery behavior is part of the claim.

1.1 Pure Logic Tests Over Tuples

Chapter 8 showed that the query engine accepts any sequence of `[e a v]` tuples as a data source. This is the fastest way to unit-test query and rule logic, because there is no store to open, transact, or clean up. The fixture data sits in the test itself. In the Clojure example, the tuple collection itself fills $, with no connection at all. The Java, Python, and JavaScript connection APIs always supply $ from a database, so those examples open a throwaway in-memory connection and bind the tuple fixture as an extra `$data` source.

```
(ns my-app.query-test
  (:require [clojure.test :refer [deftest is]]
            [datalevin.core :as d]))

(def people
  [[1 :user/name "Alice"] [1 :user/age 30]
   [2 :user/name "Bob"]   [2 :user/age 25]])

(deftest adults-are-selected
  (is (= #{["Alice"]}
         (d/q '[:find ?name
                :in $ ?min
                :where [?e :user/name ?name]
                       [?e :user/age ?age]
                       [(>= ?age ?min)]]
              people 28))))
```

Rules are data too, so a rule set can be passed straight into a tuple-backed query as the `%` input. Use this level for query shape, rule composition, predicates, and aggregation logic. It does not exercise schema, indexes, upsert, or search; those need a real store.

1.2 In-Memory Database Tests

When a test needs schema, transactions, indexes, or search, open a real in-memory connection. An in-memory database uses the same API as a persistent one, but keeps nothing on disk and is discarded when the connection closes (Chapter 2). Each test gets a fresh, isolated database with no cleanup step.

```
(deftest upsert-updates-existing-user
  (let [schema {:user/email {:db/valueType :db.type/string
                             :db/unique    :db.unique/identity}
                :user/name  {:db/valueType :db.type/string}}
        conn   (d/create-conn nil schema {:kv-opts {:inmemory? true}})]
    (try
      (d/transact! conn [{:user/email "ada@example.com" :user/name "Ada"}])
      (d/transact! conn [{:user/email "ada@example.com" :user/name "Ada L."}])
      (is (= "Ada L."
             (d/q '[:find ?name .
                    :where [?e :user/email "ada@example.com"]
                           [?e :user/name ?name]]
                  (d/db conn))))
      (finally
        (d/close conn)))))
```

Because in-memory stores are cheap to create, the simplest isolation strategy is one fresh store per test. Prefer per-test fixtures such as `:each`, `@BeforeEach`, pytest's default function-scoped fixtures, or `beforeEach` over suite-level setup, so one test cannot leave state that another test depends on. Section 1.8.1 shows complete runner-specific fixture examples.

1.3 Temporary On-Disk Tests

Some behavior only appears on disk: persistence across reopen, WAL replay, snapshots, compaction, or `copy/dump/load`. For those, open a database in a unique temporary directory and delete it afterward. The `with-conn` macro opens the connection, runs the body, and closes it, so only the directory needs explicit cleanup:

```
(ns my-app.persistence-test
  (:require [clojure.test :refer [deftest is]]
            [datalevin.core :as d]
            [datalevin.util :as u]))

(deftest data-survives-reopen
  (let [dir (u/tmp-dir (str "test-" (random-uuid)))]
    (try
      (d/with-conn [conn dir test-schema]
        (d/transact! conn [{:user/email "ada@example.com"}]))
      ;; Reopen the same path; the fact should still be there.
      (d/with-conn [conn dir test-schema]
        (is (some? (d/entity (d/db conn) [:user/email "ada@example.com"]))))
      (finally
        (u/delete-files dir)))))
```

`datalevin.util/tmp-dir` returns a platform-neutral temporary path, and `delete-files` removes the directory tree; both are utility helpers used by Datalevin's own test suite. Adding a `random-uuid` keeps parallel test runs from colliding on the same path.

1.4 Make Tests Strict to Catch Typos

Schema-on-write is convenient in production but unhelpful in a test: a misspelled attribute such as `:user/emial` is silently accepted as a new attribute, and the assertion simply fails to find the data (Chapter 5). Tests are the right place to turn that leniency off. Open test connections with `:validate-data?` and `:closed-schema?` so unknown attributes and ill-typed values fail loudly:

```
(d/create-conn nil schema
  {:kv-opts         {:inmemory? true}
   :validate-data? true
   :closed-schema? true})
```

With these options, a transaction that mentions an undeclared attribute or writes a value that does not match its declared `:db/valueType` throws instead of quietly storing surprising data. This converts a class of modeling bugs into test failures (Chapter 11).

1.5 Dry-Run Transactions With Simulated Reports

To test how transaction data resolves, including tempids, upserts, and the datoms that would be added or retracted, without mutating the database, use the simulated transaction report API. It returns the same transaction-report shape as `transact!` but leaves the connection untouched. The `:db-after` value is the would-be post-transaction database, so tests can inspect the state that would result without committing it (Appendix E).

This is useful for unit-testing transaction-function output and validation logic, where you care about the datoms or would-be state a transaction would produce rather than committing them. Section 1.8.2 shows complete examples for each language binding.

1.6 Testing Search, Vectors, and Async Indexing

Full-text, idoc, and vector indexes are maintained synchronously by default (Chapters 16-17), so an in-memory database is enough to test them: read-your-writes holds, and a `fulltext`, `idoc-match`, or `vec-neighbors` query sees data committed earlier in the same test.

Two pitfalls are worth avoiding:

- **Keep vector tests deterministic.** Prefer `:db.type/vec` with literal vectors and `vec-neighbors` rather than depending on a remote embedding provider in a unit test. A network call to an embedding API makes the test slow, nondeterministic, and dependent on credentials. Reserve provider-backed `:db/embedding` tests for a small, clearly separated integration suite.
- **Wait for async indexing before asserting.** If a domain opts into `:indexing-mode :async`, the source datoms commit before the secondary index catches up, so an immediate search may miss them. Call `wait-for-secondary-index` (or `process-secondary-index-jobs!`) before the assertion (Appendix E):

```
(let [conn (d/create-conn
             nil
             {:doc/content {:db/valueType :db.type/string
                            :db/fulltext true
                            :db.fulltext/autoDomain true}}
             {:kv-opts {:inmemory? true}
              :search-domains {"doc/content" {:indexing-mode :async}}})]
  (try
    (d/transact! conn [{:doc/content "vector search guide"}])
    (d/wait-for-secondary-index conn)
    (is (seq (d/q '[:find [?e ...]
                    :where [(fulltext $ :doc/content "vector") [[?e _ _]]]]
                  (d/db conn))))
    (finally
      (d/close conn))))
```

The same full-text, idoc, and vector queries can be tested through ordinary query calls. The wait helper is only needed for domains that explicitly choose async secondary indexing; synchronous domains are visible before `transact` returns.

1.7 Testing From Application Test Runners

Use the test runner your application already uses. Datalevin should be ordinary application code inside those tests.

The portable patterns are:

- **Tuple-backed query tests**: Bind an EAV tuple sequence as a named source such as `$data`, using a throwaway in-memory connection when the query API needs a primary database source.

- **Fresh in-memory databases**: open a fresh in-memory database and close the connection at the end of the test.
- **Per-test temporary directories**: use the test runner's temporary-directory facility, open a normal path-backed connection, and let the runner delete the directory after the test.
- **Simulated transaction reports**: dry-run transaction data against the current database state and assert on the resulting report without committing the transaction.
- **Helper APIs**: fixtures and temporary-directory helpers are conveniences; explicit connection lifecycle management works everywhere.

1.8 Testing Recipes

The following recipes turn the testing strategy into complete runner-specific fixtures and simulated-transaction examples.

1.8.1 Per-Test In-Memory Fixtures

Each test should read the connection supplied by the fixture and start from an empty database. The second test below would fail if the first test's transaction leaked into it.

```
(def test-schema
  {:user/email {:db/valueType :db.type/string
                :db/unique    :db.unique/identity}
   :user/name  {:db/valueType :db.type/string}})

(def ^:dynamic *conn* nil)

(defn with-fresh-db [f]
  (let [conn (d/create-conn nil test-schema {:kv-opts {:inmemory? true}})]
    (binding [*conn* conn]
      (try (f) (finally (d/close conn))))))

(use-fixtures :each with-fresh-db)

(deftest create-user
  (d/transact! *conn* [{:user/email "ada@example.com"
                        :user/name  "Ada"}])
  (is (= "Ada"
         (:user/name (d/entity (d/db *conn*) [:user/email
↪  "ada@example.com"])))))

(deftest each-test-starts-empty
  (is (nil? (d/q '[:find ?e .
                   :where [?e :user/email "ada@example.com"]]
                 (d/db *conn*)))))
```

1.8.2 Simulated Transaction Reports

The simulated transaction report method follows each binding's naming convention: `tx-data->simulated-report` in Clojure, `txDataToSimulatedReport` in Java and JavaScript, and `tx_data_to_simulated_report` in Python.

```
(deftest signup-adds-email-datom
  (let [conn   (d/create-conn nil schema {:kv-opts {:inmemory? true}})
        report (d/tx-data->simulated-report
                 (d/db conn)
                 [{:db/id -1
                   :user/email "ada@example.com"
                   :user/name "Ada"}])
        eid    (get (:tempids report) -1)]
    (try
      ;; The report shows the datoms that *would* be written ...
      (is (some #(and (= :user/email (d/datom-a %))
                      (= "ada@example.com" (d/datom-v %)))
                (:tx-data report)))
      ;; ... and :db-after can be read as the would-be result.
      (is (= "Ada" (:user/name (d/pull (:db-after report) [:user/name] eid))))
      ;; ... but the database itself is left untouched.
      (is (nil? (d/entity (d/db conn) [:user/email "ada@example.com"])))
      (finally
        (d/close conn)))))
```

2. Choose a Deployment Mode

Datalevin can run as an embedded library, a TCP server, a read-replica topology, a consensus-lease HA cluster, an MCP tool process, or a Babashka pod. Figure 22.1 depicts these different modes. The data model and storage format stay the same; the operational boundary changes.

Appendix A gives package coordinates, supported platforms, startup commands, and connection examples. This chapter focuses on the production decision: which operational boundary should own the database?

Mode	Best Fit	Main Trade-Off
Embedded	Single-process services, desktop apps, local tools	Fastest path, but access is local to the process and storage path
Server	Shared databases, remote clients, centralized RBAC	Adds network latency and serialization
Async read replica	Read scaling or workload isolation without failover	Eventually consistent and read-only; no automatic promotion
Consensus-lease HA	Automatic write-leader promotion and follower reads	Requires quorum discipline, fencing, and explicit membership management
MCP	AI tools that need controlled Datalevin access	Tool-call boundary; writes disabled unless explicitly allowed
Babashka pod	Scripts and maintenance tasks	IPC boundary; convenient but not the zero-copy embedded path

Embedded mode is the default when one application owns the database path. It avoids network serialization and uses the OS page cache directly, so leave RAM headroom for that cache. Server mode is the right boundary when multiple processes need the same database, when remote clients need `dtlv://` access, or when server-side RBAC is part of the deployment.

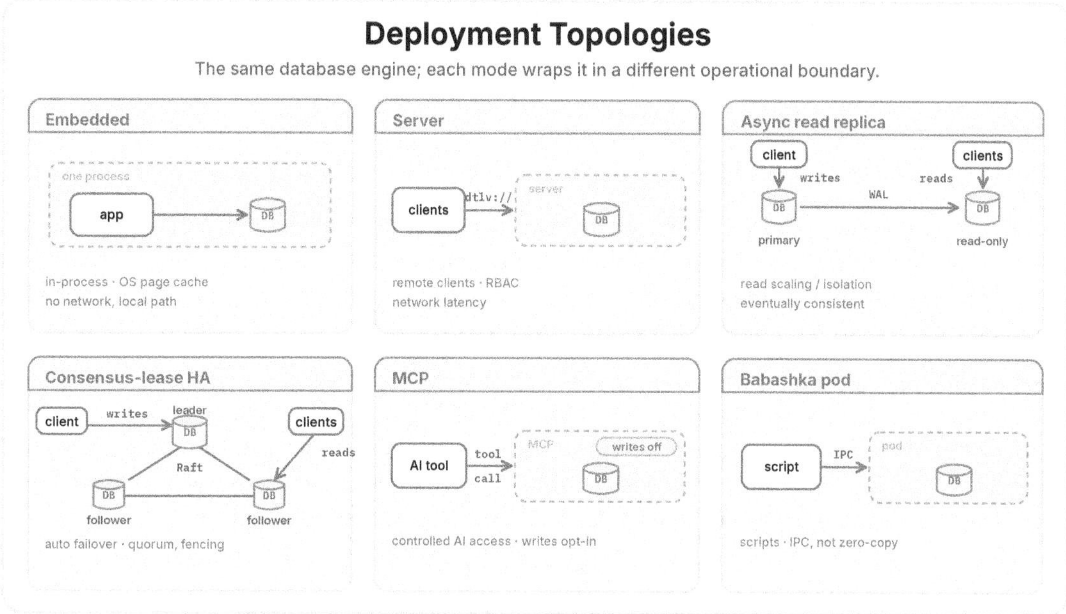

Figure 22.1: Deployment topologies around the same database: embedded keeps the app and DB in one process; server exposes the DB to remote clients over dtlv://; an async read replica ships WAL from a primary to a read-only replica; consensus-lease HA coordinates a leader and follower DBs via Raft with fencing; MCP wraps the DB in a tool-call boundary with writes off by default; a Babashka pod reaches the DB over IPC

An embedded Clojure application with the full Datalevin library, or a Java application with `org.datalevin:datalevin-java-server`, can also host a server inside its own process. Use that only when the application intentionally owns both lifecycles. Python and Node.js applications should run `dtlv`, Docker, or the standalone JVM jar as a separate server process.

Async read replicas and consensus HA both depend on WAL records, but they solve different problems:

- **Async read replicas** scale or isolate reads behind one primary writer. They do not elect a leader, fence writes, or promote automatically.
- **Consensus-lease HA** keeps one write leader per database, uses a Raft-backed control plane for leases and membership, and lets followers serve reads. Raft is a replicated-consensus algorithm for agreeing on cluster state [1].

Choose async replicas when stale-by-lag reads are acceptable. Choose HA only when automatic failover is worth the extra operational surface.

In HA discussions, **fencing** means preventing a stale or deposed writer from continuing to accept writes after another node has been allowed to lead. It is the safety mechanism that keeps failover from becoming split-brain.

`dtlv mcp` and the Babashka pod are tooling modes. They are useful for local AI tools, maintenance jobs, and ad hoc scripts, but they are not replacements for the embedded or server boundary of the main application.

Production version pinning. Pin the Datalevin version across the application, server, CLI,

Docker image, pod, and language bindings. This guide targets the Datalevin 1.1.0 release; before upgrading, read the release notes and verify the upgrade in staging. Appendix A covers runtime requirements, and Chapter 19 covers storage-format upgrade workflows.

2.1 Connection Lifecycle

A long-running application should normally keep one shared Datalog connection per Datalevin database in a process. Create it during application startup, pass it to the code that needs database access, and close it during application shutdown. Do not open and close the same local embedded database for every web request, job, or repository call.

For embedded Datalog databases, use the normal open-or-reuse path at the application boundary. In Clojure, that is `d/get-conn`; in Java, it is `Datalevin.getConn`; in Python and JavaScript, it is `connect`. Short-lived connections are fine for REPL experiments, scripts, and tests. Controlled test fixtures and specialized construction paths may deliberately create a fresh connection object, such as with `d/create-conn` in Clojure or `Datalevin.createConn` in Java.

For path-addressed embedded databases, `get-conn` keeps a process-local entry for the path. If an open connection for that path is already registered, it returns that connection; if the registered connection has been closed, it opens a new one. Treat this as connection-opening behavior, not as a reason to call `get-conn` in every function. The schema and option maps are the complicating part: the first successful open establishes the live connection for that path, and later calls may simply receive that already-open connection instead of applying a different schema or option map in the way the caller expects. Repeated calls also obscure ownership and close timing. Keep the returned connection at the application boundary and pass it to the code that needs it.

3. Multi-Tenancy Trade-Offs

Tenant shape is part of choosing the operational boundary. It affects deployment mode, RBAC, backup and restore workflows, migration strategy, monitoring cardinality, and HA topology. Choose the tenancy shape based on the isolation boundary the application really needs, not only on the number of tenants.

Pattern	Best fit	Main trade-off
Database per tenant	Strong isolation, per-tenant backup/delete/export, local-first workspaces	You operate many database paths and run migrations across a fleet.
Shared database with tenant facts	Many tiny tenants, global analytics, shared reference data	Every query and write path must preserve the tenant predicate.
Catalog database plus tenant databases	SaaS workspaces with shared identity/billing and isolated tenant data	Cross-database changes are not one Datalevin transaction.
Tiered tenancy	Free/small tenants shared, large or regulated tenants isolated	Moving a tenant between tiers becomes an operational workflow.

Database per tenant is the most Datalevin-native pattern when tenant isolation matters. Use one database path per tenant, workspace, or user. The advantages are operational: per-tenant backup, restore, migration, export, and deletion are path-level operations; the OS page cache shares memory across many idle databases efficiently; tenant isolation is physical rather than a repeated query predicate; and the same database path can often move between server and local devices. This pattern is common in personal knowledge management, hosted workspaces, and agent memory stores. A local-first per-user or per-workspace database is a deployment variant of the same pattern: the database path is a useful unit of ownership that can be backed up, moved, synced through application-specific tooling, or attached to a server later.

The trade-off is fleet management and open-store budgeting. You need a catalog that maps tenant ids to database names or paths, a migration runner that can visit every tenant database, backup and retention automation, and monitoring that can summarize many small databases. Cross-tenant analytics usually become a fan-out query over many databases or a separate reporting database fed by application jobs.

High density also does not mean every tenant database should stay open forever. Each open database consumes file descriptors and a memory-mapped virtual address reservation related to `:mapsize`. The page cache can make many idle database files cheap, but process virtual address space and OS limits are still finite. For database-per-tenant deployments, keep an explicit open-database budget, close idle tenant connections, size `:mapsize` with Chapter 19's guidance, and monitor open file counts plus process virtual size and resident memory. If the tenant database can be edited independently on several devices, conflict detection and merge policy are part of the application design; Datalevin's server replicas and HA modes are server-side operations, not a general multi-master local-first sync protocol.

Shared database with tenant facts keeps all tenants in one Datalevin database and models tenancy as ordinary data, for example with `:workspace/id`, `:account/id`, or `:entity/tenant` refs. This can be the right choice when tenants are tiny, when cross-tenant reporting is central, or when many tenants share the same reference data. Datalevin's indexes make tenant filters ordinary indexed facts, so a query can start from the tenant and join outward:

```
(d/q '[:find ?doc ?title
       :in $ ?workspace-id
       :where
       [?w :workspace/id ?workspace-id]
       [?doc :doc/workspace ?w]
       [?doc :doc/title ?title]]
     (d/db conn)
     workspace-id)
```

The risk is correctness. Server RBAC controls access to a database, not per-entity tenant predicates inside that database. If one shared database holds many tenants, the application must make tenant scoping part of every query, pull, transaction function, background job, and admin path. Consider wrapping query construction, adding transaction-time validation for required tenant refs, and testing that unscoped reads are impossible in application code.

Catalog plus tenant databases is often the best SaaS compromise. Keep a small shared catalog database for users, organizations, billing state, entitlements, and the mapping from tenant id to tenant database. Store each tenant's domain data in its own Datalevin database. The catalog answers "which database should this request open?", while the tenant database holds the application facts. This gives physical isolation for tenant data while keeping global account operations centralized.

The boundary is transactional: a Datalevin transaction is scoped to one database. If a signup updates the catalog and initializes a tenant database, that is an application-level workflow, not one atomic database transaction. Use idempotent steps, retries, and repair jobs for cross-database workflows.

Tiered tenancy mixes the first two patterns. You might place many free or trial tenants in a shared database, give enterprise tenants dedicated database paths, and place regulated tenants in separate regions or storage roots. This keeps small tenants cheap while preserving an upgrade path for isolation, custom backup schedules, or data-residency requirements. The cost is migration: tenant promotion from shared to dedicated storage needs an export/import or copy workflow, validation, and a catalog switch.

For most applications, start with the narrowest isolation boundary that matches the product and compliance requirements. If tenants must be individually exported, deleted, restored, encrypted, or hosted in different regions, prefer database-per-tenant or catalog-plus-tenant databases. If cross-tenant analytics and global queries dominate, a shared database with explicit tenant facts may be simpler.

4. Operate a Datalevin Server

The server is a non-blocking network service that exposes Datalevin APIs over a TCP connection. Appendix A shows the `dtlv`, Docker, standalone JVM jar, in-process Clojure, and in-process Java startup forms. In production, the important operational rules are the same for all of them: bind to loopback unless remote access is intended, set `DATALEVIN_DEFAULT_PASSWORD` before exposing a non-loopback address, and run the server under the platform's service manager with a service account that can read and write the data root.

The main server options are:

- **`-r <path>`**: Root directory for databases. The default is `/var/lib/datalevin` on POSIX systems and `C:\{}ProgramData\{}Datalevin` on Windows.
- **`-p <port>`**: Listening port. The default is `8898`.
- **`--host <host>`**: Listening address. Binding to a non-loopback address requires `DATALEVIN_DEFAULT_PASSWORD`.
- **`--idle-timeout <ms>`**: Disconnect inactive sessions. The default is `172800000` ms, or 48 hours.
- **`-v`**: Enable verbose server logs on stdout.

4.1 Remote Database URIs

Remote Datalog and KV databases use this URI shape:

```
dtlv://<user>:<pass>@<host>:<port>/<db-name>?store=datalog|kv
```

`store` is optional and defaults to `datalog`. Database names must be unique across the whole server, so a Datalog database and a KV database cannot share the same server-side name. Appendix A shows connection examples, and Appendix G lists the public `datalevin.client` administrative API.

4.2 Protocol and Client Tuning

The server protocol uses a TLV-style (type-length-value) message format. Clojure clients use Nippy serialization by default [3]. The wire protocol also supports Transit/JSON, Cognitect's data interchange format encoded over JSON [2], for clients that cannot use Nippy, and Datalevin also exposes a JSON command API.

Most deployments should start with defaults. Tune only when a measured workload needs it:

- **Freshness checks**: `datalevin.constants/*remote-db-last-modified-check-interval-ms*` defaults to `0`, which checks remote modification state on every call. A positive interval reduces round trips but can delay observation of remote updates.
- **Connection pool**: `datalevin.client/new-client` accepts `:pool-size` (default 3) and `:time-out` (default `60000` ms).
- **HA write retry**: `:ha-write-retry-timeout-ms` and `:ha-write-retry-delay-ms` bound automatic retries after retryable HA write rejections such as `:not-leader`. The timeout default is derived from the HA lease and promotion timings, currently `min(:time-out, 17000)`, rather than a fixed constant.
- **Wire compression**: `datalevin.constants/*wire-compression-threshold*` defaults to `8192` bytes, and `datalevin.constants/*wire-compression-level*` defaults to `3`.

For network-heavy workloads, prefer fewer larger requests. Batch writes with `transact!`, use `transact-async` when appropriate, use `pull` instead of many small entity lookups, and keep `:find` clauses specific so large unused result sets are not serialized over the wire.

4.3 Runtime UDFs in Server Mode

Descriptor-backed `:db/udf` functions resolve where the query or transaction executes. Embedded databases can receive a runtime registry in store options. Remote transactions execute in the server process, so the UDF registry or resolver must be installed on the server. Client-local registries are not consulted for remote transaction execution.

Remote queries use the server-safe resolver mode described in Chapter 8, Section 3.5. Operationally, custom query predicates, binding functions, and aggregates must be installed in the server-side UDF registry and called through `udf`; client-local functions are not callable just because the client submits the query. RBAC controls who may query or write, while the resolver boundary controls what code a permitted query may run.

In HA deployments, `:ha-require-udf-ready? true` makes a leader reject writes until installed transaction UDF descriptors can be resolved by the server runtime.

> **Server UDFs run on the server**
>
> Remote transactions and server-side query UDF calls resolve in the server process. Install the registry or resolver where the work executes; a client registry does not make client-local functions available to remote queries or writes.

5. Security and Access Control

Security has two separate layers: physical protection of database files and server authorization for remote access.

5.1 Default Admin Account

Every server has a built-in administrative user:

- **Username**: `datalevin`
- **Default password**: `datalevin`

Set `DATALEVIN_DEFAULT_PASSWORD` when starting the server. It is required when binding to a non-loopback address and should be treated as mandatory in production. Leave the `datalevin` user for administration and create separate application users with narrower roles.

> **Change the default administrator password**
> The built-in administrator is for setup and administration, not application traffic. Set a non-default password before exposing a server, then create least-privilege users for application clients.

5.2 RBAC Model

Server permissions are granted to roles, then roles are assigned to users. Every user also has a private role named `:datalevin.role/<username>`.

Each permission has three parts:

- **Action**: `:datalevin.server/view`, `:datalevin.server/alter`, `:datalevin.server/create`, or `:datalevin.server/control`.
- **Object**: `:datalevin.server/database`, `:datalevin.server/user`, `:datalevin.server/role`, or `:datalevin.server/server`.
- **Target**: A database name, username, role keyword, or `nil` for all targets of that object type.

The actions are hierarchical: `:alter` includes read access, `:create` includes lower actions, and `:control` is administrative control. The server object also subsumes database, user, and role objects, so reserve server-scoped grants for administrative roles.

5.3 Create a Least-Privilege User

This example creates a user, creates a role, grants read-only access to one database, and assigns the role to the user.

```
(c/create-user client "alice" "password123")
(c/create-role client :app.role/analyst)

(c/grant-permission client
                    :app.role/analyst
                    :datalevin.server/view
```

```
                :datalevin.server/database
                "sales-db")

(c/assign-role client :app.role/analyst "alice")
```

5.4 Encryption at Rest

Datalevin does not currently provide database-level transparent data encryption. Use encryption at the layer where it is easiest to operate and audit:

- Cloud volume encryption such as AWS EBS, GCP Persistent Disk, or Azure Disk Encryption.
- OS or filesystem encryption such as LUKS, FileVault, or `fscrypt`.
- Application-level encryption for sensitive fields such as PII or secrets.

For SaaS systems, envelope encryption is usually an application concern: keep tenant keys in a managed KMS (Key Management Service), wrap data-encryption keys with KMS-managed keys, and encrypt sensitive values before transacting them into Datalevin. NIST SP 800-57 is a useful general reference for key-management terminology and lifecycle concerns [4].

This guide does not include a hand-rolled encryption implementation. Correct field encryption depends on the runtime, threat model, KMS provider, key rotation policy, backup policy, and whether encrypted fields must be searched or joined. Use vetted libraries or managed KMS features, store enough key metadata to support rotation, and security-review the application code that encrypts and decrypts values.

6. Replication and High Availability

Replication and HA are server-mode features. Chapter 19 covers WAL durability and snapshots; this section covers the operational shape.

Both modes begin with the same physical mechanism: one server writes WAL records and another server can copy and replay them. The difference is what happens when the writer is unavailable. An async replica is a read-copy feature: it gives another server a lagging, read-only copy. Consensus-lease HA is a leadership feature: it decides which server may accept writes, rejects unsafe writes, and promotes another server when the old leader is no longer safe.

Figure 22.2 shows the two shapes side by side.

Use the smallest mode that satisfies the operational requirement:

- Use an async read replica when you want read scaling, reporting isolation, or lower-latency reads near another region, and stale-by-lag reads are acceptable.
- Use consensus-lease HA when the system must promote a new write leader after a server failure. HA needs more operational discipline: stable membership, clock discipline, tested fencing, and monitoring.

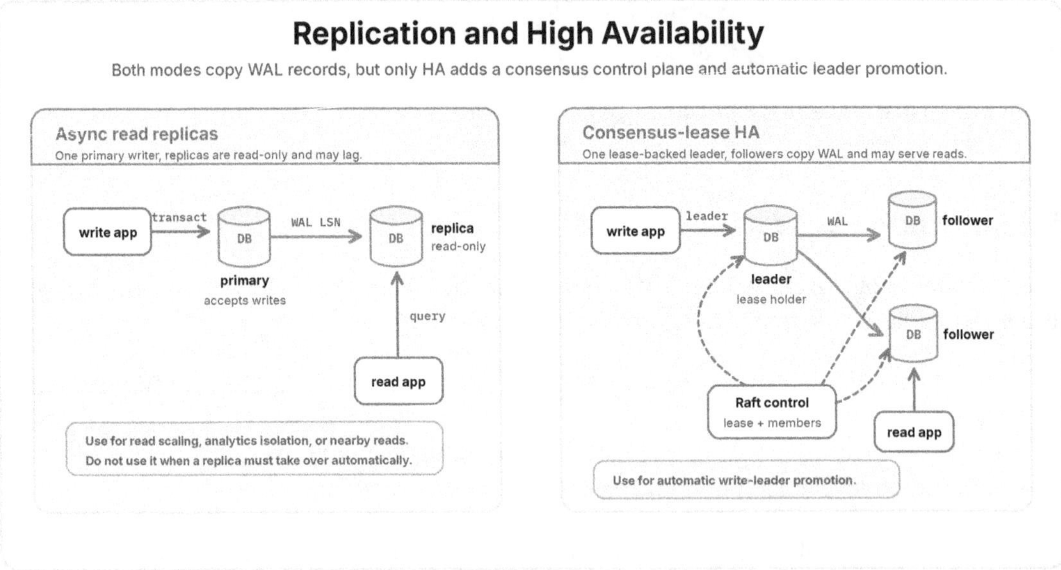

Figure 22.2: Replication and high availability guide: async read replicas have one primary writer that ships WAL records to read-only replicas; consensus-lease HA has one write leader, follower databases, and a Raft control plane that owns the lease and membership state

> **Replication is not a backup**
>
> Async replicas and HA followers copy current state, including bad writes, deletes, and application mistakes. Keep independent backup copies, off-site retention, and practiced restore procedures.

6.1 Non-HA Async Read Replicas

Use an async read replica when you want one primary writer and one or more read-only servers. The primary must have WAL enabled. The default `:strict` WAL durability profile is appropriate because the replica tails durable source records. The replica open also enables WAL so replayed records are persisted durably on the replica side.

A first async-replica rollout has six steps:

1. Start the primary Datalevin server and open the database with WAL enabled.
2. Create or choose a server user for the replica to read the primary database and WAL stream.
3. Start the replica Datalevin server with its own data root.
4. Open the database on the replica server with `:replica/read-only? true` and a `:replica/source` URI pointing at the primary.
5. Point read-only application traffic, dashboards, or batch jobs at the replica only after initial bootstrap completes.
6. Monitor `:replica-lag-lsn`, `:replica-last-error`, and degraded state.

The next example shows step 4, the replica-side database open:

```
(require '[datalevin.core :as d]
```

```
         '[datalevin.client :as cl])

(def schema
  {:name {:db/valueType :db.type/string
          :db/cardinality :db.cardinality/one}})

(def replica-conn
  (d/create-conn
   "dtlv://replica-admin:pass@replica-host:8898/app"
   schema
   {:replica/read-only? true
    :replica/source "dtlv://replicator:pass@primary-host:8898/app"
    :replica/id "app-replica-us-west"
    :replica/poll-ms 250
    :replica/report-ms 5000
    :wal? true
    :wal-durability-profile :strict}))
```

On first open, Datalevin bootstraps the replica from the primary copy interface if the local replica database does not exist. After that, the replica tails WAL records in LSN order and reports its applied LSN to the primary so WAL garbage collection preserves needed segments.

The source user in `:replica/source` must be able to open/copy the primary database, read the transaction log, and update the replica floor. In practice, grant the replicator user database `:datalevin.server/alter` permission. The user opening the local replica needs permission to create or open the database on the replica server.

Replica reads use normal APIs. User writes are rejected by the replica server:

```
(d/q '[:find [?name ...] :where [?e :name ?name]] (d/db replica-conn))

;; Throws: Replica is read-only
(d/transact! replica-conn [{:db/id -1 :name "blocked"}])
```

Replica status is available through typed admin clients:

```
(require '[datalevin.client :as cl])

(def client
  (cl/new-client "dtlv://replica-admin:pass@replica-host:8898"))

(cl/replica-status client "app")
```

The returned map includes `:replica/read-only?`, `:replica/source`, `:replica/id`, `:replica-applied-lsn`, `:replica-source-durable-lsn`, `:replica-source-committed-lsn`, `:replica-lag-lsn`, `:replica-last-sync-ms`, `:replica-degraded-reason`, and `:replica-last-error`.

Operationally, treat a replica as eventually consistent. A successful write on the primary is visible to later primary reads immediately, but a replica sees it only after it has copied and replayed the corresponding WAL record. If a user workflow needs read-your-write behavior, route that read to the primary or wait until the replica has applied the needed LSN.

6.2 Consensus-Lease HA

Consensus HA gives each HA database exactly one write leader at a time. Followers replicate from the leader using WAL records and snapshots and can serve reads.

Think of HA as an async replica topology plus a small consensus authority. The consensus group does not replicate every user transaction. User data still moves through WAL and snapshots. The consensus group decides which node owns the write lease, which membership is authoritative, and whether a candidate is allowed to become leader.

The design has three operational layers:

- **Data plane**: The leader appends writes to WAL; followers copy and replay records. A follower that falls behind retained WAL bootstraps from a snapshot.
- **Control plane**: A Raft-backed group stores the lease owner, term, leader endpoint, membership hash, and leader LSN.
- **Local state**: Each database on each node is a `:follower`, `:candidate`, `:leader`, or `:demoting`.

Writes are admitted only when the node has fresh proof that it is the current leader and is not demoting. HA forces `:wal? true` and defaults to `:wal-durability-profile :strict`.

Fencing is the external proof that the previous writer cannot keep accepting writes after leadership moves. The Raft control plane can decide that a lease has expired and that a candidate may try to lead, but it cannot by itself prove that an old process, host, load-balancer route, or storage attachment is unable to keep serving stale write traffic. That proof is deployment-specific, so Datalevin runs the configured fencing hook before a candidate admits writes.

A fencing hook should make the old writer harmless, not merely report that it looks unhealthy. Depending on the deployment, that may mean stopping or disabling the old Datalevin service, removing the old node from a write load balancer, revoking a virtual IP, detaching or remounting a writable volume, quarantining a host through a cloud or orchestration API, or acquiring a durable external lock that only one writer can hold. A health check, a failed ping, or client-side routing change is not enough by itself: the old writer might still be reachable from some clients or might recover while believing it can still write.

The hook must be idempotent and bounded. Datalevin may retry it, so running it twice should have the same effect as running it once. It should return success only after the old writer has been fenced, or after the operator's chosen lower layer already guarantees that the old writer cannot write. If fencing cannot be confirmed, the safe behavior is to remain unavailable for writes rather than risk two leaders writing divergent histories.

> **Failed fencing should stop promotion**
>
> HA is allowed to pause writes while the cluster proves safety. It should not trade a short outage for split-brain risk. Treat a failing fencing hook as an operational incident, not as a warning that can be ignored.

A first HA rollout usually looks like this:

1. Pick a stable database identity and a stable ordered member list. Do not reuse node ids for different machines.

2. Start at least three promotable nodes for normal quorum behavior. A witness voter can help quorum in some topologies, but it cannot become a data leader.
3. Open the database on every node with the same `:db-identity`, `:ha-members`, and promotable voter mapping. Each node gets its own `:ha-node-id`, control-plane `:local-peer-id`, and durable Raft directory.
4. Install and test the fencing hook before letting a node become promotable. Fencing must be idempotent because Datalevin may retry it.
5. Point write clients at the known HA endpoints and configure bounded HA write retry with the client options from Section 4.2.
6. Verify that exactly one node is leader, followers are caught up, and every node reports the same membership hash.
7. Run a failover drill before treating the cluster as production-ready. Stop or isolate the leader, watch a follower promote, verify writes succeed again, and verify the old leader rejoins as a follower.

A minimal HA option map has this shape. The exact hostnames, ports, credentials, and fencing command are deployment-specific:

```
{:db-identity "orders-prod"
 :ha-mode :consensus-lease
 :ha-node-id 1
 :ha-members
 [{:node-id 1 :endpoint "node-a:8898"}
  {:node-id 2 :endpoint "node-b:8898"}
  {:node-id 3 :endpoint "node-c:8898"}]

 :ha-client-credentials
 {:username "ha-replica"
  :password "secret"}

 :ha-fencing-hook
 {:cmd ["/usr/local/bin/dtlv-fence"]
  :timeout-ms 3000
  :retries 2
  :retry-delay-ms 1000}

 :ha-control-plane
 {:backend :sofa-jraft
  :group-id "orders-prod-ha"
  :local-peer-id "node-a:9098"
  :raft-dir "/var/lib/datalevin/orders/ha-control/node-a"
  :voters [{:peer-id "node-a:9098" :ha-node-id 1 :promotable? true}
           {:peer-id "node-b:9098" :ha-node-id 2 :promotable? true}
           {:peer-id "node-c:9098" :ha-node-id 3 :promotable? true}]}}
```

On `node-b`, use `:ha-node-id 2`, `:local-peer-id "node-b:9098"`, and a node-local `:raft-dir`. On `node-c`, use the corresponding 3 values. The membership list and voter mapping should otherwise describe the same ordered members and voters on every node.

The healthy write path is:

1. A client sends a write to a node it believes can lead.

2. The server admits the write only if its local role is `:leader`, its cached lease still names that node, its term is current, and it is not demoting.
3. The leader applies the transaction locally and appends the WAL record.
4. Followers copy and replay WAL records in LSN order.
5. If the leader fails, followers wait for the lease to expire, check membership and lag, race for the lease through the control plane, run fencing, and only then admit writes.

What HA does not mean is just as important: it is not multi-leader replication, not automatic membership discovery, not automatic failback to the old leader, and not quorum replication of every user transaction. A committed write is accepted by the current leader under a valid lease. It may not have reached a follower yet. If your risk model requires zero acknowledged-write loss after permanent destruction of the latest leader, design the deployment and client workflow around that requirement.

For follower reads, connect to a follower endpoint with the normal APIs:

```
(def follower
  (d/get-conn "dtlv://user:pass@replica-host:8898/app"))
```

Use RBAC to enforce read-only access. Grant only `:datalevin.server/view` to users that should read from followers.

6.3 HA Membership Updates

Consensus HA membership is operator managed. Datalevin does not discover data nodes automatically. Update the authoritative membership for an open HA database with the HA membership update admin call.

`ha-members` endpoints use `host:port` member endpoints, not `dtlv://` database URIs, and must be ordered by ascending `:node-id`.

Use membership updates for planned changes such as adding a node, replacing a node, or changing the control-plane voter set. A typical replacement procedure is:

1. Start the new server with the target Datalevin version, database identity, credentials, and node-local Raft directory.
2. Let it bootstrap or catch up as a follower before it is relied on for promotion.
3. Submit one desired membership spec that includes the full replacement member list, not only the changed node.
4. Keep `:clear-leases? true` unless you have a tested reason not to; leadership should be reacquired under the new membership.
5. Apply the same desired configuration to surviving or newly staged servers so their local options match the authoritative membership hash.
6. Verify one leader, matching membership hashes, follower lag, and fencing health after the update.

```
(require '[datalevin.client :as cl])

(def client
  (cl/new-client "dtlv://admin:pass@node-a:8898"))

(cl/ha-update-membership!
 client
```

```
"app"
{:ha-members
 [{:node-id 1 :endpoint "node-a:8898"}
  {:node-id 2 :endpoint "node-b:8898"}
  {:node-id 3 :endpoint "node-c-new:8898"}]
 :ha-control-plane
 {:voters [{:peer-id "node-a:9098" :promotable? true :ha-node-id 1}
           {:peer-id "node-b:9098" :promotable? true :ha-node-id 2}
           {:peer-id "node-c-new:9098" :promotable? true :ha-node-id 3}]}})
```

This is an operator API; keep it in deployment tooling that is tested against the target Datalevin version. The binding names are `ha-update-membership!`, `haUpdateMembership`, and `ha_update_membership`.

The request validates the member and voter lists, persists the new HA options on the target server, optionally replaces the Raft voter set, CAS-updates the authoritative membership hash, clears existing leases by default, and restarts that server's local HA runtime.

Useful request keys:

- **`:ha-members`**: Replacement data-node member list.
- **`:ha-control-plane {:voters [...]}`** or **`:ha-control-plane-voters`**: Replacement control-plane voters.
- **`:expected-membership-hash`**: Optional CAS guard.
- **`:replace-voters?`**: Defaults to `true`.
- **`:clear-leases?`**: Defaults to `true`, forcing leadership to be reacquired.
- **`:timeout-ms`**: Optional control-plane operation timeout.

Run the same desired update on surviving or newly staged servers so local persisted options match the authoritative membership hash. A node with a membership mismatch fails closed and will not promote or admit writes until the mismatch is resolved.

7. Monitoring and Diagnostics

Datalevin relies heavily on the operating system for memory mapping and disk I/O. Production monitoring should combine host metrics with Datalevin-specific checks.

> **Monitor the host and the database together**
>
> Datalevin depends on memory-mapped files, the operating system page cache, and storage flush behavior. Pair Datalevin signals with disk latency, page faults, swap, free space, and process residency metrics.

7.1 Host Metrics

Watch these first:

- **Disk latency and `iowait`**: `iowait` is CPU time spent waiting for storage I/O. Durable direct-LMDB commits are bounded by synchronous flush latency. WAL commit latency depends on the chosen durability profile.

- **Free disk space:** LMDB copy-on-write and WAL files need enough room for new pages, snapshots, and retained WAL segments.
- **Page cache and major page faults:** Datalevin's effective buffer pool is the OS page cache. Major page faults indicate that the active working set is not resident.
- **RSS vs virtual size:** RSS is resident set size, the memory pages actually resident in RAM. A large virtual mapping reflects `:mapsize`; it is not the same as resident memory.

Chapter 19 shows concrete Linux and macOS commands for checking page-cache residency, major page faults, swap activity, and file-backed cache usage.

7.2 Database Statistics

Use `stat` to inspect LMDB B+Tree statistics. In Clojure, `stat` operates on a KV handle. For a Datalog database directory, the `dtlv stat` command opens the store and reports the same class of LMDB statistics.

```
(def kv (d/open-kv "/data/companydb"))

(d/stat kv)

(d/open-dbi kv "datalevin/eav")
(d/stat kv "datalevin/eav")
```

```
$ dtlv -d /data/companydb stat
$ dtlv -d /data/companydb stat datalevin/eav
```

Useful fields include `:branch-pages`, `:leaf-pages`, `:overflow-pages`, and `:entries`. LMDB reuses deleted pages rather than shrinking the file in place. After large deletions, use the compact-copy workflow from Chapter 19 during controlled maintenance.

7.3 Query and Write Diagnostics

- Use `explain` from Chapter 21 to inspect the selected physical plan and, with `{:run? true}`, confirm the executed root and compare estimates with actual result sizes.
- If writes are slow and disk is busy, check storage latency and the selected durability profile.
- If writes are slow while disk is idle, look for writer contention, many small transactions, or WAL group-commit settings that do not fit the workload.
- Log transaction latency, query latency, and result sizes at the application boundary. Datalevin cannot know which latency budget matters to your service.

7.4 Replication and HA Checks

For async replicas, monitor `replica-status` (`replicaStatus` in Java and JavaScript, `replica_status` in Python), especially `:replica-lag-lsn`, `:replica-degraded-reason`, and `:replica-last-error`.

For HA clusters, monitor leader identity, follower lag, membership hash agreement, clock skew, and fencing health. Keep each node's `:ha-members` and promotable control-plane voters aligned with the authoritative membership hash.

7.5 Health Checks

A service health check should prove that the application can reach the database path it depends on. For a Datalog database, use a scalar probe:

```
(d/q '[:find ?e . :where [?e _ _]] (d/db conn))
```

This is cheaper than returning a relation with `:limit 1`; the health check only needs one scalar value or `nil`.

For server deployments, check both the server process and the application-level operation that your service needs, such as a read-only query for read endpoints or a small write in a dedicated health database for write-path checks.

8. Concurrency and Failure Handling

Datalevin's normal transaction examples focus on successful writes. Production code also needs a failure policy. Separate failures into three groups:

1. **Semantic transaction failures**: the transaction data is invalid for the current database state or schema.
2. **Logical conflicts**: the transaction was well-formed, but a condition such as `:db/cas` did not hold.
3. **Operational failures**: the local store, WAL, remote server, or HA topology could not complete the write.

Handle those groups differently, as shown in Figure 22.3.

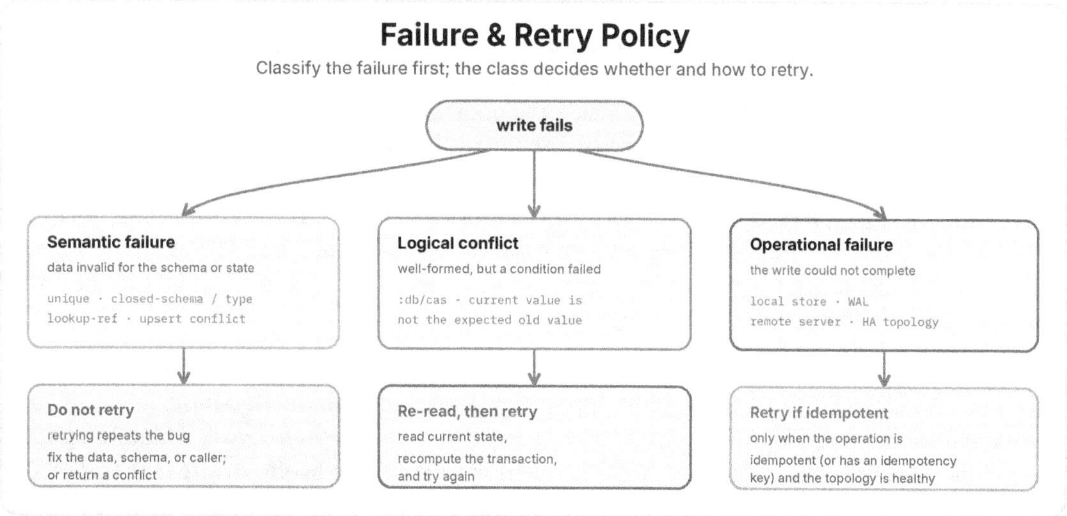

Figure 22.3: Failure and retry policy: a failed write is classified into a semantic failure (invalid data for the schema or state - do not retry, fix the cause), a logical conflict (a well-formed transaction whose condition such as :db/cas did not hold - re-read current state, recompute, and retry), or an operational failure (the store, WAL, remote server, or HA topology could not complete the write - retry only when the operation is idempotent and the topology is healthy)

Retrying a validation error just repeats a bug. Retrying a CAS conflict can be correct if the operation re-reads current state on every attempt. Retrying an operational failure is only safe when the application operation is idempotent or has an idempotency key.

> **Retry by failure class**
>
> Do not retry semantic transaction failures blindly. Retry logical conflicts only by re-reading current state and recomputing the transaction. Retry operational failures only for idempotent operations and only when the topology is healthy enough to accept writes safely.

8.1 The Write Serialization Model

Datalevin inherits LMDB's MVCC read model: readers do not block writers, and writers do not block readers. A reader sees a stable read view for its current operation. A writer commits a new version that later reads can observe.

Writes are different. For a single Datalevin store, the durable B+Tree commit path has one writer at a time. In embedded Clojure, `transact!` serializes writes through the connection and the underlying LMDB write transaction. WAL mode can queue and batch concurrent callers, but the committed writes still have a single order. The caller either receives a transaction report or an exception.

In SQL isolation terminology, committed Datalevin write transactions are serializable within that store because they are physically ordered by the single writer. Ordinary read-only queries use MVCC read views: the view is stable and isolated from concurrent writes, but it is not an application-level read/write transaction unless the read and write are placed inside one transaction callback or encoded as one conditional transaction.

`with-transaction` opens one read/write transaction around its body. Reads and writes inside the body observe the same write transaction, so no other write can interleave between the read and the write within that store. A query inside the body also reads writes already made earlier in the same body, which gives `with-transaction` the usual read-your-writes behavior. If the body throws, the transaction is aborted. Explicit Datalog and KV transaction callbacks can also be given a timeout; when it expires, Datalevin aborts the transaction and reports `:type :transaction/timeout`. `with-transaction` does not retry logical conflicts such as CAS failures or unique-constraint violations. It only retries Datalevin's internal map-resize condition, where the LMDB map was grown and the write can be replayed safely.

Normal Datalevin writes should serialize rather than deadlock. If a write path appears stuck, look for a long-running transaction function, application locks held around database calls, callbacks that synchronously wait on other writes, slow storage flushes, or HA/network waits. Keep transaction functions small and side-effect-free, and perform external side effects after the transaction commits.

8.2 Common Transaction Failures

Most Datalevin-originated transaction failures are `clojure.lang.ExceptionInfo`. When Datalevin supplies structured `ex-data`, use that for application branching and treat the message as human diagnostics. Native storage and remote failures may wrap lower-level exceptions, so log the full exception and its cause chain at the service boundary.

Failure	When it happens	Typical `ex-data`	Retry policy
Unique constraint violation	A transaction adds a value already present for a `:db.unique/value` attribute, or creates an inconsistent unique assertion. `:db.unique/identity` normally upserts when used as identity.	`{:error :transact/unique, :attribute attr, ...}`	Do not retry blindly. Return a conflict or use identity upsert intentionally.
Closed schema or type validation	`:closed-schema?` rejects an undeclared attribute, or `:validate-data?` rejects a value that does not match `:db/valueType`.	Often `:transact/syntax` for syntax errors; type and closed-schema checks may only include input context.	Do not retry. Fix schema, data, or caller validation.
Lookup-ref syntax or non-unique attribute	A lookup ref is not `[attr value]`, or `attr` is not unique.	`:lookup-ref/syntax` or `:lookup-ref/unique`	Do not retry. Fix the transaction shape or schema.
Lookup-ref miss in a transaction	A transaction form requires an existing entity and `[attr value]` resolves to nothing.	`{:error :entity-id/missing, :entity-id [...]}`	Usually do not retry. Create the referenced entity first, use identity upsert, or treat it as a domain conflict.
`:db/cas` or `:db.fn/cas` failure	The current value is not the expected old value.	`{:error :transact/cas, :old old, :expected old-value, :new new-value}`	Retry only by re-reading current state and recomputing the new transaction.
Upsert conflict	Multiple identity assertions in one entity resolve to different existing entities.	`{:error :transact/upsert, ...}`	Do not retry. The input identifies two different entities.
Transaction function failure	An installed transaction function or descriptor-backed UDF throws, cannot be resolved, or returns invalid transaction data.	Depends on the function or UDF validation.	Depends on the function. Keep transaction functions deterministic and side-effect-free.
Explicit transaction timeout	A `with-transaction` or `with-transaction-kv` body exceeds its per-call timeout or the configured default explicit transaction timeout.	`{:type :transaction/timeout, :timeout-ms n}`	Treat as an aborted transaction. Retry only if the application operation is idempotent and still useful. Shorten or move slow work out of the transaction.

Failure	When it happens	Typical `ex-data`	Retry policy
LMDB map full / map resize	The write exceeds the current memory map size.	Normally handled internally as `"DB resized"` with `{:resized true}` and retried.	Usually no application action. If it happens often, set a larger `:mapsize` up front.
WAL, remote, or HA write failure	WAL sync fails or times out, a remote server returns an error, or HA rejects a write because leadership or fencing changed.	Often `:type :txlog/...` or `:error :ha/...`	Retry only if the operation is idempotent and the topology is healthy. For HA, use the HA retry settings and still design writes to be replay-safe.

The important distinction is that Datalevin does not turn logical conflicts into transparent retries. If the old value for `:db/cas` is wrong, the transaction fails because the caller's premise is false. The application must decide whether to report the conflict or compute a new transaction against current state.

8.3 Recommended Retry Patterns

In embedded Clojure, Java, or Python, prefer the Datalog transaction callback (`with-transaction`, `withTransaction`, or `with_transaction`) for read-modify-write logic that must be atomic within one store:

```
(defn increment-counter! [conn counter-id]
  (d/with-transaction [cn conn]
    (let [db  (d/db cn)
          n   (or (d/q '[:find ?n .
                          :in $ ?id
                          :where [?e :counter/id ?id]
                                 [?e :counter/value ?n]]
                        db counter-id)
                   0)]
      (d/transact! cn
        [{:counter/id counter-id
          :counter/value (inc n)}]))))
```

This pattern does not need a CAS retry loop because the read and write happen inside the same write transaction. Use it only for short critical sections. Do not perform HTTP calls, file uploads, email delivery, or long CPU work inside the transaction body.

If the host API does not expose the Datalog transaction callback, keep the read-modify-write operation as a single transaction when possible, use `:db/cas` for optimistic updates, or move the serialized command into the Datalevin-hosting process.

When a write is intentionally optimistic, use CAS and retry at the application operation boundary. Re-read inside every attempt:

Listing 22.1. Optimistic CAS retry pattern

```
(defn cas-conflict?
  [e]
  (= :transact/cas (:error (ex-data e))))

(defn retry-cas
  [f]
  (loop [attempt 1]
    (let [result (try
                   {:ok? true :value (f)}
                   (catch clojure.lang.ExceptionInfo e
                     {:ok? false :error e}))]
      (if (:ok? result)
        (:value result)
        (let [e (:error result)]
          (if (and (cas-conflict? e) (< attempt 5))
            (do
              (Thread/sleep (* 25 attempt))
              (recur (inc attempt)))
            (throw e)))))))

(defn deposit! [conn email amount]
  (retry-cas
    #(let [db      (d/db conn)
           balance (or (d/q '[:find ?balance .
                              :in $ ?email
                              :where [?e :account/email ?email]
                                     [?e :account/balance ?balance]]
                            db email)
                       0)]
       (d/transact! conn
         [[:db/cas [:account/email email]
           :account/balance balance (+ balance amount)]]))))
```

The retry wrapper does not retry unique, lookup-ref, schema, or validation failures. It retries only CAS conflicts, and each retry re-reads the balance from the current database state before preparing a new transaction.

For operational retries, apply the same rule at a higher level:

- Retry whole application operations, not half-built transaction data.
- Make every retried operation idempotent. Use a request id, event id, or unique identity attribute so a replay updates the same logical fact instead of creating a duplicate.
- Keep external side effects outside the transaction, and make listeners idempotent because they observe committed transactions after the write.
- Use bounded retries with backoff. Infinite retries can hide a broken schema, a missing entity, full disk, or an HA cluster that cannot safely accept writes.

8.4 Recovery Strategy Ladder

A full Erlang/OTP supervision treatment is outside this guide's scope, but the supervision idea is useful for production Datalevin services: when a component fails, recover at the smallest boundary that can safely repair the problem, and escalate only when that boundary is no longer credible. The database should not silently retry every error, and transaction functions should not supervise the rest of the application. Instead, the application or service manager should decide whether the right response is a local retry, a component restart, or failover.

Figure 22.4 shows the basic ladder.

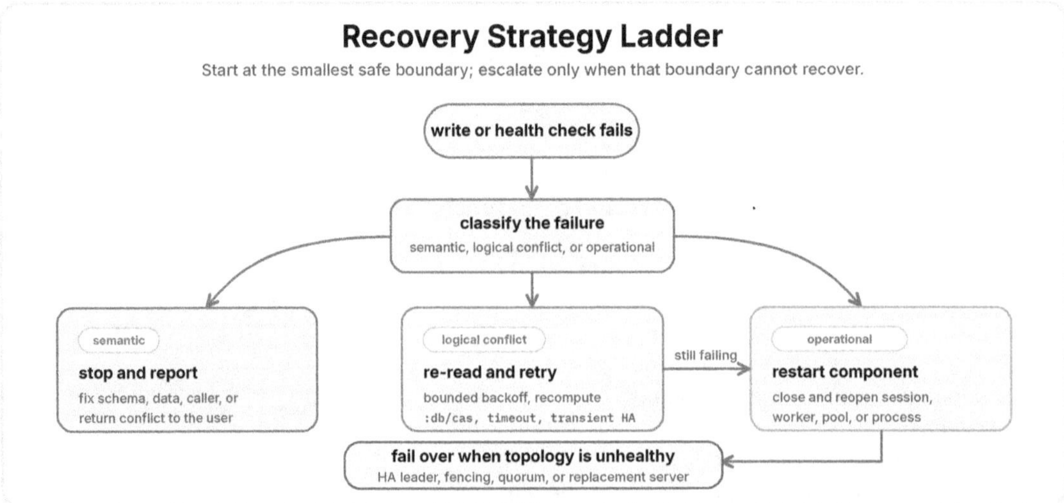

Figure 22.4: Recovery strategy ladder: classify the failure first, stop semantic failures, use bounded local retry for logical conflicts and transient operational failures, restart one database-facing component when local state may be stale, and fail over only when the topology is unhealthy

Strategy	Use when	What to do	Avoid
Local retry	The operation is still valid and the failure may clear quickly, such as a CAS conflict, transaction timeout, transient network failure, or retryable HA write rejection.	Re-read current state, rebuild the transaction, and retry with a small bounded backoff. Require idempotency for writes that may have reached the database.	Retrying validation, schema, unique-conflict, or lookup-ref bugs.
Component restart	One local database-facing component appears unhealthy: a worker, connection pool, remote client session, queue consumer, or process keeps failing after bounded retries.	Stop accepting new work for that component, close Datalevin connections or clients, create fresh ones, and replay only idempotent queued work.	Restarting the whole service for every logical conflict, or hiding a persistent data bug behind repeated restarts.

Strategy	Use when	What to do	Avoid
Failover	The database topology is unhealthy: the write leader is unreachable, fencing or membership changed, a host failed, or a server cannot safely accept writes.	Route writes to the current HA leader after the cluster converges, or let the deployment supervisor replace the failed server. Continue only when health checks and HA status agree.	Inventing a second writer in application code, bypassing fencing, or using replication as a backup substitute.

The CAS helper above is a local retry example. A component restart helper is usually not a Datalevin function; it belongs in the worker framework, connection pool, job runner, service manager, or orchestration system that owns the failed component. Failover belongs to the deployment topology. In a non-HA deployment, failover is manual repair or restore. In consensus-lease HA, the cluster must settle on a safe leader before write traffic resumes.

8.5 Production Policy

A useful HTTP or job-worker policy is:

- Return `400` for transaction syntax, closed-schema, and type-validation failures.
- Return `404` or `409` for lookup-ref misses, depending on whether the missing entity is a resource path or a domain precondition.
- Return `409` for unique conflicts and CAS conflicts unless the operation has a specific retry loop.
- Return `503` for operational storage, remote, WAL, or HA failures that may clear after repair or failover.
- Log the exception class, message, `ex-data`, and cause chain. Do not log full transaction data if it can contain secrets.

This policy gives production readers a consistent mental model: Datalevin serializes writes for a store, exposes logical conflicts as exceptions, retries only safe internal resize events, and leaves application-level retries to the code that understands the business operation.

Summary

Production Checklist

Use this as a final review before production:

- Cover schema, queries, rules, and transaction logic with tests against in-memory or temporary Datalevin databases, and run tests with `:validate-data?` and `:closed-schema?` enabled (Section 1).
- Choose the simplest deployment mode that satisfies the availability and access requirements.
- Choose a tenancy shape deliberately: shared database, database per tenant, catalog-plus-tenant databases, or tiered tenancy; treat local-first per-user databases as a database-per-tenant variant (Section 3).

- Run the server under a service manager with a dedicated data directory and restricted filesystem permissions.
- Set `DATALEVIN_DEFAULT_PASSWORD`; do not expose a server using the built-in default password.
- Create application users and roles with least-privilege RBAC permissions.
- Use disk or volume encryption for the data directory, and application-level encryption for sensitive fields.
- Follow Chapter 19 for `:mapsize`, `:max-readers`, WAL, durability profile, copy, dump/load, snapshots, and compaction.
- Define an application recovery policy for CAS conflicts, uniqueness conflicts, validation failures, lookup-ref misses, operational write failures, component restarts, and failover boundaries (Section 8).
- Automate transactionally consistent backups and test restores.
- Monitor disk latency, free space, page-cache behavior, query latency, write latency, and result sizes.
- For async replicas, remember that reads are eventually consistent; monitor replica lag, last errors, and degraded state.
- For HA, keep membership, voter mapping, clock discipline, fencing checks, and failover drills part of normal operations. HA promotes a write leader; it is not a backup system or quorum replication of every user transaction.
- Install server-side UDF registries before relying on descriptor-backed UDFs in remote queries or transactions. Remote queries can call built-ins and server-registered UDFs, not arbitrary host functions.

Datalevin is operationally quiet when the deployment boundary is chosen correctly. Most production issues come from choosing a topology that is more complex than the application needs, starving the OS page cache, or failing to test backup and restore paths.

With the operational base in place, Part VI builds application architectures on top of these guarantees, especially intelligent-memory systems.

References

[1] Diego Ongaro and John Ousterhout, "In Search of an Understandable Consensus Algorithm," USENIX ATC 2014. URL: https://raft.github.io/raft.pdf.

[2] Cognitect, "Transit Format." URL: https://github.com/cognitect/transit-format.

[3] Peter Taoussanis, "Nippy: High-performance serialization library for Clojure." URL: https://github.com/ptaoussanis/nippy.

[4] Elaine Barker, "Recommendation for Key Management: Part 1 - General," NIST Special Publication 800-57 Part 1 Revision 5, 2020. DOI: https://doi.org/10.6028/NIST.SP.800-57pt1r5.

PART VI

Datalevin for Intelligent Systems

CHAPTER 23

Agent Memory Architectures

A central limitation of large language models (LLMs), when used as application components, is that they do not provide durable application state on their own. The moment an agent must continue work across turns, remember a user's durable preferences, resume a goal after a restart, reason over factual knowledge, audit why it believed a fact, or retract an obsolete conclusion, more prompt context is not enough. The application needs a memory layer that is durable, queryable, transactional, and explicit about provenance and authority.

A prompt can carry context, and a product may add session memory, but the model call itself should not be treated as the system of record. Stateful intelligent agents need persistent memory: a place to record episodes, extracted claims, curated factual knowledge, goals, tasks, evidence, and the working set assembled for the next model call. This chapter shows how Datalevin can serve as that unified substrate, bridging unstructured semantic search and structured logical reasoning.

1. Part VI Contract: What Datalevin Provides

Before going further, set the boundary. Part VI of this guide is about application architecture on top of Datalevin. The intent is not to describe a bundled agent framework, but to give guidance on how to use Datalevin's capabilities to build persistent agent memory. Figure 23.1 shows the boundary.

Datalevin gives you the database capabilities needed to build durable agent memory:

- ACID transactions over facts, documents, indexes, and task state.
- Datalog queries, rules, lookup refs, schema, and graph traversal.
- Full-text search, vector search, embedding indexes, and idoc fields.
- Transaction reports, listeners, WAL, replication, server access, and MCP tooling for operational integration.

You build the agent-specific machinery:

- The ingestion loop that turns messages, tool calls, and observations into episode records.
- The consolidation job that extracts candidate facts from episodes.
- The working-memory refresh policy, slot scoring, and eviction rules.
- The apperception or truth-maintenance layer that decides which facts become current, superseded, rejected, or pending review.
- The operating envelope, tool authorization checks, budgets, and model-calling policy.

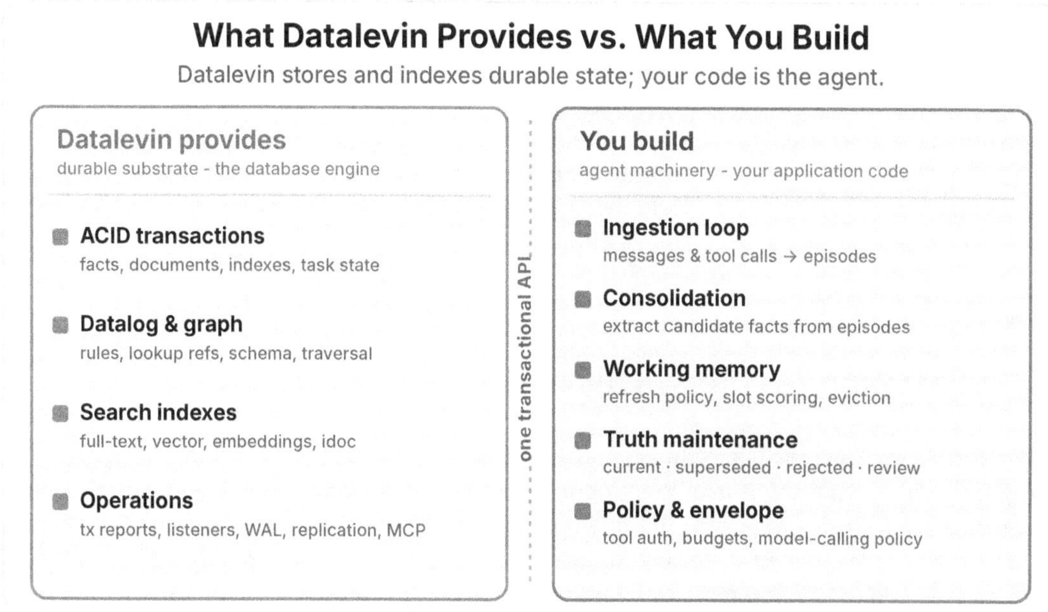

Figure 23.1: The Datalevin/application boundary: Datalevin provides ACID transactions, Datalog and graph, search indexes, and operations, while your application code builds the ingestion loop, consolidation, working memory, truth maintenance, and policy

> **Datalevin is the substrate, not the agent runtime**
>
> Datalevin stores and indexes the facts, documents, tasks, evidence, and working-memory projections. The application still owns ingestion policy, consolidation, tool authorization, model calls, truth maintenance, and prompt assembly.

The examples in this part model those pieces as ordinary entities, references, attributes, and idocs. When a snippet demonstrates a basic Datalevin API call, it may use the same multi-language style as earlier chapters. When a snippet shows an end-to-end application loop, it is Clojure-first so the control flow, transactions, and data shapes stay runnable in one place. The same model can be implemented from Java, Python, JavaScript, or over the server API by composing the same `transact`, `query`, `pull`, full-text, vector, and embedding operations shown earlier in the guide. For exact language coverage, use the compatibility matrix linked in Appendix A. The notable gaps for Part VI examples are the same as elsewhere: Clojure, Java, and Python expose the Datalog transaction callback, but JavaScript does not; staged mutation of existing Entity objects is Clojure-only.

Before the chapters move into concrete records, keep the main memory families in view. Chapter 24 defines the record shapes and examples; here the important point is how the pieces fit together.

Record family	Role in Part VI
`:user/*`	Scope memory to a person, workspace, or account.
`:episode/*`	Preserve timestamped interactions, tool calls, observations, and summaries.

Record family	Role in Part VI
`:fact/*`	Store extracted semantic claims with source, confidence, and review status.
`:goal/*`, `:task/*`, and `:session/*`	Track durable objectives, bounded task state, structured contracts, and interaction channels.
`:wm/*` and `:wm.slot/*`	Hold the current prompt-facing projection and links back to source entities.

2. The Human Memory Analogy

When people say "memory" in ordinary conversation, they often mean one thing: keeping something for later, like putting it in a box. Cognitive science uses a more useful distinction. Human memory is not a single box where all remembered material has the same shape and lifetime. Remembering what happened yesterday, knowing that Paris is a city, and keeping a phone number in mind long enough to type it are different operations. They interact, but they are not the same operation.

This distinction is useful for agent design because LLM applications fail in similar ways when all context is treated as one undifferentiated pile of text. A chat transcript, a durable user preference, a current task constraint, and a retrieved document excerpt should not have the same authority. Some information is evidence. Some is current state. Some is background knowledge. Some is only a temporary focus for the next model call.

Cognitive science provides a useful vocabulary here: episodic memory for event records, semantic memory for durable facts and concepts [1], and working memory for the bounded context active in the current task [2]. Chapter 24 turns these categories into concrete Datalevin records.

The analogy should not be taken too literally. A database is not a brain, and an LLM does not have human memory. The value of the analogy is architectural: do not store everything as one transcript, and do not paste everything into one prompt. Episodes, semantic facts, and working-memory projections have different lifetimes, different query patterns, and different authority. Those lifetimes also imply deletion and consolidation: Chapter 15 explained why Datalevin makes forgetting cheap, and Chapters 24 and 26 use that property for agent memory.

> **Use the memory analogy architecturally**
>
> The point is not that a database or LLM has human memory. The useful design rule is to give episodes, semantic facts, and working-memory projections different lifetimes, authority, and query patterns.

By using Datalevin, an agent can query past experiences, organized world knowledge, and current task state through multiple lenses at the same time. That is the foundation for the rest of Part VI.

3. Perception Is Not Organized Knowledge

Modern generative models are very strong perceptual systems: they map text, images, audio, and other raw inputs into labels, concepts, and high-dimensional vector representations. This is enormously useful, but it is not the same thing as an organized world model. A model can recognize that two passages are similar without connecting them to semantic knowledge, such as which facts are current, which goal they support, who is allowed to see them, or which old belief they supersede.

Agent memory needs both bottom-up and top-down machinery. Bottom-up perception turns messy input into candidates: entities, summaries, embeddings, extracted facts, classifications, and possible next actions. Top-down knowledge gives those candidates an explicit ontology, knowledge-graph relationships, domain rules, goals, and policies. This is the constructive role long associated with knowledge representation and expert systems: deciding what distinctions matter and making them inspectable and executable. The two approaches are complements, not rivals. Datalevin can store perceptual outputs and symbolic knowledge and control state under one transactional model, so learned evidence can be tested against explicit meaning, authority, and constraints. Chapters 5 and 9 develop the ontology and rule foundations.

This distinction also explains why a plain vector database is not enough for long-lived agents. Vectors help recall things that feel similar. They do not, by themselves, maintain propositions, graph structure, task state, authority, or evidence. Those are explicit knowledge-organization problems.

> **Similarity is not organized knowledge**
> Use vectors and embeddings to find candidates, not to decide which facts are current, authorized, supported, or superseded. Those decisions need explicit facts, graph links, status fields, and application policy.

4. Why a Unified Memory Engine?

Many agent architectures attempt to combine a vector database for similarity with a SQL database for state. This polyglot approach creates significant friction:

- **Consistency**: The vector index and the relational database can get out of sync.
- **Complexity**: The agent must manage two different APIs and data models.
- **Latency**: Multiple network hops to different services slow down the "think-act" loop.

Datalevin's advantage is that the source text, optional user-supplied vector, embedding-index metadata or jobs, and logical facts about an episode live in one database. The source datoms and durable secondary-index work are committed together, and Datalevin maintains the index. That removes a common source of inconsistency between separately updated stores. It does not decide what an episode means; your ingestion and consolidation code does that. At retrieval time, a Datalog query can combine the relevant pieces of information in one semantically coherent fashion.

```
;; An episode entity in Datalevin
{:episode/id      #uuid "00000000-0000-0000-0000-000000000001"
```

```
:episode/text     "I helped the user debug a Clojure macro today."
:episode/vector   [0.12 -0.03 0.44] ; Optional user-supplied vector
:episode/user     [:user/id "alice"]
:episode/success true
:episode/topics   [:clojure :macros]}
```

For many text-memory workloads, set `:db/embedding` to `true` on the text attribute instead of storing a separate vector datom. Datalevin keeps the original text as the source of truth and maintains the embedding index as a secondary index.

5. The "Context Graph" Pattern

True agency requires understanding the relationships between things. Instead of storing memories as isolated silos, Datalevin allows you to build a **Context Graph**.

- **Nodes**: Memorable events, users, concepts, and goals.
- **Edges**: How these things relate (e.g., "Goal A was superseded by Goal B", "User X is an expert in Topic Y").

Because Datalevin supports recursive graph traversal (Chapter 13), application code can navigate this graph to find non-obvious connections in stored memory. The graph supplies explainable associations; the agent runtime decides how those associations affect a turn.

A small graph is enough to show the pattern. A goal owns a task, a task owns a session, the session records episodes, and extracted facts point back to their source episode. For legibility, this example uses human-readable keyword ids; production schemas, like the ones in Section 6 and Chapter 24, typically declare these id attributes as UUIDs, and one database must pick one value type per attribute:

```
[{:goal/id     :goal/release-readiness
  :goal/status :goal.status/active}
 {:task/id   :task/check-blockers
  :task/goal [:goal/id :goal/release-readiness]}
 {:session/id   :session/morning-standup
  :session/task [:task/id :task/check-blockers]}
 {:episode/id      #uuid "00000000-0000-0000-0000-000000000201"
  :episode/session [:session/id :session/morning-standup]
  :episode/text    "Backup verification still blocks release M184."}
 {:fact/id             #uuid "00000000-0000-0000-0000-000000000202"
  :fact/subject        [:goal/id :goal/release-readiness]
  :fact/kind           :fact.kind/blocker
  :fact/content        "Release M184 is blocked on backup verification."
  :fact/source-episode [:episode/id #uuid
   ↪ "00000000-0000-0000-0000-000000000201"]}]
```

With those links, the application can ask for the facts that support an active goal and keep the evidence path available:

```
[:find ?content ?episode-text
 :in $ ?goal-id
 :where
 [?g :goal/id ?goal-id]
 [?f :fact/subject ?g]
```

```
[?f :fact/content ?content]
[?f :fact/source-episode ?ep]
[?ep :episode/text ?episode-text]]
```

Working-memory slots can then point to the fact or episode they selected. That makes prompt assembly explainable: a sentence in the prompt is not anonymous text, but a projection from a fact, which came from an episode, which happened inside a session, task, and goal.

6. Memory Scopes: Goal, Task, Session, Turn

Real agents need more than a bag of memories. They need scoped state with clear lifetimes:

- **Goal**: the durable contract with the user: objective, constraints, budget, and success criteria.
- **Task**: an execution unit under a goal, with its own plan, runtime state, checkpoints, and boundary summaries.
- **Session**: an interaction channel with transient scratch context, recent turns, approvals, and working memory.
- **Turn**: the smallest reasoning/action cycle, useful for audit, token accounting, tool calls, and replay.

Keeping these scopes separate prevents a common failure mode in agent systems: temporary context gets promoted into durable truth, while durable instructions get buried in chat history. A task may need a retry counter, a partially completed plan, or a tool result cache; those are not the same thing as the user's long-term goal. A session may contain copied notes, a recap, or a pending approval; those should influence the next turn, but they should not silently rewrite the user's preferences or the goal contract.

> **Separate temporary focus from durable truth**
>
> A session note, pending approval, or copied scratch value can guide the next turn without becoming a long-term preference. A goal contract or durable policy should remain visible even when recent chat history is noisy.

Model these scopes as ordinary entities and references. The records below are schematic transactable maps, not a complete schema declaration; the exact attributes belong to the application schema.

```
{:goal/id       #uuid "00000000-0000-0000-0000-000000000101"
 :goal/status   :goal.status/active
 :goal/contract {:objective "Monitor release readiness"
                 :success   "Open blockers are summarized before 09:00"
                 :budget    {:max-llm-calls 12}}}

{:task/id       #uuid "00000000-0000-0000-0000-000000000102"
 :task/goal     [:goal/id #uuid "00000000-0000-0000-0000-000000000101"]
 :task/state    :task.state/running
 :task/contract {:kind :task
                 :version 1
                 :steps [{:id :collect :kind :tool}
```

```
                        {:id :summarize :kind :llm}]}}

{:session/id       #uuid "00000000-0000-0000-0000-000000000103"
 :session/task     [:task/id #uuid "00000000-0000-0000-0000-000000000102"]
 :session/scratch {:notes ["User prefers short deployment summaries."]}}
```

When the schema declares them with `:db/valueType :db.type/idoc`, the `:goal/contract`, `:task/contract`, and `:session/scratch` fields are good idoc attributes: they are structured documents that need to move transactionally with the graph, but do not need every nested field promoted to a top-level attribute. In the Python and JavaScript forms, `tx.entity` converts the outer attribute names to Datalevin keywords, while keys inside those idoc values remain native strings. Chapter 14 explains why an idoc query path must use the same key type as the stored document.

Use top-level attributes for fields you will join, filter, or index frequently: status, owner, timestamps, parent references, and current lifecycle state. Use `idoc` fields for nested data that is usually loaded as a whole, such as a task spec or an operating envelope. This gives the agent a durable object model without forcing every prompt-facing detail into schema.

Looking Ahead

Chapters 24 through 26 cover concrete memory records, retrieval and prompt assembly, and long-term consolidation, respectively.

Part VI draws on the author's work on Juji and the human-centered agentic AI perspective in [3], as well as Xia, a separate open-source assistant [4]. Xia is not part of Datalevin; it is a reference application for building long-term memory, scoped task state, tool boundaries, and scheduled work on ordinary database capabilities.

Summary

Datalevin is a database substrate for AI memory, not an agent runtime. It keeps episodic records, semantic facts, scoped task state, and working-memory projections together; learning, consolidation, and action selection remain application responsibilities.

References

[1] Endel Tulving, "Episodic and Semantic Memory," in *Organization of Memory*, Academic Press, 1972, pp. 381-403.

[2] Alan D. Baddeley and Graham Hitch, "Working Memory," in *Psychology of Learning and Motivation*, vol. 8, Academic Press, 1974, pp. 47-89.

[3] Michelle X. Zhou, Jie Lu, Huahai Yang, and Wenxi Chen, *Human-Centered Agentic AI: Fundamentals, Practice, Applications, and Future Directions*, ACM Press, in press.

[4] Huahai Yang, "Xia," GitHub repository. URL: https://github.com/huahaiy/xia.

CHAPTER 24

Episodic, Semantic, and Working Memory

Chapter 23 described the overall architecture: a persistent agent needs scoped state, a context graph, and a transactional memory substrate. This chapter defines the memory records themselves. The goal is to separate kinds of memory by lifetime and purpose, while keeping them queryable together.

As in Chapter 23, the records below are schematic transactable maps, not a complete schema declaration. The application schema should make lookup-ref targets such as `:episode/id`, `:fact/id`, `:user/id`, and `:migration/id` unique, define relationships such as `:episode/session`, `:fact/subject`, and `:fact/source-episode` as refs, and index text or vector attributes as needed.

Figure 24.1 shows the three categories of memory used in this book.

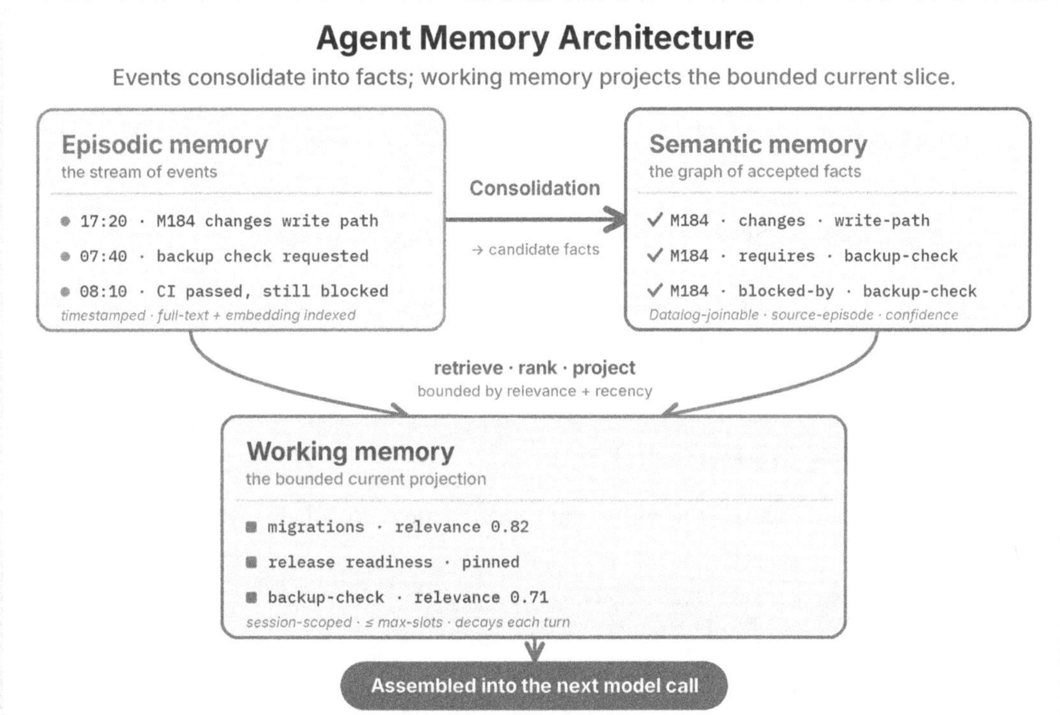

Figure 24.1: Agent memory architecture: episodic memory holds timestamped events, a consolidation step extracts candidate facts into semantic memory, and working memory combines perceived environment with selected long-term memory for the next model call

1. **Episodic memory** records what happened. For an agent, this is the chronological stream of user turns, assistant responses, tool calls, observations, approvals, failures, and task steps. It answers questions such as "what happened?", "when did it happen?", and "what

evidence explains a later decision?" In Datalevin, episodic memory is usually represented as timestamped episode entities linked to users, sessions, tasks, tools, and source artifacts.

2. **Semantic memory** records what the system currently knows or believes. It contains generalized facts, concepts, profiles, preferences, policies, and graph relationships derived from episodes or curated by application code. It answers questions such as "what is true now?", "which fact supersedes an older claim?", and "which concepts or entities are related?" In Datalevin, semantic memory is represented as ordinary facts with identity, status, confidence, provenance, and relationships.
3. **Working memory** records what should be active right now. It is the bounded prompt-facing context for a task or turn, combining perceived environment with selected episodes, semantic facts, external chunks, constraints, and pending tool results. It answers "what should the model pay attention to for this step?" In Datalevin, working memory is usually represented as scoped slots that point back to the source entities they summarize.

Datalevin can store all three in one database without forcing them into one shape. Episodes can be full-text and embedding indexed. Semantic facts can be joined and checked with Datalog. Working-memory slots can point back to the entities they came from, so a prompt can carry provenance instead of anonymous text.

External documents appear later as environment records, not as a fourth memory category. They are inputs from outside the agent that can be indexed and pulled into the perceived-environment side of working memory. If their contents become durable knowledge, promote extracted claims into semantic memory instead of treating the document itself as agent memory.

1. Episodic Memory: The Stream of Experience

Episodic memory is a chronological record of specific events. For an AI agent, this usually means user messages, assistant responses, task executions, tool calls, observations, approvals, and failures.

1.1 Implementing the Episode Schema

An episode should be a lightweight entity that captures the who, what, when, and where:

```
{:episode/id        #uuid "00000000-0000-0000-0000-000000000301"
 :episode/timestamp #inst "2026-06-08T10:00:00Z"
 :episode/type      :episode.type/chat
 :episode/summary   "User Alice asked about vector search."
 :episode/vector    [0.12 -0.03 0.44] ; Optional user-supplied vector
 :episode/content   "..." ; Raw log; full-text indexed if schema declares it
 :episode/session   [:session/id #uuid "00000000-0000-0000-0000-000000000103"]}
```

If Datalevin should compute embeddings for you, define `:episode/summary` or `:episode/content` as a string attribute with `:db/embedding true`. Query it with `embedding-neighbors` using text input. Use `:db.type/vec` and `vec-neighbors` only when your application supplies vectors directly.

Keep raw content and summaries separate. Raw content is useful for audit and reprocessing.

Summaries are useful for retrieval and prompt assembly. If a model or background worker creates the summary, store enough metadata to know which episode version, model, or extraction policy produced it.

> **Keep raw episodes separate from summaries**
> Raw content is the replay and audit source. Summaries are derived retrieval material. If a model creates the summary, record the source episode version, model, and extraction policy so later jobs can reprocess or challenge it.

1.2 Temporal Retrieval

The primary way to query episodic memory is by time. Because Datalevin indexes every attribute in AVE, range queries on `:episode/timestamp` are efficient.

```
;; Find all chat episodes from the last 24 hours.
(d/q '[:find ?summary
       :in $ ?since
       :where [?e :episode/type :episode.type/chat]
              [?e :episode/timestamp ?ts]
              [(> ?ts ?since)]
              [?e :episode/summary ?summary]]
     db twenty-four-hours-ago)
```

The time cutoff should be computed by the application, not hidden inside the query. That keeps recency policy testable: one agent may need the last hour, another may need the last week, and a scheduled report may need everything since the previous run.

1.3 Retention and Downsampling

Episodic memory grows without bound unless you design a lifecycle. A common policy is:

1. Keep recent episodes at full resolution.
2. Mark episodes as processed after consolidation.
3. Retain older episodes by importance, recency, and diversity.
4. Preserve compact summaries even when raw detail is archived or deleted.

Chapter 15 explained why Datalevin's current-state model makes this kind of forgetting a normal database operation rather than a special history rewrite.

Represent the policy as data so maintenance jobs can be tested:

```
{:episode/id          #uuid "00000000-0000-0000-0000-000000000302"
 :episode/summary     "Release discussion about migration lock timing."
 :episode/timestamp   #inst "2026-06-08T17:20:00Z"
 :episode/importance  0.85
 :episode/processed?  true
 :episode/retention   :episode.retention/full-resolution}
```

The important point is not the exact formula. It is that retention should depend on explicit fields, not on hidden prompt history. Datalevin can then query, copy, compact, and audit old memory with ordinary database operations.

> **Make retention policy queryable**
> Retention should depend on explicit facts such as timestamp, importance, processed status, and retention class. Hidden prompt history cannot be audited, compacted, retried, or tested with ordinary database operations.

This also keeps consolidation (Section 3) honest. An episode should be marked processed only after the semantic facts, summaries, or task updates derived from it have been written. If consolidation fails, record the error and leave enough metadata for a later maintenance run to retry or skip the episode deliberately. Otherwise, agents accumulate invisible gaps: the raw conversation exists, but the knowledge graph was never updated from it.

2. Semantic Memory: The Graph of Knowledge

Semantic memory stores generalized knowledge: facts that are useful beyond the single episode where they were learned. This is where user profiles, preferences, domain concepts, policies, and concept graphs live.

A knowledge graph is more than a bag of extracted facts. Its ontology supplies a top-down account of the concepts and relationships the application considers meaningful; episodes, documents, models, and people supply bottom-up evidence that populates and revises the graph. The ontology may be a formal shared vocabulary, or a lightweight application model expressed with Datalevin attributes, reference relationships, enum entities, and Datalog rules. Either way, the explicit structure complements data-driven AI by making knowledge queryable, governable, and explainable. Chapters 5 and 9 cover these modeling choices in more detail.

2.1 Implementing the Knowledge Graph

Semantic memory can use idents (Chapter 11) and graph relationships (Chapter 13):

```
;; Defining a concept.
{:db/ident :concept/clojure
 :concept/description "A modern, dynamic Lisp for the JVM."
 :concept/related-to [:concept/lisp :concept/jvm :concept/datalog]}
```

Facts extracted from episodes should carry evidence and confidence, not only content:

The user record behind these examples can be as small as `{:user/id "alice"}`. In application schema, `:user/id` should be unique so lookup refs such as `[:user/id "alice"]` resolve consistently.

```
{:fact/id             #uuid "00000000-0000-0000-0000-000000000303"
 :fact/subject        [:user/id "alice"]
 :fact/content        "Alice is evaluating Datalevin for agent memory."
 :fact/source-episode [:episode/id #uuid "00000000-0000-0000-0000-000000000301"]
 :fact/confidence     0.8
 :fact/utility        0.5}
```

Chapter 26 explains the distinction between confidence and utility and how those scores drive long-term maintenance.

Fact records usually have a lifecycle. A newly extracted claim starts as `:fact.status/pending` until a rule, review step, or confidence policy promotes it. Claims that pass validation become `:fact.status/current` semantic memory, often with a `:fact/kind` and a structured `:fact/object` that make them easy to join and summarize. Later evidence may mark that assertion `:fact.status/superseded`, reject a pending claim, or lower confidence instead of deleting the evidence trail. Chapter 26 develops that truth maintenance flow.

The vocabulary in this part is: `:fact.status/pending` for newly extracted candidate claims, `:fact.status/current` for the maintained assertion the system currently uses, `:fact.status/superseded` for an older assertion replaced by newer evidence, and `:fact.status/rejected` when an application chooses to keep a discarded candidate for audit instead of omitting it.

Treat these facts as claims about the world, not as the world itself. Multiple facts can refer to the same entity, disagree with each other, or describe different time periods. The entity reference gives Datalog something stable to join on, while the content and evidence fields preserve the natural-language claim that the agent may later quote, summarize, or challenge.

> **Facts are claims, not unquestioned truth**
>
> Store source, confidence, status, and subject links with extracted facts. Current semantic memory is a maintained view over claims; later evidence may make a claim current, supersede it, reject it, or lower confidence without erasing the trail.

2.2 Inference over Semantic Data

Use Datalog rules to navigate the graph. For example, an agent can infer that a user interested in Clojure may also be interested in Datalog if the concept graph records that Clojure is related to Datalog and the user has repeatedly asked about Clojure.

Semantic inference should remain explicit. If a rule derives `:user/interested` from several episodes, write the derived fact with a source rule, supporting evidence, and confidence. This lets the apperception layer in Chapter 26 later revise or retract the derived fact without losing the original episodes.

3. The Consolidation Loop: From Episode to Knowledge

A powerful agent architecture needs a background consolidation loop, shown in Figure 24.2, to move data from episodic memory into semantic memory:

1. **Ingest Episodes**: Store every interaction as a raw episode.
2. **Analyze**: Periodically use an LLM or Datalog rule set to scan recent episodes.
3. **Abstract**: Identify recurring patterns, preferences, or durable facts.
4. **Update Semantic State**: Update user profiles, concept graphs, or task state with derived facts.
5. **Refresh Working Memory**: Combine the current environment with the relevant memory subset for the next turn.

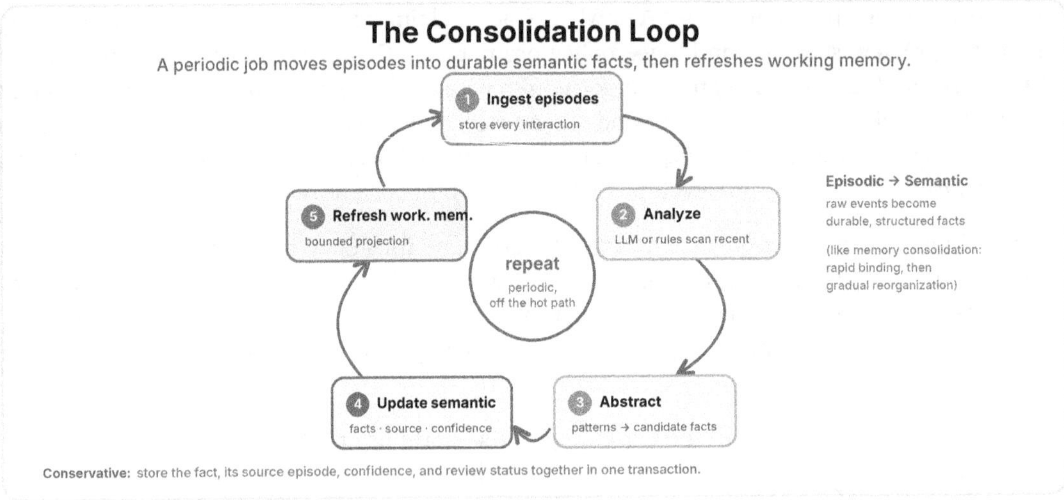

Figure 24.2: The consolidation loop: a periodic job, run off the hot path, moves episodes into durable semantic facts and then refreshes working memory in five steps - (1) ingest episodes by storing every interaction, (2) analyze recent episodes with an LLM or rules, (3) abstract patterns into candidate facts, (4) update semantic memory with the fact plus its source and confidence, and (5) refresh working memory by combining current environment with selected memory - turning raw events into durable, structured facts much like memory consolidation

By doing this, you prevent episodic memory from becoming a dumping ground and ensure the agent's core knowledge stays relevant and structured.

This is similar in function, though not in biological mechanism, to the role often assigned to the hippocampus in systems-consolidation theories of human memory. In that view, the hippocampus rapidly binds new experiences and supports their later reactivation, while longer-term knowledge is gradually reorganized into broader cortical networks [1], and this often happens during sleep! In an agent architecture, the episode store plays the rapid binding role, and the consolidation job turns selected episodes into durable semantic facts, profiles, and graph relationships.

Good consolidation is conservative. Store the extracted fact, the source episode, the confidence, and any review status in one transaction. If the extraction came from an LLM, consider leaving lower-confidence facts out of working memory until they are reinforced by later evidence or reviewed by a human. Datalevin makes that easy because raw episodes, candidate facts, review state, and working-memory slots can live in the same database.

Here is the simplest ingestion-time extraction variant using the episode and fact shapes defined above. The function records one episode and one candidate fact atomically, so it collapses ingest, analyze, abstract, and update into one transaction. The background-job shape in Section 3.1 generalizes this pattern to episodes that were stored earlier. The extraction policy is passed in as data; Datalevin only commits the records and keeps the links queryable.

Listing 24.1. Recording an episode and candidate fact atomically

```
(defn record-episode-and-candidate-fact!
  [conn {:keys [user-id session-id summary fact-kind fact-content confidence]}]
  (let [episode-id (random-uuid)
        fact-id    (random-uuid)
```

```
        now         (java.util.Date.)]
    (d/transact! conn
      [{:episode/id         episode-id
        :episode/user       [:user/id user-id]
        :episode/session [:session/id session-id]
        :episode/timestamp now
        :episode/summary summary
        :episode/processed? true}

       {:fact/id                fact-id
        :fact/subject           [:user/id user-id]
        :fact/kind              fact-kind
        :fact/source-episode [:episode/id episode-id]
        :fact/status            :fact.status/pending
        :fact/content           fact-content
        :fact/confidence        confidence
        :fact/created-at        now}])
    {:episode/id episode-id
     :fact/id    fact-id}))

(record-episode-and-candidate-fact!
  conn
  {:user-id "alice"
   :session-id #uuid "00000000-0000-0000-0000-000000000103"
   :summary "Alice asked for release notes to emphasize migration risk."
   :fact-kind :preference/release-notes
   :fact-content "Alice prefers release notes that call out migration risk."
   :confidence 0.74})

(d/q '[:find ?content ?confidence
       :where [?f :fact/status :fact.status/pending]
              [?f :fact/content ?content]
              [?f :fact/confidence ?confidence]]
     (d/db conn))
```

3.1 Example: Consolidating Episodes into a Knowledge Graph

Recorded Episodes. Suppose an agent records three release-planning episodes:

```
[{:episode/id         #uuid "00000000-0000-0000-0000-000000000311"
  :episode/timestamp #inst "2026-06-08T17:20:00Z"
  :episode/user       [:user/id "alice"]
  :episode/summary    "Alice said migration M184 changes the write path."
  :episode/processed? false}

 {:episode/id         #uuid "00000000-0000-0000-0000-000000000312"
  :episode/timestamp #inst "2026-06-09T07:40:00Z"
  :episode/user       [:user/id "alice"]
  :episode/summary    "The DBA asked for backup verification before M184 ships."
  :episode/processed? false}

 {:episode/id         #uuid "00000000-0000-0000-0000-000000000313"
```

```
  :episode/timestamp #inst "2026-06-09T08:10:00Z"
  :episode/user      [:user/id "alice"]
  :episode/summary   "CI passed, but M184 is still blocked on backup
   ↪ verification."
  :episode/processed? false}]
```

Candidate Facts. A consolidation job can extract candidate semantic facts from those episodes. The extraction might be performed by an LLM, by rules, or by application code. The transaction should preserve the source episode and review status, and mark the source episodes processed once their derived facts have been written.

One transactional detail matters here. The fact records point at graph entities that do not exist yet: the migration, the component, and the check. A lookup ref used as a ref value must resolve to an existing entity or to one upserted earlier in the same transaction, so the transaction data starts with stub identity maps for those entities. The promotion step below then fills in their remaining attributes:

```
[{:migration/id "M184"}
 {:component/id :component/write-path}
 {:check/id :check/backup-verification}

 {:fact/id              #uuid "00000000-0000-0000-0000-000000000321"
  :fact/subject         [:migration/id "M184"]
  :fact/kind            :fact.kind/changes-component
  :fact/object          [:component/id :component/write-path]
  :fact/content         "Migration M184 changes the write path."
  :fact/source-episode [:episode/id #uuid
   ↪ "00000000-0000-0000-0000-000000000311"]
  :fact/confidence      0.86
  :fact/status          :fact.status/pending}

 {:fact/id              #uuid "00000000-0000-0000-0000-000000000322"
  :fact/subject         [:migration/id "M184"]
  :fact/kind            :fact.kind/requires-check
  :fact/object          [:check/id :check/backup-verification]
  :fact/content         "Migration M184 requires backup verification before
   ↪ shipping."
  :fact/source-episode [:episode/id #uuid
   ↪ "00000000-0000-0000-0000-000000000312"]
  :fact/confidence      0.92
  :fact/status          :fact.status/current}

 {:fact/id              #uuid "00000000-0000-0000-0000-000000000323"
  :fact/subject         [:migration/id "M184"]
  :fact/kind            :fact.kind/blocked-by
  :fact/object          [:check/id :check/backup-verification]
  :fact/content         "Migration M184 is blocked on backup verification."
  :fact/source-episode [:episode/id #uuid
   ↪ "00000000-0000-0000-0000-000000000313"]
  :fact/confidence      0.94
  :fact/status          :fact.status/current}
```

```
 {:episode/id          #uuid "00000000-0000-0000-0000-000000000311"
  :episode/processed? true}
 {:episode/id          #uuid "00000000-0000-0000-0000-000000000312"
  :episode/processed? true}
 {:episode/id          #uuid "00000000-0000-0000-0000-000000000313"
  :episode/processed? true}]
```

Promoted Graph State. Facts promoted to current can then update the knowledge graph in the same transaction or in a second promotion step:

```
[{:component/id   :component/write-path
  :component/name "Write path"}

 {:migration/id         "M184"
  :migration/status     :migration.status/blocked
  :migration/affects    [:component/id :component/write-path]
  :migration/blocked-by [:check/id :check/backup-verification]}

 {:check/id     :check/backup-verification
  :check/type   :check.type/manual
  :check/status :check.status/pending}]
```

Querying the Result. Now the agent can answer operational questions from the graph instead of re-reading the whole transcript:

```
(d/q '[:find ?migration ?component ?check
       :where [?m :migration/id ?migration]
              [?m :migration/status :migration.status/blocked]
              [?m :migration/affects ?c]
              [?c :component/id ?component]
              [?m :migration/blocked-by ?b]
              [?b :check/id ?check]]
     db)
;; => #{["M184" :component/write-path :check/backup-verification]}
```

The episode store still matters. If the agent needs to justify the graph fact, audit a mistaken extraction, or lower confidence after contradictory evidence, the `:fact/source-episode` links lead back to the original episodes. The knowledge graph gives the agent compact operational state; the episodes provide the evidence trail.

Consolidation is not retrieval. Consolidation changes the long-term memory model. Retrieval selects the small slice of that model needed right now. Section 4 describes external documents as retrieval material from the environment; Section 5 shows working-memory slots as bounded current context built from environment and memory inputs.

> **Consolidation changes memory**
>
> Retrieval chooses context for a turn. Consolidation writes long-term semantic state, source links, review status, and task updates. Treat consolidation as a database maintenance operation, not as a prompt-formatting step.

4. External Context: Documents and Chunks

External documents can feed working memory. Users may upload or paste PDFs, notes, spreadsheets, logs, and drafts that matter to the current task or session. Store a local reference or copy of the parent document and its chunks separately. Retrieve chunks, but keep the parent summary nearby for prompt assembly.

The parent document is the stable object: it has a name, media type, checksum, summary, and ownership or session scope. Chunks are retrieval units. They should be sized around natural boundaries when possible, such as headings, paragraphs, or spreadsheet sections, because those boundaries preserve enough local context for the model to understand an excerpt.

```
{:local.doc/id            #uuid "00000000-0000-0000-0000-000000000304"
 :local.doc/session       [:session/id #uuid
  ↪ "00000000-0000-0000-0000-000000000305"]
 :local.doc/name          "release-notes-draft.md"
 :local.doc/media-type    "text/markdown"
 :local.doc/summary       "Draft release notes for Datalevin 1.1.0."
 :local.doc/chunk-count 12}

{:local.doc.chunk/id        #uuid "00000000-0000-0000-0000-000000000306"
 :local.doc.chunk/doc       [:local.doc/id #uuid
  ↪ "00000000-0000-0000-0000-000000000304"]
 :local.doc.chunk/index     3
 :local.doc.chunk/heading "Migration notes"
 :local.doc.chunk/text      "The migration changes the write path ..."}
```

Declare summaries and chunk text as string attributes. Add `:db/fulltext true` or `:db/embedding true` when the agent needs lexical, semantic, or hybrid recall; Datalevin computes embeddings for these text values and lets you query them with text input. Chunk-level retrieval keeps large documents from crowding out precise evidence, while parent summaries keep document context nearby.

Scope matters here. A document uploaded into one session may be temporary environment context, while a curated artifact or knowledge-base page may be global external context. It is not agent memory merely because Datalevin indexes it for retrieval. When the agent should remember something from the document, extract that claim into episodic or semantic memory with evidence and status.

> **Indexed documents are environment context**
>
> A retrieved chunk can help a task without becoming agent memory. Promote only claims the agent should remember into episodic or semantic records, with source links and status.

5. Working Memory: The Current Context

Working memory should not be the whole database. Treat it as bounded, session-scoped current context with two inputs: perceived environment and a projection from agent memory. The perceived environment includes current user input, tool results, and external document chunks.

The memory projection includes selected episodes, semantic facts, and new information just learned in the session:

```
{:wm/id       #uuid "00000000-0000-0000-0000-000000000104"
 :wm/session [:session/id #uuid "00000000-0000-0000-0000-000000000103"]
 :wm/topics  "Release readiness and database migration risk"
 :wm/config  {:max-slots 15
              :decay-factor 0.85
              :eviction-threshold 0.1}}

{:wm.slot/wm          [:wm/id #uuid "00000000-0000-0000-0000-000000000104"]
 :wm.slot/entity      [:db/ident :concept/migrations]
 :wm.slot/relevance 0.82
 :wm.slot/pinned?     false
 :wm.slot/reason      "Current release has pending schema changes."}
```

Each turn can refresh this context by combining current environmental evidence with retrieved concepts, episodes, and external chunks, expanding one hop through the context graph, then decaying entries that were not refreshed. Pinned slots preserve user-important or application-critical context without making every retrieved fact permanent prompt material.

This is a prompt-management pattern as much as a storage pattern. The database can hold years of facts, but a model call needs a small, current, explainable context. Working memory gives that context a shape: each slot has a reason to be present, a relevance score, and an entity it came from. If the process crashes, the current context can be restored from its last snapshot or rebuilt from long-term memory; it is not the only copy of the knowledge.

> **Working memory is a projection**
>
> Working memory should be small, current, and explainable. It points back to source entities so prompt text can be audited, but it is not the only copy of the knowledge and should be rebuildable from long-term memory.

Working memory is also where goal/task/session boundaries matter most. A fact may be true globally but irrelevant to the current task. A session note may be important right now but should not become a durable preference. Keeping working-memory slots separate from long-term semantic facts prevents temporary context from silently becoming permanent truth.

Summary

Episodic, semantic, and working memory solve different problems. Episodes preserve history. Semantic facts organize current claims and durable knowledge. A consolidation loop transforms selected episodes into facts. External documents add scoped retrieval material from the environment. Working memory combines the perceived environment with the projected subset of memory that should matter now.

Datalevin's advantage is that memory records and external context references are not separate silos. They can be queried, updated, audited, and joined in one transactional system, which gives the next stage, recall and context assembly, a coherent substrate to work from.

References

[1] James L. McClelland, Bruce L. McNaughton, and Randall C. O'Reilly, "Why There Are Complementary Learning Systems in the Hippocampus and Neocortex: Insights from the Successes and Failures of Connectionist Models of Learning and Memory," *Psychological Review* 102(3):419-457, 1995. DOI: https://doi.org/10.1037/0033-295X.102.3.419.

CHAPTER 25

Recall and Context Assembly

Chapter 24 defined the records an agent stores: episodes, semantic facts, working-memory slots, and session documents. This chapter explains how those records become useful to an LLM at inference time.

The key distinction is that recall is not just search. Search finds candidates. Recall selects, ranks, filters, explains, and formats the small context packet that should guide the next model call. Datalevin helps because symbolic constraints, full-text matching, vector similarity, graph expansion, and task state all live in the same queryable substrate. Figure 25.1 shows the recall pipeline.

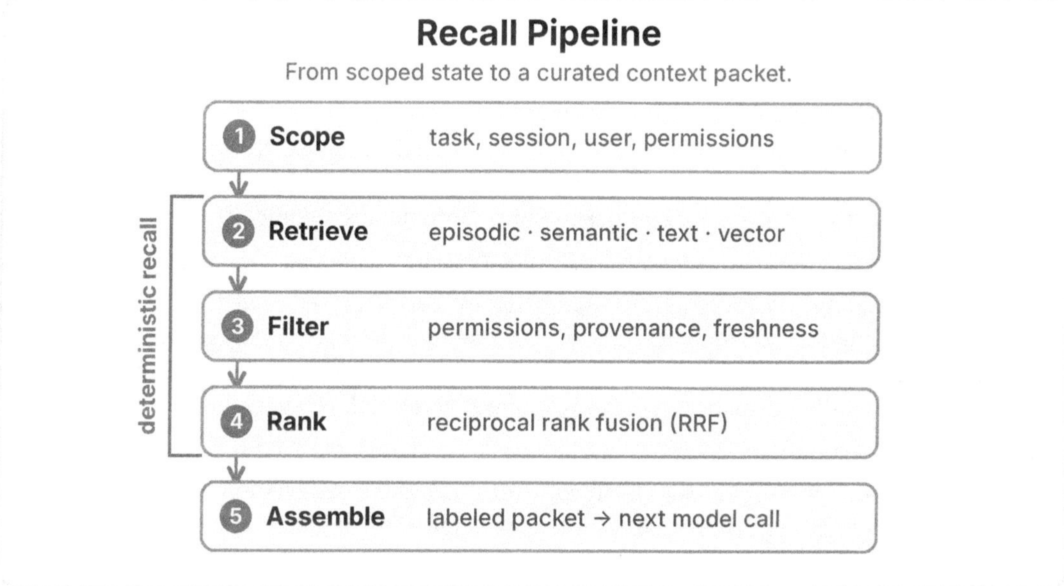

Figure 25.1: The recall pipeline: scoped state feeds candidate retrieval across lenses, then permission and provenance filtering, ranking with reciprocal rank fusion, and finally assembly into a labeled context packet for the next model call

1. Recall Is a Control Problem

In many AI architectures, Retrieval-Augmented Generation (RAG) is treated as a vector search step: embed the user question, fetch similar chunks, paste them into the prompt. That can work for simple question answering, but agent memory needs more control.

An agent usually needs to ask:

1. Which goal and task are active?
2. Which user, session, document scope, or permission boundary applies?

3. Which recent episodes matter?
4. Which semantic facts are current and reliable?
5. Which working-memory slots are pinned or stale?
6. Which evidence should be shown to the LLM, and in what order?

Those are database questions as much as model questions. If retrieval ignores goal state, policy, permissions, or evidence quality, the model receives a plausible-looking but unsafe context. Good recall makes the boundary explicit: deterministic code retrieves and organizes candidate context; the LLM reasons over the curated packet.

> **Recall is policy-controlled selection**
>
> Recall applies scope, permissions, freshness, evidence quality, ranking, and formatting policy before anything reaches the model. Keep that control path in deterministic application code.

2. Why Vectors Are Not Enough

Vector embeddings are good at capturing semantic similarity, but they are fuzzy. A search for "Datalevin 1.0.15" might return results for "Datomic 1.0.1015" because the texts are semantically related, even though the software and versions are wrong. Product IDs, exact names, incident numbers, schema attributes, dates, and permissions often need symbolic precision.

The useful mental model is symbolic structure as the skeleton and data-driven models as flexible functions. Embeddings, classifiers, and LLM calls are excellent at broad matching, similarity, extraction, and judgment under uncertainty. They should produce candidate facts, candidate entities, candidate answers, or scores. Datalog then gives those candidates a place in a structured world: permissions, identities, timestamps, task state, and graph relationships.

This keeps model calls from becoming unreviewable authority. A model may say "this message appears to be about billing," but Datalevin can store that as a classification with a score, join it with the user's account permissions, and route it through explicit task logic. If the classification was wrong, the stored score and source give you something to inspect and correct, and the transactions and queries give you a concrete trail to inspect.

> **Exact constraints stay symbolic**
>
> Use embeddings for fuzzy candidate generation, not for identifiers, permissions, dates, policy, or task state. Those constraints should remain ordinary facts and Datalog joins so the recall path can be audited and tested.

3. The Multi-Lens Retrieval Pattern

A production-grade agent usually combines several retrieval lenses:

1. **Episodic lens**: "Have I talked about this specific topic with this user or task recently?"
2. **Semantic lens**: "What durable facts or concepts are known about this topic?"

3. **Textual lens**: "Which documents contain the exact terms, identifiers, or phrases?"
4. **Vector lens**: "Which records are semantically similar even if they use different wording?"
5. **Logical lens**: "Which candidates are allowed, current, scoped to this session, and relevant to the active goal?"

A simple double-lens query can combine recent episodic context with semantic knowledge. Here the semantic lens is a named rule:

```
(def memory-rules
  '[[(semantic-memory ?topic ?fact)
     [?c :concept/name ?topic]
     [?c :concept/description ?fact]]])

(d/q '[:find ?fact
       :in $ % ?topic ?user ?since
       :where (or-join [?fact ?topic ?user ?since]
               ;; Lens 1: recent episodic context.
               (and [?e :episode/user ?user]
                    [?e :episode/topic ?topic]
                    [?e :episode/timestamp ?ts]
                    [(> ?ts ?since)]
                    [?e :episode/summary ?fact])

               ;; Lens 2: semantic knowledge.
               (semantic-memory ?topic ?fact))]
     db memory-rules "Vector Search" user-id one-hour-ago)
```

This section is the conceptual pattern: retrieval is a controlled combination of lenses, not a single search call. The next section turns that pattern into a bounded pipeline that decides which lenses run, how many candidates each lens may return, how results are fused, and what becomes working memory.

4. One Concrete Pipeline Implementation

Agent applications usually need a retrieval pipeline, not simply an unbounded query. A practical pipeline looks like this:

1. Extract cheap lexical hints from the user turn.
2. Search knowledge graph nodes, facts, episodes, and uploaded document chunks with small `:top` limits.
3. Merge lexical and semantic hits with Reciprocal Rank Fusion (RRF) or another deterministic ranker.
4. Expand graph hits by one hop to activate adjacent concepts.
5. Enforce permissions, freshness, and task scope.
6. Materialize only the best references into working memory.

Section 5 explains the candidate limits that keep step 2 bounded. Section 6 shows one implementation of the search, filtering, and rank-fusion stages as a hybrid Datalevin query.

The important design choice is the boundary: the model should receive the final curated context, not every intermediate hit from every index. After all, agent context is at a premium.

Once deterministic retrieval has selected the best references, materialize that selection as ordinary working-memory entities. The database does not decide the ranking policy; it stores the result so later prompt assembly, audit, and refresh logic can inspect it.

```
(defn materialize-working-memory!
  [conn session-id selected-facts]
  (let [wm-id (random-uuid)]
    (d/transact! conn
      (into
        [{:wm/id      wm-id
          :wm/session [:session/id session-id]}]
        (map (fn [{:keys [fact-id relevance reason pinned?]}]
               {:wm.slot/wm          [:wm/id wm-id]
                :wm.slot/entity      [:fact/id fact-id]
                :wm.slot/relevance relevance
                :wm.slot/reason      reason
                :wm.slot/pinned?     (boolean pinned?)}))
        selected-facts))
    wm-id))

(materialize-working-memory!
  conn
  #uuid "00000000-0000-0000-0000-000000000103"
  [{:fact-id (d/q '[:find ?id .
                    :where [?f :fact/status :fact.status/current]
                           [?f :fact/id ?id]]
                  (d/db conn))
    :relevance 0.91
    :reason "The current task asks for release-note guidance."
    :pinned? true}])
```

In production code, `selected-facts` would come from fused and filtered retrieval results, such as the hybrid query in Section 6. The example is deliberately small: it shows the boundary between retrieval policy and durable state.

This is especially important when the agent can call tools repeatedly. If every full-text hit, embedding neighbor, and document chunk enters the transcript, the model spends its context budget sorting retrieval artifacts instead of solving the task. A bounded pipeline lets deterministic code do the repetitive work: search several domains, merge rankings, remove duplicates, enforce permissions, and return only the selected evidence.

```
{:retrieval.run/id        #uuid "00000000-0000-0000-0000-000000000401"
 :retrieval.run/session   [:session/id #uuid
 ↪  "00000000-0000-0000-0000-000000000402"]
 :retrieval.run/query     "migration lock timing"
 :retrieval.run/terms     ["migration" "lock" "timing"]
 :retrieval.run/limits    {:nodes 10
                           :facts 15
                           :episodes 5
                           :doc-chunks 8}
 :retrieval.run/result    {:facts []
                           :episodes []
                           :chunks []}}
```

Storing the run record is optional for latency-sensitive paths, but it is useful for audit, debugging, and prompt-quality evaluation. It also gives you a place to record why a fact entered working memory.

The run record can be small. You do not need to store every rejected candidate forever. In production systems, it is often enough to store the query, limits, selected references, ranking method, and the versions or epochs of the indexed state that were searched. That gives you reproducibility without turning retrieval logs into another unbounded memory store.

> **Audit the decision, not every discarded hit**
>
> For many systems, storing the query, limits, selected references, ranking method, and index epochs is enough to reproduce why context was chosen. Avoid turning retrieval logs into another unbounded memory stream.

5. Performance: Candidate Control

Keep candidate generation bounded. Use domains and `:top` first. Use application-specific `:vec-filter` or `:doc-filter` functions only when the predicate can be evaluated from the index reference itself.

General Datalog constraints such as `:doc/status :published` are still valuable, but they are joins over the candidates returned by the search function. If a hard constraint is very selective, consider using separate search, vector, or embedding domains for that slice, or overfetch with a larger `:top` and let Datalog remove unauthorized or stale candidates.

There is a practical tradeoff here. Separate domains make candidate generation cheap and precise, but they increase schema and indexing decisions. Overfetching keeps the schema simpler, but wastes work when most candidates are filtered out after retrieval. Start with simple domains and explicit `:top` limits, then add specialized domains only for constraints that are both common and selective.

> **Control candidates before fusion**
>
> Use domains and `:top` limits before relying on downstream filters. Overfetching is useful, but if most candidates are discarded after retrieval, create a more selective domain or retrieval path for that slice.

6. Implementing a Hybrid Query

This section implements the candidate-search, permission-filtering, and rank-fusion stages from Section 4. In Datalevin, full-text search, vector search, and logical constraints share the same Datalog query model. For recall, it is often clearer to run one bounded query per lens and then fuse the ordered lists in application code.

The practical pattern is not to subtract a vector distance from a text score. Those numbers come from different scoring systems. Instead, run bounded queries for each lens, apply the same logical constraints to each candidate set, and fuse the resulting ranked lists with reciprocal rank fusion (RRF).

> **Apply hard filters in every lens**
>
> When separate lens queries feed rank fusion, each lens should enforce the same mandatory scope and permission constraints. Fusion is for ranking evidence, not for admitting unauthorized or stale candidates.

Chapter 18 introduced RRF for hybrid database queries. The version here extends that application-side pattern by carrying evidence rows for provenance, so prompt assembly can explain why each item was selected.

The examples below assume `:doc/content` has `:db/fulltext true` and `:db.fulltext/autoDomain true`; the vector lens uses a `:doc/vec` attribute declared with `:db.type/vec` and vector index options. The `?user-id` input represents the user entity id or lookup ref whose `:user/permissions` refs point to authorized document entities.

Listing 25.1. Hybrid recall with full-text, vector search, authorization, and RRF

```
(def text-hits
  (d/q '[:find ?e ?content ?score
         :in $ ?q-text ?user-id
         :where
         [(fulltext $ :doc/content ?q-text {:top 50
                                             :display :refs+scores})
          [[?e _ ?content ?score]]]
         [?e :doc/status :published]
         [?user-id :user/permissions ?e]
         :order-by [?score :desc]]
       db
       "performance tuning"
       user-id))

(def vector-hits
  (d/q '[:find ?e ?content ?dist
         :in $ ?q-vec ?user-id
         :where
         [(vec-neighbors $ :doc/vec ?q-vec {:top 50
                                             :display :refs+dists})
          [[?e _ _ ?dist]]]
         [?e :doc/content ?content]
         [?e :doc/status :published]
         [?user-id :user/permissions ?e]
         :order-by [?dist :asc]]
       db
       query-embedding
       user-id))

(defn rrf-fuse
  "Fuse ordered result rows. The first column must be the entity id."
```

```
  [ranked-lists]
  (let [k 60]
    (->> ranked-lists
         (mapcat
           (fn [rows]
             (map-indexed
               (fn [idx row]
                 (let [rank (inc idx)
                       e    (first row)]
                   [e (/ 1.0 (+ k rank)) row]))
               rows)))
         (reduce
           (fn [acc [e score row]]
             (-> acc
                 (update-in [e :rrf] (fnil + 0.0) score)
                 (update-in [e :evidence] (fnil conj []) row)))
           {})
         (map (fn [[e result]] (assoc result :e e)))
         (sort-by :rrf >))))

(take 10 (rrf-fuse [text-hits vector-hits]))
```

RRF uses only each candidate's rank in each list. The top full-text hit contributes `1 / (60 + 1)`, the second contributes `1 / (60 + 2)`, and so on. If the same entity appears in both the full-text and vector lists, its contributions are added. This makes RRF robust when lexical scores and vector distances have incompatible scales.

When Datalevin owns the text embedding index with `:db/embedding true` and `:db.embedding/autoDomain true`, replace `vec-neighbors` with `embedding-neighbors`. The query input is text, not a vector:

```
(d/q '[:find ?content ?dist
       :in $ ?q
       :where
       [(embedding-neighbors $ :doc/content ?q {:top 20
                                                 :display :refs+dists})
        [[?e _ ?content ?dist]]]
       [?e :doc/status :published]]
     db
     "performance tuning")
```

7. Retrieval Epochs and Working-Memory Refresh

The `:retrieval.run/*` record in Section 4 is an audit record for one retrieval decision: what query ran, what limits applied, and which references were selected. Retrieval epochs are different: they are compact scope state used by a running turn to decide whether search-visible memory has changed enough to refresh working memory.

Long-running agent turns may write new facts, documents, or episodes before the final response. Keep a cheap retrieval epoch so the agent can make that refresh decision without rerunning every search on every step.

```
{:retrieval.scope/id         #uuid "00000000-0000-0000-0000-000000000403"
 :retrieval.scope/session    [:session/id #uuid
 ↪ "00000000-0000-0000-0000-000000000402"]
 :retrieval/knowledge-epoch 42
 :retrieval/local-doc-epoch 7}
```

When a transaction changes an indexed knowledge fact, episode summary, or local document chunk, increment the relevant epoch in the same transaction. A running turn can compare its baseline epoch with the current epoch before deciding whether to run retrieval again. This keeps prompt context fresh without turning every tool call into a full memory refresh.

Epochs are not a replacement for database consistency. They are a cheap signal for prompt assembly. The database is already consistent after each transaction; the question is whether the agent's current working-memory projection is stale enough to rebuild. Keeping that decision explicit avoids both extremes: stale context during long tasks and wasteful retrieval after every harmless state change.

> **Epochs decide refresh, not consistency**
>
> Datalevin transactions keep the database consistent. Retrieval epochs are application state that tells a running turn whether its current working-memory projection is stale enough to rebuild.

8. Prompt Assembly from Memory

The final model prompt should not be a raw dump of retrieved memories. Raw concatenation is hard for an LLM to use reliably: old facts, current task state, user preferences, retrieved episodes, and tool policy all compete in the same flat text. A better pattern is to assemble a labeled context packet from the different memory sources, then give the model an explicit output contract.

Use this assembly policy before constructing the packet:

1. Put durable instructions, safety policy, and goal contracts near the beginning.
2. Put the current user request, fresh tool results, and required answer format near the end.
3. Keep supporting evidence in the middle, but label it clearly and keep it compact.
4. If a fact is both safety-critical and easy to overlook, summarize it at one boundary and include the evidence in the body.
5. Keep stable material in a stable order when prompt caching matters, and measure the exact layout with the models you deploy.

With that policy in mind, suppose the refresh step has produced the following Datalevin-backed context packet. The packet uses `:fact/ref` for prompt-facing labels; stored semantic facts still use UUID `:fact/id` values, as shown in Chapter 24.

> **Assemble a labeled packet, not a dump**
>
> The model should see a compact packet with labeled policy, goal state, evidence, tool results, and output expectations. Datalevin keeps the source records and provenance; the assembler decides the small text shape.

Listing 25.2. Datalevin-backed context packet for prompt assembly

```
{:goal
 {:id        #uuid "00000000-0000-0000-0000-000000000101"
  :objective "Monitor release readiness"
  :success   "Summarize open blockers before 09:00"
  :policy    {:do-not-deploy-with-open-migrations true}}

 :task
 {:id     #uuid "00000000-0000-0000-0000-000000000102"
  :state  :task.state/running
  :step   :summarize
  :user   [:user/id "alice"]}

 :working-memory
 [{:entity    :concept/migrations
   :reason    "Current release has pending schema changes."
   :relevance 0.82
   :pinned?   false}
  {:entity    :preference/release-summary-style
   :reason    "User asked for short deployment summaries."
   :relevance 1.0
   :pinned?   true}]

 :semantic-facts
 [{:fact/ref     :fact/release-2026-06-09-has-migration
   :statement    "Release 2026-06-09 includes database migration M184."
   :confidence   0.98
   :source       [:episode/id #uuid "00000000-0000-0000-0000-000000000211"]}
  {:fact/ref     :fact/m184-blocked
   :statement    "Migration M184 is blocked until backup verification finishes."
   :confidence   0.93
   :source       [:tool.call/id #uuid "00000000-0000-0000-0000-000000000212"]}]

 :episodic-evidence
 [{:episode/id #uuid "00000000-0000-0000-0000-000000000213"
   :when       #inst "2026-06-09T07:40:00Z"
   :summary    "CI passed, but the DBA asked for manual backup verification."
   :score      0.76}
  {:episode/id #uuid "00000000-0000-0000-0000-000000000214"
   :when       #inst "2026-06-08T18:15:00Z"
   :summary    "A previous migration was rolled back because verification was
   ↪  skipped."
   :score      0.68}]

 :recent-turn
 {:user-request "Can we ship this morning?"
  :tool-results [{:tool :deploy/status
                  :result {:ci :passed
                           :open-blockers ["backup verification"]}}]}}
```

This assembly step can be mechanical. Deterministic application code can take the ranked and filtered retrieval results, enforce a context budget, group rows under fixed headings, and render a prompt template. An LLM may be useful for compressing long evidence into summaries, but it is not required to preserve the structure shown here.

For the packet above, the prompt sent to the LLM can preserve the distinction between durable instructions, current state, evidence, and the requested answer:

Listing 25.3. Mechanically assembled prompt from the context packet

```
You are assisting with a release-readiness task.

Priority order:
1. Follow the Goal Contract and Safety Policy.
2. Use Current Task State and Recent Tool Results as the freshest operational
   state.
3. Treat Semantic Facts as durable knowledge, but prefer higher-confidence facts
   when facts conflict.
4. Treat Episodic Evidence as supporting evidence, not as policy.
5. Do not invent deployment status. If required data is missing, say what must
   be checked.

Goal Contract:
- Objective: Monitor release readiness.
- Success condition: Summarize open blockers before 09:00.
- Safety policy: Do not recommend deployment while open migrations or backup
  verification blockers remain.

Current Task State:
- Task state: running.
- Current step: summarize.
- User: alice.
- User request: "Can we ship this morning?"

Working Memory:
- :concept/migrations, relevance 0.82
  Reason: Current release has pending schema changes.
- :preference/release-summary-style, pinned
  Reason: User asked for short deployment summaries.

Semantic Facts:
- [fact/release-2026-06-09-has-migration, confidence 0.98]
  Release 2026-06-09 includes database migration M184.
- [fact/m184-blocked, confidence 0.93]
  Migration M184 is blocked until backup verification finishes.

Episodic Evidence:
- [episode 00000000-0000-0000-0000-000000000213, score 0.76, 2026-06-09 07:40Z]
  CI passed, but the DBA asked for manual backup verification.
- [episode 00000000-0000-0000-0000-000000000214, score 0.68, 2026-06-08 18:15Z]
  A previous migration was rolled back because verification was skipped.

Recent Tool Results:
```

```
- deploy/status: CI passed; open blocker is backup verification.

Answer Format:
- Start with one sentence: "Ship?" followed by yes/no and the main reason.
- Then list blockers, if any.
- Then list the next concrete check or action.
- Keep the answer under 120 words.
```

This structure makes the prompt less error-prone in three ways. First, it keeps policy and goal state above retrieved evidence, so a similar old episode cannot override the current contract. Second, it labels provenance and confidence, so the model can distinguish durable facts from supporting anecdotes. Third, it turns working memory into a compact agenda rather than a second chat history. The LLM receives enough context to reason, but the database remains the system of record for state, evidence, and later audit.

8.1 Context Position Matters

The ordering policy above is part of recall policy, not formatting decoration. Long-context models do not always use all positions equally well. In controlled question-answering and key-value retrieval experiments, the study found that models were often best at using relevant information placed near the beginning or end of the input, while performance degraded when the relevant information was buried in the middle [1]. This is often called the "lost in the middle" effect.

The pattern has a useful analogy in human memory. Classic free-recall experiments found a serial-position effect: people tend to recall early items and recent items better than middle items [2]. The mechanisms are different, so do not take the analogy literally. The practical lesson is the same: placement affects use.

Treat this as operational guidance, not a permanent law of prompt design. Long-context retrieval, attention mechanisms, prompt caching, and model serving systems are active areas of research and engineering. A future model may use long context more evenly, while a production runtime may reward stable, deterministic prompt prefixes because they cache better. The durable rule is to make assembly explicit and measure the exact layout with the models you deploy.

This is another reason not to dump retrieval results directly into a prompt. A Datalevin-backed assembly step can rank, label, group, and position memory records deliberately: pinned policy first, current operational state last, supporting semantic facts and episodes in between. The database stores the evidence and provenance; the context assembler decides the small amount of text the model should see and where it should appear.

Summary

Recall is the bridge between stored memory and model reasoning. The database can hold years of episodes, facts, documents, and state, but the LLM needs a compact packet with clear priority, provenance, and output expectations.

Datalevin's advantage is that the same engine can retrieve fuzzy semantic matches, exact text matches, graph relationships, scoped state, permissions, and evidence links. That makes context assembly inspectable instead of a pile of prompt strings.

References

[1] Nelson F. Liu, Kevin Lin, John Hewitt, Ashwin Paranjape, Michele Bevilacqua, Fabio Petroni, and Percy Liang, "Lost in the Middle: How Language Models Use Long Contexts," arXiv:2307.03172, 2023. URL: https://arxiv.org/abs/2307.03172.

[2] Bennet B. Murdock, "The Serial Position Effect of Free Recall," *Journal of Experimental Psychology* 64(5):482-488, 1962. DOI: https://doi.org/10.1037/h0045106.

CHAPTER 26

Apperception and Truth Maintenance

The previous chapters built the memory substrate and the recall pipeline. This chapter covers the next problem: what happens when new information should change the agent's long-term model?

Memory is about storage and retrieval. **Apperception** is about integrating new information into a coherent internal model of the world [1]. In an AI application, the apperception layer does not just record events; it applies rules, model outputs, review policy, and domain logic to decide how those events should change current assertions, goal hierarchy, operating envelope, and future behavior.

Datalevin is suited for storing and checking the state behind this layer because it combines Datalog's logical rigor, graph relationships, full-text and vector evidence, and transactional updates in one engine.

As in the previous agent-memory chapters, the examples below are schematic data shapes, not complete schema declarations. Application schema should declare lookup-ref targets such as `:fact/id`, `:episode/id`, `:agent/id`, and `:strategy/id` as unique identities; relationship attributes as refs; and nested document-shaped attributes as `:db.type/idoc` only when the application needs path-level idoc indexing.

1. What Is an Apperception Engine?

In philosophy, apperception is the process of perceiving new things through the lens of past experience to form a unified sense of self. In AI architecture, an apperception engine is application code that:

1. **Validates**: Ensures new information is logically consistent with what is already known.
2. **Synthesizes**: Uses inference to derive new facts from raw input.
3. **Updates**: Modifies the agent's world model, identity, and goal hierarchy.
4. **Maintains**: Archives stale claims, preserves evidence, and tracks why current facts are current.

This chapter deliberately comes after recall and context assembly. A retrieval pipeline can decide what to show the LLM for this turn. Apperception decides what should become part of long-term state after the turn.

Datalevin does not ship this engine. It supplies the transaction boundary, indexes, query language, and constraints that the engine uses.

> **Apperception is application policy**
>
> Datalevin can make candidate facts, current assertions, evidence, and supersession links transactional and queryable. A supersession link is an explicit ref from a newer assertion to an older assertion it replaces. The application still decides which observations to accept, reject, review, or use to change behavior.

The term is also useful because of recent AI work on the Apperception Engine, which frames "making sense" as constructing an interpretable symbolic theory that explains observations while satisfying coherence or unity conditions [1]. This chapter does not implement that system directly; it borrows the same design ideas. New observations should be integrated into an explicit, coherent, inspectable model rather than left as disconnected prompt fragments.

A practical apperception pass follows a simple lifecycle:

1. **Observe and extract**: Turn episodes, tool results, or model outputs into candidate facts, usually stored as `:fact.status/pending`.
2. **Validate**: Check the candidates against current facts, policy, and domain constraints.
3. **Accept or supersede**: Mark validated facts `:fact.status/current`, reject weak candidates, and mark older replaced facts `:fact.status/superseded`.
4. **Score evidence**: Track confidence, utility, provenance, and freshness.
5. **Synthesize insights**: Derive higher-level interests, plans, or summaries from repeated evidence.
6. **Update identity and policy**: Propose or commit long-term changes to the agent's strategy, operating envelope, and goal hierarchy.

The rest of this chapter reuses that loop as its organizing spine:

Loop phase	Where it appears	What Datalevin stores or checks
Observe and extract	Chapter 24 and Section 2	Episodes, tool results, and candidate facts.
Validate	Section 2.1	Datalog checks, constraints, and policy-specific predicates.
Accept or supersede	Section 2.2	Current facts, retired facts, source links, and supersession links.
Score evidence	Section 3	Confidence, utility, provenance, and freshness attributes.
Synthesize insights	Section 4	Derived insight records backed by source episodes and rule versions.
Update identity and policy	Sections 5 and 6	Strategy proposals, approved current strategies, and resolved operating envelopes.

In an application these phases usually run as one pipeline, but separating them in the chapter makes each maintenance decision easier to inspect.

2. Maintaining a Coherent World Model

A major challenge in AI is hallucination or drift. An agent may accept contradictory information, over-promote a temporary user statement, or let old facts linger after they have been superseded.

In this application layer, use Datalevin constraints, Datalog checks, and explicit evidence records to make the world model easier to keep coherent.

This section covers the validation and accept-or-supersede phases of the apperception loop.

2.1 Logical Validation with Datalog

Before committing a new claim to memory, the engine can run a Datalog query to check for contradictions.

```
;; Find a current fact with the same subject and kind but different content.
(d/q '[:find ?fact-id ?existing-content
       :in $ ?user-id ?kind ?candidate-content
       :where [?user :user/id ?user-id]
              [?fact :fact/subject ?user]
              [?fact :fact/kind ?kind]
              [?fact :fact/status :fact.status/current]
              [?fact :fact/content ?existing-content]
              [?fact :fact/id ?fact-id]
              [(not= ?existing-content ?candidate-content)]]
     db
     "alice"
     :preference/deployment-window
     "Alice's deployment window is Thursday.")
```

This simple check treats "same subject, same kind, different content" as a conflict. Richer domains may also compare time ranges, mutually exclusive statuses, or policy scopes. Keep those checks explicit and repeatable. In server deployments, put richer custom logic behind a registered descriptor-backed UDF or transaction function, following the server-mode rules in Chapter 22; do not rely on arbitrary host functions or a prompt instruction such as "do not contradict yourself" as the only guardrail.

> **Validate before accepting long-term claims**
>
> Prompt instructions are not a truth-maintenance policy. Check candidate facts against current assertions, policy scopes, time ranges, and domain constraints before making them current semantic memory.

2.2 Truth Maintenance

If new data supersedes old data, application code can use Datalevin's atomic transactions to update the state. Because Datalevin is fact-centric, you can retire or replace specific old facts while keeping the rest of the model intact.

This is the practical version of a classic AI idea: a **truth-maintenance system** (TMS) records beliefs together with the reasons or dependencies that support them, then updates the current belief set when assumptions or evidence change [2]. In Datalevin, those dependencies can be ordinary facts: source episodes, rule identifiers, supersession links, confidence scores, and review status.

In this chapter, a **supersession link** is an ordinary application-level reference that records replacement history: the newer assertion points to the older assertion it replaces. For facts, the examples use `:fact/supersedes` on the new current fact. Datalevin does not give this attribute special built-in meaning; the application schema and queries make the relationship useful.

For agent systems, truth maintenance should usually preserve evidence even when the current belief changes. If a user says "my deployment window is Tuesday" and later says "move it to Thursday," the current scheduling fact should change, but the old episode remains historically true. This distinction matters when the agent explains a decision, audits a mistake, or reconstructs why a task was scheduled at a particular time.

> **Retire assertions, preserve evidence**
>
> When a current fact changes, mark the older assertion superseded rather than deleting the episode or source link. The old belief may no longer be current, but the evidence remains part of the audit trail.

Figure 26.1 shows this status lifecycle, in which evidence links survive every status change.

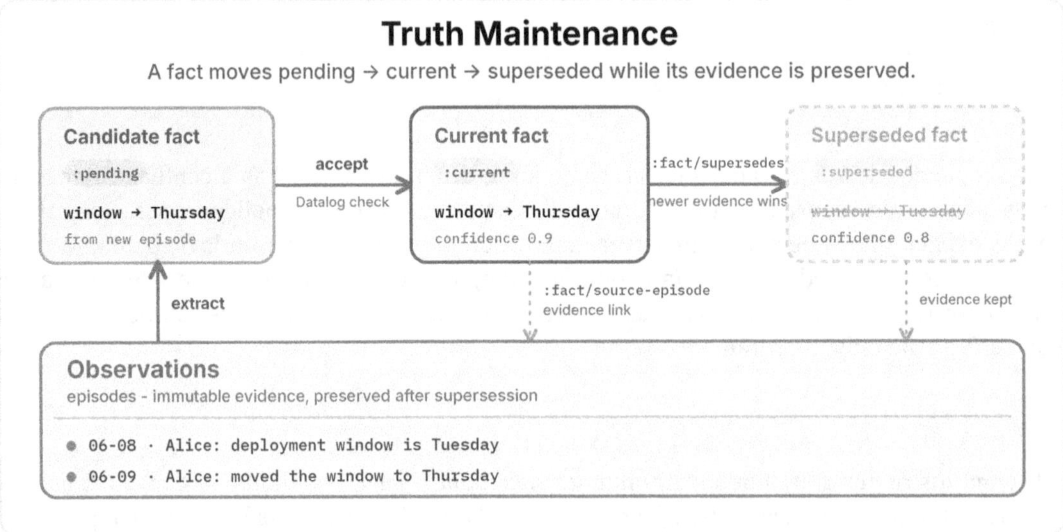

Figure 26.1: Truth maintenance: an observation is extracted into a candidate fact, validated into a current fact, and superseded by newer evidence, while source-episode evidence links are preserved through every status change

A practical pattern is to separate evidence from current assertions:

```
{:episode/id        #uuid "00000000-0000-0000-0000-000000000701"
 :episode/summary "Alice moved the deployment window from Tuesday to Thursday."
 :episode/timestamp #inst "2026-06-09T11:00:00Z"}

{:fact/id             #uuid "00000000-0000-0000-0000-000000000702"
 :fact/subject        [:user/id "alice"]
 :fact/kind           :preference/deployment-window
 :fact/content        "Alice's deployment window is Thursday."
 :fact/source-episode [:episode/id #uuid "00000000-0000-0000-0000-000000000701"]
```

```
  :fact/status          :fact.status/current
  :fact/supersedes      [:fact/id #uuid "00000000-0000-0000-0000-000000000699"]}
```

The older fact can be marked superseded rather than deleted. Retrieval can then prefer current facts while audit and explanation paths can still follow the evidence chain.

The intended invariant in this example is one current fact per `(:fact/subject, :fact/kind)` pair, not one current fact in the whole database. For Alice's deployment-window preference, accepting a newer current fact retires any older current facts for that same user and kind. Other users and other fact kinds can still have their own current facts. The read step collects a sequence of old ids rather than a single one, so the helper can also clean up any duplicates that already exist inside that one equivalence class, for example after an import or an earlier bug.

> **Define the current-fact equivalence class**
>
> The application must decide what "one current fact" means. In this example it means one current assertion for a given subject and kind, while other subjects and fact kinds keep their own current assertions.

The helper is a read-retire-write transaction:

Phase	Operation	Result
Read	Find current facts for the same subject and kind.	The transaction knows which facts must stop being current.
Retire	Mark those facts `:fact.status/superseded`.	Old assertions remain available for audit and explanation.
Write	Add the new `:fact.status/current` fact with `:fact/supersedes` links.	The equivalence class has one intended current assertion.
Return	Return the new fact id.	Callers can attach follow-up evidence, review state, or logs.

The helper also stores `:fact/confidence` on the new current fact. Section 3 explains how confidence and utility are maintained after the fact becomes part of the current model.

Here is a small runnable helper using the Chapter 24 episode and fact record shapes. The policy decision to accept a fact remains outside Datalevin; this function only makes the state transition atomic and inspectable:

Listing 26.1. Accepting a current fact with supersession tracking

```
(defn current-fact-ids
  [db user-id kind]
  (d/q '[:find [?fact-id ...]
         :in $ ?user-id ?kind
         :where [?user :user/id ?user-id]
                [?fact :fact/subject ?user]
                [?fact :fact/kind ?kind]
                [?fact :fact/status :fact.status/current]
                [?fact :fact/id ?fact-id]]
       db user-id kind))
```

```
(defn accept-current-fact!
  [conn {:keys [user-id source-episode-id kind content confidence]}]
  (let [fact-id (random-uuid)
        now     (java.util.Date.)]
    (d/with-transaction [cn conn]
      (let [old-ids (current-fact-ids (d/db cn) user-id kind)
            retired (mapv (fn [old-id]
                            {:fact/id old-id
                             :fact/status :fact.status/superseded})
                          old-ids)
            current (cond-> {:fact/id fact-id
                             :fact/subject [:user/id user-id]
                             :fact/kind kind
                             :fact/content content
                             :fact/source-episode [:episode/id
                              ↪  source-episode-id]
                             :fact/status :fact.status/current
                             :fact/confidence confidence
                             :fact/created-at now}
                      (seq old-ids)
                      (assoc :fact/supersedes
                             (mapv (fn [old-id] [:fact/id old-id]) old-ids)))]
        (d/transact! cn (conj retired current))))
    fact-id))

(let [old-episode-id (random-uuid)
      new-episode-id (random-uuid)
      session-id     #uuid "00000000-0000-0000-0000-000000000103"]
  (d/transact! conn
    [{:user/id "alice"}
     {:session/id session-id}
     {:episode/id old-episode-id
      :episode/user [:user/id "alice"]
      :episode/session [:session/id session-id]
      :episode/timestamp #inst "2026-06-08T10:00:00Z"
      :episode/summary "Alice said her deployment window is Tuesday."}
     {:episode/id new-episode-id
      :episode/user [:user/id "alice"]
      :episode/session [:session/id session-id]
      :episode/timestamp #inst "2026-06-09T11:00:00Z"
      :episode/summary "Alice moved the deployment window to Thursday."}])

  (accept-current-fact!
    conn
    {:user-id "alice"
     :source-episode-id old-episode-id
     :kind :preference/deployment-window
     :content "Alice's deployment window is Tuesday."
     :confidence 0.8})

  (accept-current-fact!
    conn
```

```
    {:user-id "alice"
     :source-episode-id new-episode-id
     :kind :preference/deployment-window
     :content "Alice's deployment window is Thursday."
     :confidence 0.9}))

(d/q '[:find ?status ?content
       :in $ ?kind
       :where [?fact :fact/kind ?kind]
              [?fact :fact/status ?status]
              [?fact :fact/content ?content]]
     (d/db conn) :preference/deployment-window)
```

JavaScript does not expose the explicit Datalog transaction callback described in Chapter 6, so the same read-retire-write logic should run as a server-side command, transaction function, or Clojure/Java/Python service method when concurrent updates must be serialized. From JavaScript, call that command or use a single `conn.transact` when the old fact ids are already known.

3. Confidence, Utility, and Evidence

Long-lived agents should not treat every extracted fact as equally durable. A practical world model records where a fact came from, how confident the system is, and how useful that fact has been during later turns.

This is the scoring phase of the apperception loop: after a fact becomes part of the model, maintenance jobs keep its confidence, utility, provenance, and freshness explicit.

Confidence and utility answer different questions. Confidence asks "how likely is this fact to be true?" Utility asks "how often does this fact help the agent?" A low-confidence fact should not be promoted aggressively into prompts even if it sounds relevant. A high-confidence but low-utility fact can stay in the database without taking up working-memory space. Keeping these scores separate prevents retrieval from becoming a popularity contest over whatever was mentioned most recently.

> **Separate truth confidence from usefulness**
>
> Confidence controls how strongly the system believes a claim. Utility controls whether the claim helps future tasks. A useful-sounding but low-confidence claim should still be gated before it reaches working memory.

```
{:fact/id                #uuid "00000000-0000-0000-0000-000000000201"
 :fact/content           "Alice prefers release notes with risk summaries."
 :fact/subject           [:user/id "alice"]
 :fact/source-episode    [:episode/id #uuid
 ↪  "00000000-0000-0000-0000-000000000202"]
 :fact/confidence        0.74
 :fact/utility           0.6
 :fact/created-at        #inst "2026-06-01T10:00:00Z"
 :fact/last-confirmed-at #inst "2026-06-08T09:30:00Z"}
```

Confidence can be reinforced when new evidence agrees with an existing fact, lowered when the fact is contradicted, and decayed when it has not been confirmed for a long time. Utility captures a different signal: whether the fact actually helped answer questions, choose tools, or personalize behavior.

```
(d/q '[:find ?fact ?confidence ?utility
       :in $ ?min-confidence ?min-utility
       :where [?f :fact/content ?fact]
              [?f :fact/confidence ?confidence]
              [?f :fact/utility ?utility]
              [(< ?confidence ?min-confidence)]
              [(< ?utility ?min-utility)]]
     db 0.3 0.2)
```

This makes maintenance explicit. A background job can archive low-confidence, low-utility facts while preserving the original episodes as historical evidence. It can also promote repeatedly confirmed facts by increasing confidence or utility, which gives the agent a controlled path from raw episode to durable semantic memory.

Chapter 15's semiring provenance plan is the more general version of this idea. In this chapter, confidence and utility are ordinary attributes that application jobs update directly. With provenance annotations, a system could also propagate confidence through derived facts: combine supporting evidence when a rule joins facts, merge alternative derivations when several sources support the same claim, and keep the explanation trail that says why a confidence score changed. Until that machinery exists, model confidence, utility, source episodes, and supersession links explicitly as application data.

4. Synthesizing Higher-Level Insights

Sensory data such as chat logs, tool results, and pasted documents is low-level. Apperception requires higher-level abstractions.

Datalog rules and aggregate queries allow you to define these abstractions explicitly:

- **Raw data**: "User Alice asked about Clojure five times this week."
- **Derived fact**: "Alice is an active Clojure learner."
- **Inference**: "The tutor should prioritize advanced Clojure resources."

By running these rules periodically or during ingestion, the application compresses many episodic memories into a smaller set of high-value semantic facts.

Derived insights are still maintained facts. They need the same machinery as the current assertions in Section 2: status, explicit evidence links, confidence, and a way to supersede the insight when later evidence changes the conclusion. The difference is that the source is a synthesis step over many observations, not a single observation.

4.1 Example: Deriving an Active Interest

Suppose the episode store contains recent questions from Alice:

```
[{:episode/id          #uuid "00000000-0000-0000-0000-000000000801"
  :episode/user        [:user/id "alice"]
```

```
  :episode/type      :episode.type/question
  :episode/topic     :concept/clojure
  :episode/timestamp #inst "2026-06-02T10:00:00Z"
  :episode/summary   "Alice asked how Clojure protocols differ from
  ↪  interfaces."}

 {:episode/id        #uuid "00000000-0000-0000-0000-000000000802"
  :episode/user      [:user/id "alice"]
  :episode/type      :episode.type/question
  :episode/topic     :concept/clojure
  :episode/timestamp #inst "2026-06-04T14:30:00Z"
  :episode/summary   "Alice asked about transducer performance."}

 {:episode/id        #uuid "00000000-0000-0000-0000-000000000803"
  :episode/user      [:user/id "alice"]
  :episode/type      :episode.type/question
  :episode/topic     :concept/clojure
  :episode/timestamp #inst "2026-06-08T09:15:00Z"
  :episode/summary   "Alice asked for macro debugging examples."}]
```

A synthesis job can query for repeated interest in a topic over a time window:

```
(d/q '[:find ?user (count ?episode)
       :keys user question-count
       :in $ ?topic ?since
       :where [?episode :episode/type :episode.type/question]
              [?episode :episode/topic ?topic]
              [?episode :episode/user ?user]
              [?episode :episode/timestamp ?ts]
              [(> ?ts ?since)]]
     db :concept/clojure #inst "2026-06-01T00:00:00Z")
```

If the count crosses the application's threshold, the job writes a derived insight and keeps the evidence explicit:

```
{:insight/id         #uuid "00000000-0000-0000-0000-000000000811"
 :insight/kind       :insight.kind/active-interest
 :insight/subject    [:user/id "alice"]
 :insight/topic      :concept/clojure
 :insight/content    "Alice is actively learning Clojure."
 :insight/rule       :rule/active-interest-v1
 :insight/window     {:since #inst "2026-06-01T00:00:00Z"
                      :until #inst "2026-06-09T00:00:00Z"}
 :insight/evidence [[:episode/id #uuid "00000000-0000-0000-0000-000000000801"]
                    [:episode/id #uuid "00000000-0000-0000-0000-000000000802"]
                    [:episode/id #uuid "00000000-0000-0000-0000-000000000803"]]
 :insight/confidence 0.78
 :insight/status     :insight.status/current}
```

The application can also project the current insight into the user knowledge graph:

```
{:user/id              "alice"
 :user/interested-in   :concept/clojure
 :user/learning-state  {:topic       :concept/clojure
```

```
:state       :learning.state/active
:evidence    [:insight/id #uuid
 ↪  "00000000-0000-0000-0000-000000000811"]
:updated-at #inst "2026-06-09T00:00:00Z"}}
```

The important property is explainability. The agent can later retrieve the compact fact "Alice is actively learning Clojure" for personalization, but the system can still inspect the episodes, rule version, time window, and confidence that produced it. If Alice stops asking about Clojure or starts focusing on a different topic, a later apperception pass can supersede the insight rather than silently rewriting history.

The compression should remain inspectable. A derived fact should carry the rule or model version that produced it, the source episodes or facts, and the time range over which the pattern was observed. This prevents "insight" from becoming an unexplained assertion.

5. Identity and Long-Term State

For an agent to have a persistent style, role, or strategy, it must have a stable model of itself.

This is the same apperception loop applied to the agent's own strategy state: performance evidence is observed, the application validates whether a change is justified, a proposed strategy update is written for review, and an approved proposal becomes the new current strategy while the old strategy is superseded.

- **Identity store**: system prompt fragments, behavioral constraints, role definitions, and self-description entities.
- **Performance history**: records of which strategies led to successful outcomes.
- **Goal hierarchy**: a graph of active, pending, blocked, superseded, and completed goals.

Unlike a static config file, this state is dynamic. As the application reviews the agent's performance, it can update strategy entities in Datalevin to improve over time. That does not mean the model should rewrite its own policy at will. It means the application can record evidence about strategy performance, then use rules, review workflows, or explicit approvals to update the long-term agent configuration.

5.1 Example: Updating an Agent Strategy

Consider a release-writing agent whose identity and strategy are stored as data.

This example assumes `:agent/id` and `:strategy/id` are unique identity attributes. Relationships to agents and strategies therefore use lookup refs such as `[:strategy/id :strategy/release-summary-v1]`. If the application needs path-level queries over nested maps such as `:agent/style` and `:strategy/constraints`, declare those attributes as `:db.type/idoc`, following Chapter 23's guidance for nested data that is usually loaded as a whole.

```
{:agent/id         :agent/release-writer
 :agent/role       "Prepare release-readiness summaries."
 :agent/style      {:brevity :short
                    :tone    :direct}
 :agent/strategy [:strategy/id :strategy/release-summary-v1]}
```

```
{:strategy/id          :strategy/release-summary-v1
 :strategy/status      :strategy.status/current
 :strategy/instruction "Summarize CI, migrations, blockers, and next action."
 :strategy/constraints {:max-words 120
                        :include-risk? true}}
```

After several runs, the agent has performance evidence:

```
[{:run/id          #uuid "00000000-0000-0000-0000-000000000901"
  :run/agent       [:agent/id :agent/release-writer]
  :run/strategy    [:strategy/id :strategy/release-summary-v1]
  :run/outcome     :run.outcome/revised-by-user
  :run/feedback    "Summary omitted migration risk."
  :run/timestamp   #inst "2026-06-02T09:10:00Z"}

 {:run/id          #uuid "00000000-0000-0000-0000-000000000902"
  :run/agent       [:agent/id :agent/release-writer]
  :run/strategy    [:strategy/id :strategy/release-summary-v1]
  :run/outcome     :run.outcome/revised-by-user
  :run/feedback    "Need explicit backup verification status."
  :run/timestamp   #inst "2026-06-05T09:05:00Z"}]
```

An apperception pass can detect the pattern:

```
(d/q '[:find ?strategy (count ?run)
       :keys strategy revision-count
       :in $ ?since
       :where [?run :run/strategy ?strategy]
              [?run :run/outcome :run.outcome/revised-by-user]
              [?run :run/timestamp ?ts]
              [(> ?ts ?since)]]
     db #inst "2026-06-01T00:00:00Z")
```

If the revision count crosses a threshold, the system should not let the model silently rewrite its own identity. Instead, write a proposed strategy update with evidence and review state:

```
{:strategy.proposal/id       #uuid "00000000-0000-0000-0000-000000000911"
 :strategy.proposal/agent    [:agent/id :agent/release-writer]
 :strategy.proposal/replaces [:strategy/id :strategy/release-summary-v1]
 :strategy.proposal/body     {:instruction "Summarize CI, migrations, backup
  ↪  verification, blockers, and next action."
                              :constraints {:max-words 120
                                            :include-risk? true
                                            :include-backup-status? true}}
 :strategy.proposal/evidence [[:run/id #uuid
  ↪  "00000000-0000-0000-0000-000000000901"]
                              [:run/id #uuid
 ↪  "00000000-0000-0000-0000-000000000902"]]
 :strategy.proposal/status   :review.status/pending}
```

After approval, the update becomes a normal graph transition. The new strategy record comes first in the transaction data so that the lookup refs to it from the superseded strategy and the agent resolve within the same transaction:

```
[{:strategy/id          :strategy/release-summary-v2
```

```
 :strategy/status      :strategy.status/current
 :strategy/instruction "Summarize CI, migrations, backup verification,
  ↪ blockers, and next action."
 :strategy/constraints {:max-words 120
                        :include-risk? true
                        :include-backup-status? true}
 :strategy/derived-from [:strategy.proposal/id #uuid
  ↪ "00000000-0000-0000-0000-000000000911"]}

{:strategy/id     :strategy/release-summary-v1
 :strategy/status :strategy.status/superseded
 :strategy/superseded-by [:strategy/id :strategy/release-summary-v2]}

{:agent/id       :agent/release-writer
 :agent/strategy [:strategy/id :strategy/release-summary-v2]}]
```

This is long-term state evolution without prompt drift. The agent's identity and strategy changed, but the change is explainable: there is performance evidence, a proposal, an approval state, and a supersession link from the old strategy to the new one.

6. Operating Envelopes

Do not hide hard constraints in prompt prose. Store them as data, resolve them before each turn, and attach the resolved envelope to the turn record.

This closes the loop from after-turn maintenance to before-turn control. The state maintained by apperception shapes what the next turn is allowed to perceive, retrieve, and do; Chapter 27 builds the full stateful-agent loop on top of that boundary.

An operating envelope is the effective set of constraints for a turn: tool permissions, model budgets, output style, privacy limits, retry policy, and any domain-specific guardrails. The model may see a summarized version of the envelope, but enforcement should happen in application code before tools run and before expensive model calls are made.

This is the agentic version of top-down attention. A pure bottom-up system waits for input and reacts to whatever looks salient. A goal-directed system asks a different question: given the current objective, which inputs deserve attention, which memories should be recalled, and which tools are even eligible? The operating envelope gives that top-down pressure a concrete representation.

A useful precedence order is:

```
organization policy > goal contract > task constraints > user preferences >
 ↪ session scratch
```

> **Hard constraints belong outside prompt prose**
>
> Resolve tool permissions, budgets, privacy limits, retry policy, and output constraints as data before each turn. The model may see a summary, but application code should enforce the resolved envelope.

Each layer can live in Datalevin as an `idoc` value when the corresponding attribute is declared with `:db/valueType :db.type/idoc`:

```
{:policy/id       :policy/org
 :policy/source   :policy.source/organization
 :policy/envelope {:tools {:email-send :require-approval}
                   :limits {:max-llm-calls 20}}}

{:goal/id       #uuid "00000000-0000-0000-0000-000000000203"
 :goal/contract {:objective "Prepare a weekly account update"
                 :envelope {:style {:brevity :short}
                            :limits {:max-llm-calls 8}}}}

{:task/id          #uuid "00000000-0000-0000-0000-000000000204"
 :task/goal        [:goal/id #uuid "00000000-0000-0000-0000-000000000203"]
 :task/constraints {:tools {:calendar-event-create :deny}}}
```

The resolved envelope is ordinary data, so it can be audited, tested, replayed, and queried. This is especially important for autonomous or scheduled work, where the agent may run without a human watching every turn.

Precedence is important because the sources will sometimes disagree. A user preference may ask for terse replies, while a task constraint may require a detailed audit note. An organization policy may deny a tool that a task would otherwise allow. Resolving these conflicts outside the prompt makes the behavior predictable and gives tests a stable target.

The same envelope can also drive retrieval. For example, a task that is waiting for user approval should retrieve the pending decision and the evidence behind it, while a task that is executing a scheduled report should retrieve the last successful run and the configured reporting interval. In both cases, the goal and task state shape what the agent perceives as relevant.

Summary

An apperception layer is application code that turns stored memory into maintained long-term state. Datalevin supplies the durable facts, indexes, constraints, and transaction boundary that make that layer auditable.

- **Consistency**: Datalog checks make conflicts explicit.
- **Inference**: Rules derive knowledge that is not explicitly stated.
- **Evidence**: Episodes and source links explain why current facts exist.
- **Evolution**: Graphs allow the world model, goals, and identity to grow over time.
- **Policy**: Operating envelopes keep behavioral constraints inspectable and separate from prompt text.

By building this layer on Datalevin, you give the AI application a substrate for coherent long-term state without hiding the maintenance policy inside prompt text.

References

[1] Richard Evans, Jose Hernandez-Orallo, Johannes Welbl, Pushmeet Kohli, and Marek Sergot,

"Making Sense of Sensory Input," *Artificial Intelligence* 293, 2021, article 103438. DOI: https://doi.org/10.1016/j.artint.2020.103438.

[2] Jon Doyle, "A Truth Maintenance System," *Artificial Intelligence* 12(3):231-272, 1979.

CHAPTER 27

Stateful AI Applications

Many useful AI applications are not isolated chat sessions. They are systems that maintain persistent state. A "Stateful AI" application is one where facts, goals, and behavior evolve alongside users under explicit application policy.

This chapter synthesizes the Part VI arc around one loop and one running scenario: a documentation-feedback assistant that turns failed searches into reviewable documentation tasks. The scenario uses memory architecture, concrete memory records, recall and context assembly, truth maintenance, and application-level task control. The capstone then shows the whole loop end to end as a small runnable program.

As in the preceding chapters, compact record maps outside the capstone are schematic examples, not complete schema declarations. Lookup-ref targets should be unique identities, relationship attributes should be refs, and nested document-shaped values such as task contracts, board metadata, boundaries, and tool arguments should be declared as `:db.type/idoc` when the application needs path-level idoc queries.

1. Pattern: The "Database-as-Environment"

In most applications, the database is treated as a passive storage layer. In a Stateful AI application, a better pattern is **database-as-environment-state**: the runtime treats durable database facts as the inspectable state of the application environment.

- The agent runtime reads and writes durable environment state in Datalevin.
- Every interaction can update episode records, task state, audit records, or semantic facts.
- Persona, strategy, permissions, and style are entities interpreted by application code, not hidden prompt text.

An agent runtime also needs a control loop. In the standard agent model, an agent receives percepts from its environment, updates internal state, and chooses actions [1]. In application terms, perceptors convert external input into structured observations, actuators call tools or services, and a central controller decides which task, agenda item, or exception should run next. Memory keeps the loop from starting over every turn, and plugins extend the loop with domain-specific skills.

Datalevin is helpful in this control layer because the agenda can be data. For example, Juji's agent-platform has a main stack for the active task, an agenda queue for planned work, an ad-lib queue for handling opportunistic contingencies, and an exception queue as the last resort of guardrails. In Datalevin, these can be task entities with different state, priority, and scheduling attributes. The important modeling point is that each control surface is durable, queryable state, regardless of the names your application uses.

> **The environment is durable state**
>
> In a stateful AI application, persona, strategy, agenda, permissions, task state, and audit records should be inspectable data. Hidden prompt text is not a reliable environment model.

2. Pattern: The Memory Loop

A typical persistent agent memory loop follows these steps:

1. **Perceive**: The agent receives input and records enough metadata to replay the turn later.
2. **Resolve Scope and Envelope**: The application resolves the active task, permissions, budgets, and operating envelope for the turn.
3. **Recall**: The agent performs bounded hybrid retrieval (Chapter 25) to find relevant experiences, documents, and structured facts.
4. **Refresh Working Memory**: The selected recall results are materialized into a bounded working-memory projection.
5. **Reason**: The agent uses the assembled context to decide on an action.
6. **Act and Audit**: Tool calls, approvals, and model usage are recorded as turn data.
7. **Consolidate**: After the action, the agent transacts the new experience back into Datalevin, updating graph relationships and any configured search, vector, or embedding indexes.

The loop is deliberately transactional at the boundaries. A turn should not write a new episode without the tool calls, approvals, and task state that explain it. Likewise, consolidation should preserve a link back to the source episode so later maintenance jobs can explain why a semantic fact exists. This is what makes memory useful for debugging and not only for retrieval.

> **Transact the turn boundary together**
>
> When a turn changes durable state, write the episode, tool calls, approvals, usage records, and task-state transition together where possible. Memory without its evidence is hard to debug and unsafe to replay.

In application code, this loop is also where policy enforcement belongs. The model can propose an action, but the system should check the operating envelope, tool permissions, budgets, and task state before executing it. The database is the shared environment that makes the loop resumable and inspectable.

A minimal turn boundary can reuse the helper functions introduced in Chapters 24 and 25. This is the persistence boundary for one observed turn, not the whole agent loop. It does not call a model or execute tools; it only shows the durable state transition around the observation:

```
(defn memory-turn!
  [conn {:keys [user-id session-id summary fact-kind fact-content confidence]}]
  (let [{episode-id :episode/id
         fact-id    :fact/id}
        (record-episode-and-candidate-fact!
          conn
```

```
          {:user-id user-id
           :session-id session-id
           :summary summary
           :fact-kind fact-kind
           :fact-content fact-content
           :confidence confidence})

        wm-id
        (materialize-working-memory!
          conn
          session-id
          [{:fact-id fact-id
            :relevance 1.0
            :reason "Newly observed fact from this turn."
            :pinned? false}])]
    {:episode/id episode-id
     :fact/id fact-id
     :wm/id wm-id}))

(memory-turn!
  conn
  {:user-id "reader-1"
   :session-id "docs-session-1"
   :summary "Reader searched for backup and restore and marked the result not
   ↪  helpful."
   :fact-kind :fact.kind/docs-gap
   :fact-content "Readers may need clearer backup and restore guidance."
   :confidence 0.62})
```

The application still decides when a candidate fact is promoted, rejected, or sent for review. Datalevin stores the records, links, and working-memory projection that make that decision auditable.

The rest of this chapter keeps that loop in one domain. First it defines the documentation-feedback scenario, then it zooms into the review-task and tool-audit records, and finally the capstone runs the whole loop end to end.

3. Main Scenario: Documentation Feedback

Imagine a documentation site that records search sessions. A reader searches for "backup and restore," opens the storage chapter, and marks the result as not helpful. One event is just feedback. Several similar events become evidence that the docs may have a gap.

- **Episodes** preserve the raw search-feedback events.
- **Semantic facts** summarize current documentation gaps, such as "readers are struggling to find backup and restore guidance."
- **Tasks** turn current gaps into reviewable work with explicit approval steps.
- **Working memory** keeps the current gap, evidence, and task state in a bounded context for the reviewer or agent.

The system needs stateful logic, but not hidden state. If ingestion stores a normalized topic for each search, a query can count topics with repeated unhelpful feedback:

```
;; Find query topics with enough unhelpful feedback to review.
(d/q '[:find ?topic (count ?episode)
       :where [?episode :episode/type :episode.type/search-feedback]
              [?episode :episode/topic ?topic]
              [?episode :episode/helpful? false]]
     db)
```

The query is not the whole application. The application still decides the threshold, how to group similar queries, whether a human should review the candidate, and whether any generated draft is allowed to publish. Datalevin's role is to keep the raw evidence, current facts, task state, and policy records in one queryable environment.

4. The Loop in the Scenario

The documentation-feedback assistant follows the same loop from Section 2:

1. **Perceive**: Record a search-feedback episode with the query, page, user, helpfulness flag, and timestamp.
2. **Resolve Scope and Envelope**: Decide which documentation workspace, budget, and review policy apply.
3. **Recall**: Retrieve related feedback episodes, existing docs-gap facts, and relevant pages.
4. **Refresh Working Memory**: Materialize a bounded packet containing the current candidate gap and its evidence.
5. **Reason**: Decide whether the candidate is strong enough to become a current fact and task.
6. **Act and Audit**: Open a review task, record tool authorization, and write a sanitized usage entry.
7. **Consolidate**: Promote or reject the gap, supersede stale current facts, and preserve evidence links.

The next two sections zoom in on the two control records that make the scenario operational: durable review tasks and tool-audit records. The capstone then puts the full loop together.

5. Pattern: Durable Tasks and Board Projections

Tasks are the control state of the memory loop. They let a stateful application pause, resume, delegate, and audit long-running work without relying on the model's chat history.

For long-running work, store the plan as data. A task contract should be stable and inspectable, while runtime state changes independently as the task runs.

The contract is what the system intends to do. Runtime state is what has happened so far. Keeping them separate makes resumption and review much easier: the agent can retry a failed step without rewriting the plan, and a human can inspect whether the task is still following the original objective. Stable step ids are important because later outputs, approvals, and boundary summaries need to point back to the same step even after display text changes.

> **Keep task contracts stable**
>
> Store the intended plan separately from runtime state. Stable step ids let approvals, outputs, retries, and handoff summaries refer to the same step even when labels or display text change.

```
{:task/id        #uuid "00000000-0000-0000-0000-000000000501"
 :task/type      :task.type/task
 :task/state     :task.state/ready
 :task/title     "Review documentation gap: backup and restore"
 :task/contract {:kind :task
                 :version 1
                 :goal "Decide how to fix the backup and restore gap"
                 :steps [{:id :inspect-feedback
                          :kind :tool
                          :tool :docs-feedback-query}
                         {:id :draft-change
                          :kind :llm
                          :depends-on :inspect-feedback}
                         {:id :approve
                          :kind :approval}
                         {:id :publish
                          :kind :tool
                          :tool :docs-publish
                          :depends-on :approve}]}
 :task/runtime  {:current-step :inspect-feedback
                 :attempts {}
                 :outputs {}}}
```

A Kanban board, scheduler, or branch-worker queue should be a projection over tasks, not a separate store. For example:

> **Boards and queues are projections**
>
> The task table is the source of truth. Kanban boards, schedulers, and worker queues should project task state rather than maintain independent status that can drift from the durable task records.

```
{:task/id            #uuid "00000000-0000-0000-0000-000000000502"
 :task/board         {:visible? true
                      :lane :board.lane/open
                      :priority :normal}
 :task/claim-token   "opaque-worker-token"
 :task/claimed-by    [:agent/id :agent/docs-reviewer]}
```

Workers can claim and update tasks transactionally. If a claimed task requires a token for later updates, enforce the token check in the same transaction that changes the task. That gives the application an optimistic ownership protocol without a separate lock table.

This keeps coordination boring in the right way. The board is a view of durable work, not another source of truth that can drift from the task table. A scheduler can create or wake tasks by changing task state. A worker can claim a task by adding claim fields. A UI can render lanes by querying task metadata. All of those views use the same facts.

Task boundaries should also be durable. When a task pauses, completes, or hands off to another agent, write a small boundary document:

```
{:task/id       #uuid "00000000-0000-0000-0000-000000000501"
 :task/boundary {:summary "Feedback evidence was inspected; approval is
  ↪  pending."
                 :resume-hint "Resume at :approve."
                 :next-step :approve
                 :open-questions ["Should this become a guide or a
                  ↪  troubleshooting note?"]}}
```

This makes continuation safer than relying on chat history alone.

Boundary documents are small on purpose. They are not a full transcript; they are the minimum state needed for another turn, worker, or human to continue without guessing. A good boundary states what was done, why the task stopped, what should happen next, and which questions remain open.

6. Pattern: Tool Authority, Audit, and Usage Ledgers

Autonomous systems should not rely on the model to remember which tools are allowed. Store tool authority and audit records as data.

The permission check should run before the tool handler sees sensitive arguments or credentials. In the documentation-feedback scenario, the detector may inspect aggregated search feedback, but the system still decides whether that tool is allowed for the workspace and task. The audit record then captures the decision and the actor without exposing secrets to the model.

> **Authorize before tool execution**
>
> Check the operating envelope before a tool handler receives sensitive arguments or credentials. The model can request a tool, but application policy decides whether the call is allowed and records the audit event.

```
{:tool.call/id       #uuid "00000000-0000-0000-0000-000000000503"
 :tool.call/task     [:task/id #uuid "00000000-0000-0000-0000-000000000501"]
 :tool.call/tool     :tool/docs-gap-detector
 :tool.call/decision :permission/allowed
 :tool.call/args     {:topic "backup and restore"
                      :evidence-count 3}}

{:audit.event/id      #uuid "00000000-0000-0000-0000-000000000504"
 :audit.event/type    :audit.type/tool-authorized
 :audit.event/task    [:task/id #uuid "00000000-0000-0000-0000-000000000501"]
 :audit.event/tool-call [:tool.call/id #uuid
  ↪  "00000000-0000-0000-0000-000000000503"]
 :audit.event/actor   :actor/background-job
 :audit.event/message "Policy allowed docs-gap-detector for backup and
  ↪  restore."}
```

The same principle applies to LLM budgets. Record sanitized usage entries by scope so goals, tasks, sessions, and schedules can share enforcement:

```
{:limit.usage/id                      #uuid "00000000-0000-0000-0000-000000000505"
 :limit.usage/scope                   :limit.scope/task
 :limit.usage/task                    [:task/id #uuid
 ↪  "00000000-0000-0000-0000-000000000501"]
 :limit.usage/provider                :provider/openai
 :limit.usage/model                   "doc-review-assistant"
 :limit.usage/prompt-tokens           1800
 :limit.usage/completion-tokens 420
 :limit.usage/cost-micros             890}
```

With these records in Datalevin, the inputs and evidence for policy enforcement become queryable state. The enforcement decision still belongs in application code, before tools run or sensitive payloads leave the boundary.

Budgets become more useful when they are scoped. A session-level ceiling protects interactive use, a goal-level ceiling protects long-running work, and a schedule-level ceiling prevents recurring jobs from quietly consuming unlimited tokens. Because the ledger is sanitized, it can be used for policy and reporting without storing prompt text or secret-bearing tool payloads.

> **Ledger usage without storing secrets**
>
> Record provider, model, token counts, cost, and scope for budget enforcement. Keep prompt text and secret-bearing tool payloads out of the usage ledger unless the application has a separate retention and privacy policy for them.

7. Capstone: A Documentation Feedback Memory Loop

The documentation-feedback scenario now comes together in one small, runnable program. It is not production code for a documentation site. Instead, it models the same application shape in miniature: readers search the docs, leave feedback, a background job detects a documentation gap, and the system creates a reviewable task with working memory. The memory loop is illustrated in Figure 27.1.

The important boundary remains the same as in Chapter 23. Datalevin stores the episodes, facts, task state, tool-call audit records, usage ledger, and working-memory projection. The application implements the extraction policy, thresholds, review workflow, and publication rules.

> **Datalevin stores the loop state**
>
> The capstone keeps episodes, current facts, review tasks, working memory, tool audit, and usage records in Datalevin. Thresholds, extraction, review, and publishing remain application policy.

The full capstone is written as one Clojure program so it can be read and run as a single listing. It uses the same record families shown above, with three scenario-specific adaptations: `:fact/subject` is a normalized topic string, `:fact/evidence` is a many-ref because several episodes

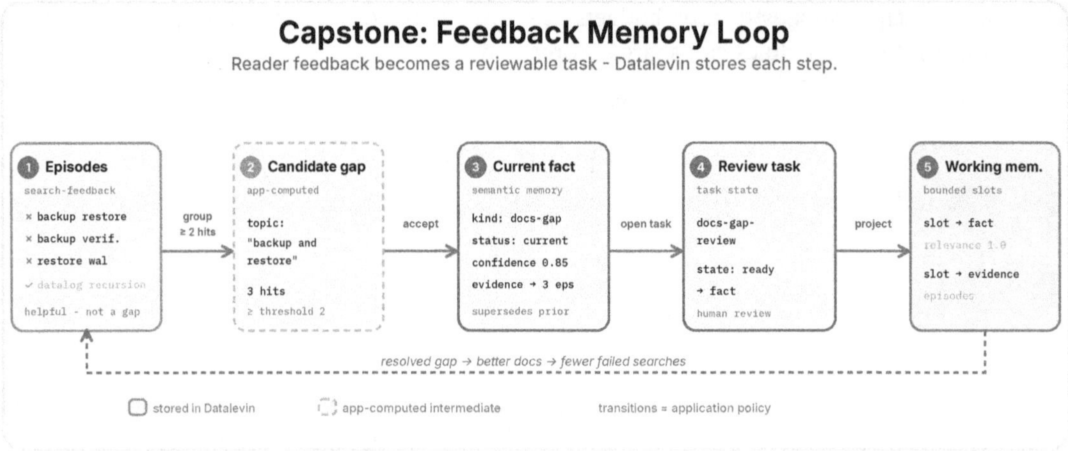

Figure 27.1: Capstone feedback memory loop: not-helpful search-feedback episodes are grouped into a candidate gap once they meet a threshold; the app promotes a current docs-gap fact (with evidence and supersession), opens a docs-gap-review task against it, and projects the fact and its evidence into working memory; resolving the task improves the docs and closes the loop with fewer failed searches. Boxes are stored in Datalevin; the dashed candidate-gap box is application-computed; transitions are application policy

support one gap, and working memory is task-scoped with `:wm/task` rather than session-scoped with `:wm/session`. In Java and Python, the serialized read-modify-write portions map to `withTransaction` / `with_transaction`; in JavaScript, put those serialized commands behind a Datalevin-hosting service or transaction function and call that boundary from the client.

Read the listing in phases:

1. Define the schema and seed a reader session.
2. Record feedback episodes from unsuccessful documentation searches.
3. Detect a candidate documentation gap from repeated feedback.
4. Promote the gap to a current fact, superseding older current facts for the same topic.
5. Create a review task, including tool authorization, audit, and usage records.
6. Materialize working memory for the task from the current fact and its evidence; phases 5 and 6 are written in one transaction.
7. Query the resulting semantic memory, recall context, task state, working memory, tool audit, and usage ledger.

The program below implements this loop end to end.

Listing 27.1. Documentation feedback memory loop capstone

```
(require '[clojure.string :as str])
(require '[datalevin.core :as d])

(def capstone-schema
  {:user/id          {:db/valueType :db.type/string
                      :db/unique    :db.unique/identity}
   :session/id       {:db/valueType :db.type/string
                      :db/unique    :db.unique/identity}
   :session/user     {:db/valueType :db.type/ref}
```

```
:episode/id        {:db/valueType :db.type/uuid
                    :db/unique    :db.unique/identity}
:episode/user      {:db/valueType :db.type/ref}
:episode/session   {:db/valueType :db.type/ref}
:episode/type      {:db/valueType :db.type/keyword}
:episode/query     {:db/valueType :db.type/string
                    :db/fulltext  true}
:episode/page      {:db/valueType :db.type/string}
:episode/helpful?  {:db/valueType :db.type/boolean}
:episode/summary   {:db/valueType :db.type/string
                    :db/fulltext  true}
:episode/timestamp {:db/valueType :db.type/instant}

:fact/id           {:db/valueType :db.type/uuid
                    :db/unique    :db.unique/identity}
:fact/kind         {:db/valueType :db.type/keyword}
:fact/subject      {:db/valueType :db.type/string}
:fact/content      {:db/valueType :db.type/string
                    :db/fulltext  true}
:fact/status       {:db/valueType :db.type/keyword}
:fact/confidence   {:db/valueType :db.type/double}
:fact/evidence     {:db/valueType :db.type/ref
                    :db/cardinality :db.cardinality/many}
:fact/supersedes   {:db/valueType :db.type/ref
                    :db/cardinality :db.cardinality/many}
:fact/created-at   {:db/valueType :db.type/instant}

:task/id           {:db/valueType :db.type/uuid
                    :db/unique    :db.unique/identity}
:task/type         {:db/valueType :db.type/keyword}
:task/state        {:db/valueType :db.type/keyword}
:task/title        {:db/valueType :db.type/string}
:task/fact         {:db/valueType :db.type/ref}
:task/contract     {:db/valueType :db.type/idoc}
:task/created-at   {:db/valueType :db.type/instant}

:tool.call/id      {:db/valueType :db.type/uuid
                    :db/unique    :db.unique/identity}
:tool.call/task    {:db/valueType :db.type/ref}
:tool.call/tool    {:db/valueType :db.type/keyword}
:tool.call/decision {:db/valueType :db.type/keyword}
:tool.call/args    {:db/valueType :db.type/idoc}

:audit.event/id    {:db/valueType :db.type/uuid
                    :db/unique    :db.unique/identity}
:audit.event/type {:db/valueType :db.type/keyword}
:audit.event/task {:db/valueType :db.type/ref}
:audit.event/tool-call {:db/valueType :db.type/ref}
:audit.event/actor {:db/valueType :db.type/keyword}
:audit.event/message {:db/valueType :db.type/string}
```

```
   :limit.usage/id    {:db/valueType :db.type/uuid
                       :db/unique    :db.unique/identity}
   :limit.usage/scope {:db/valueType :db.type/keyword}
   :limit.usage/task {:db/valueType :db.type/ref}
   :limit.usage/provider {:db/valueType :db.type/keyword}
   :limit.usage/model {:db/valueType :db.type/string}
   :limit.usage/prompt-tokens {:db/valueType :db.type/long}
   :limit.usage/completion-tokens {:db/valueType :db.type/long}
   :limit.usage/cost-micros {:db/valueType :db.type/long}

   :wm/id             {:db/valueType :db.type/uuid
                       :db/unique    :db.unique/identity}
   :wm/task           {:db/valueType :db.type/ref}
   :wm.slot/wm        {:db/valueType :db.type/ref}
   :wm.slot/entity    {:db/valueType :db.type/ref}
   :wm.slot/relevance {:db/valueType :db.type/double}
   :wm.slot/reason    {:db/valueType :db.type/string}})

(def conn
  (d/create-conn nil capstone-schema {:kv-opts {:inmemory? true}}))

(d/transact! conn
  [{:user/id "reader-1"}
   {:session/id "docs-session-1"
    :session/user [:user/id "reader-1"]}])

(defn record-feedback!
  [conn {:keys [user-id session-id query page helpful?]}]
  (let [episode-id (random-uuid)
        summary    (format "Search for '%s' opened %s and was marked %s."
                           query
                           page
                           (if helpful? "helpful" "not helpful"))]
    (d/transact! conn
      [{:episode/id        episode-id
        :episode/user      [:user/id user-id]
        :episode/session   [:session/id session-id]
        :episode/type      :episode.type/search-feedback
        :episode/query     query
        :episode/page      page
        :episode/helpful?  helpful?
        :episode/summary   summary
        :episode/timestamp (java.util.Date.)}])
    episode-id))

(def feedback-events
  [{:query "datalevin backup restore"
    :page "/docs/19-storage-tuning-durability"
    :helpful? false}
   {:query "backup verification"
    :page "/docs/22-production-operations"
    :helpful? false}
```

```
   {:query "restore backup wal"
    :page "/docs/19-storage-tuning-durability"
    :helpful? false}
   {:query "datalog recursion"
    :page "/docs/09-rules-recursion"
    :helpful? true}])

(doseq [event feedback-events]
  (record-feedback!
    conn
    (assoc event
           :user-id "reader-1"
           :session-id "docs-session-1")))

(defn failed-searches
  [db]
  (d/q '[:find ?episode-id ?query ?page
         :where [?e :episode/type :episode.type/search-feedback]
                [?e :episode/helpful? false]
                [?e :episode/id ?episode-id]
                [?e :episode/query ?query]
                [?e :episode/page ?page]]
       db))

(defn gap-topic
  [query]
  (let [q (str/lower-case query)]
    (cond
      (re-find #"backup|restore|wal" q) "backup and restore"
      :else q)))

(defn candidate-gaps
  [db]
  (->> (failed-searches db)
       (map (fn [[episode-id query page]]
              {:topic (gap-topic query)
               :episode-id episode-id
               :query query
               :page page}))
       (group-by :topic)
       (keep (fn [[topic rows]]
               (when (>= (count rows) 2)
                 {:topic topic
                  :count (count rows)
                  :evidence (mapv :episode-id rows)
                  :pages (vec (distinct (map :page rows)))})))
       vec))

(defn current-gap-fact-ids
  [db topic]
  (d/q '[:find [?fact-id ...]
         :in $ ?topic
```

```
        :where [?f :fact/kind :fact.kind/docs-gap]
               [?f :fact/subject ?topic]
               [?f :fact/status :fact.status/current]
               [?f :fact/id ?fact-id]]
      db topic))

(defn accept-gap-fact!
  [conn {:keys [topic count evidence pages]}]
  (let [fact-id    (random-uuid)
        confidence (min 0.95 (+ 0.55 (* 0.1 count)))
        content    (format "Readers are struggling to find documentation about
        ↪ %s."
                           topic)]
    (d/with-transaction [cn conn]
      (let [old-ids (current-gap-fact-ids (d/db cn) topic)
            retired (mapv (fn [old-id]
                            {:fact/id old-id
                             :fact/status :fact.status/superseded})
                          old-ids)
            current (cond-> {:fact/id fact-id
                             :fact/kind :fact.kind/docs-gap
                             :fact/subject topic
                             :fact/content content
                             :fact/status :fact.status/current
                             :fact/confidence confidence
                             :fact/evidence (mapv (fn [episode-id]
                                                    [:episode/id episode-id])
                                                  evidence)
                             :fact/created-at (java.util.Date.)}
                      (seq old-ids)
                      (assoc :fact/supersedes
                             (mapv (fn [old-id] [:fact/id old-id])
                                   old-ids)))]
        (d/transact! cn (conj retired current))))
    fact-id))

(defn recall-gap-context
  [db topic]
  (d/q '[:find ?query ?page ?summary
         :in $ ?topic
         :where [?f :fact/kind :fact.kind/docs-gap]
                [?f :fact/subject ?topic]
                [?f :fact/status :fact.status/current]
                [?f :fact/evidence ?episode]
                [?episode :episode/query ?query]
                [?episode :episode/page ?page]
                [?episode :episode/summary ?summary]]
       db topic))

(defn create-doc-review-task!
  [conn fact-id]
  (let [db          (d/db conn)
```

```
topic         (d/q '[:find ?topic .
                     :in $ ?fact-id
                     :where [?f :fact/id ?fact-id]
                            [?f :fact/subject ?topic]]
                   db fact-id)
evidence-ids (d/q '[:find [?episode-id ...]
                     :in $ ?fact-id
                     :where [?fact :fact/id ?fact-id]
                            [?fact :fact/evidence ?episode]
                            [?episode :episode/id ?episode-id]]
                   db fact-id)
task-id       (random-uuid)
wm-id         (random-uuid)
tool-call-id (random-uuid)
audit-id      (random-uuid)
usage-id      (random-uuid)
task          {:task/id task-id
               :task/type :task.type/docs-gap-review
               :task/state :task.state/ready
               :task/title (str "Review documentation gap: " topic)
               :task/fact [:fact/id fact-id]
               :task/created-at (java.util.Date.)
               :task/contract {:kind :docs-gap-review
                               :topic topic
                               :publish? false
                               :steps [{:id :inspect-evidence
                                         :kind :human-review}
                                        {:id :draft-change
                                         :kind :llm-assisted
                                         :depends-on :inspect-evidence}
                                        {:id :approve
                                         :kind :human-approval
                                         :depends-on :draft-change}]}}
wm            {:wm/id wm-id
               :wm/task [:task/id task-id]}
fact-slot     {:wm.slot/wm [:wm/id wm-id]
               :wm.slot/entity [:fact/id fact-id]
               :wm.slot/relevance 1.0
               :wm.slot/reason "Current documentation-gap fact."}
tool-call     {:tool.call/id tool-call-id
               :tool.call/task [:task/id task-id]
               :tool.call/tool :tool/docs-gap-detector
               :tool.call/decision :permission/allowed
               :tool.call/args {:topic topic
                                :evidence-count (count evidence-ids)}}
audit-event   {:audit.event/id audit-id
               :audit.event/type :audit.type/tool-authorized
               :audit.event/task [:task/id task-id]
               :audit.event/tool-call [:tool.call/id tool-call-id]
               :audit.event/actor :actor/background-job
               :audit.event/message
               (str "Policy allowed docs-gap-detector for " topic ".")}
```

```
        usage          {:limit.usage/id usage-id
                        :limit.usage/scope :limit.scope/task
                        :limit.usage/task [:task/id task-id]
                        :limit.usage/provider :provider/local
                        :limit.usage/model "docs-gap-detector"
                        :limit.usage/prompt-tokens 0
                        :limit.usage/completion-tokens 0
                        :limit.usage/cost-micros 0}
        evidence-slots
        (map-indexed
          (fn [idx episode-id]
            {:wm.slot/wm [:wm/id wm-id]
             :wm.slot/entity [:episode/id episode-id]
             :wm.slot/relevance (- 0.9 (* idx 0.05))
             :wm.slot/reason "Failed-search evidence for the review task."})
          evidence-ids)]
    (d/transact! conn
      (into [task wm fact-slot tool-call audit-event usage] evidence-slots))
    {:task/id task-id
     :wm/id wm-id
     :tool.call/id tool-call-id
     :audit.event/id audit-id
     :limit.usage/id usage-id}))

(def candidate (first (candidate-gaps (d/db conn))))
(def fact-id (accept-gap-fact! conn candidate))
(def task-ref (create-doc-review-task! conn fact-id))

(def results
  {;; Semantic memory: current documentation-gap fact.
   :semantic-memory
   (d/q '[:find ?status ?content ?confidence
          :where [?f :fact/kind :fact.kind/docs-gap]
                 [?f :fact/status ?status]
                 [?f :fact/content ?content]
                 [?f :fact/confidence ?confidence]]
        (d/db conn))

   ;; Recall: evidence that explains why the fact exists.
   :recall-context
   (recall-gap-context (d/db conn) "backup and restore")

   ;; Task state: reviewable work item, not an automatic publication.
   :task-state
   (d/q '[:find ?title ?state ?contract
          :where [?task :task/type :task.type/docs-gap-review]
                 [?task :task/title ?title]
                 [?task :task/state ?state]
                 [?task :task/contract ?contract]]
        (d/db conn))

   ;; Working memory: bounded context selected for this task.
```

```
   :working-memory
   (d/q '[:find ?reason ?summary
          :where [?wm :wm/id _]
                 [?slot :wm.slot/wm ?wm]
                 [?slot :wm.slot/reason ?reason]
                 [?slot :wm.slot/entity ?entity]
                 [?entity :episode/summary ?summary]]
        (d/db conn))

   ;; Tool authority and audit: the detector ran as an allowed internal tool.
   :tool-audit
   (d/q '[:find ?tool ?decision ?message
          :where [?task :task/type :task.type/docs-gap-review]
                 [?call :tool.call/task ?task]
                 [?call :tool.call/tool ?tool]
                 [?call :tool.call/decision ?decision]
                 [?event :audit.event/tool-call ?call]
                 [?event :audit.event/message ?message]]
        (d/db conn))

   ;; Usage ledger: sanitized accounting for the background detector.
   :usage-ledger
   (d/q '[:find ?model ?prompt-tokens ?completion-tokens ?cost-micros
          :where [?task :task/type :task.type/docs-gap-review]
                 [?usage :limit.usage/task ?task]
                 [?usage :limit.usage/model ?model]
                 [?usage :limit.usage/prompt-tokens ?prompt-tokens]
                 [?usage :limit.usage/completion-tokens ?completion-tokens]
                 [?usage :limit.usage/cost-micros ?cost-micros]]
        (d/db conn))})

(d/close conn)

results
```

The example covers the whole Part VI arc in one place:

- **Episodes** preserve the raw feedback events.
- **Consolidation** groups failed searches into a candidate documentation-gap fact.
- **Truth maintenance** marks any older current fact for the same topic as superseded before writing a new current fact.
- **Recall** follows the evidence links from the current fact back to the episodes that justify it.
- **Working memory** records a bounded context for the review task.
- **Task state** creates a contract that requires review and approval before publication.
- **Tool authority and ledgers** record that the detector ran as an allowed internal tool, with an audit event and sanitized usage entry.

Replacing the pure `gap-topic` function with an LLM extractor would not change the database model. It would only change the application policy that decides which candidate facts deserve review.

Summary: The Path to Stateful AI

Building stateful AI is about building systems that accumulate useful, inspectable state over time.

- **Memory architecture** (Chapter 23) provides the durable substrate and scope boundaries.
- **Episodic, semantic, and working memory** (Chapter 24) provide the records that history and knowledge live in.
- **Recall and context assembly** (Chapter 25) provide the model with a bounded, organized view of relevant state.
- **Apperception and truth maintenance** (Chapter 26) keep long-term knowledge coherent as new evidence arrives.
- **Tool authority and usage records** provide a controlled surface for AI actions.
- **Durable task contracts** keep long-running work resumable, auditable, and safe to hand off.

By choosing Datalevin, you are choosing a database substrate for stateful AI applications: exact structured queries, fuzzy semantic retrieval, durable task state, controlled tool use, and long-term knowledge maintenance. The agent policies and control loops remain application code built on top of those database capabilities.

References

[1] Stuart Russell and Peter Norvig, *Artificial Intelligence: A Modern Approach*, 4th US ed., Pearson, 2020. URL: https://aima.cs.berkeley.edu/.

Afterword: Where Datalevin Is Going

This guide began with a practical claim: SQL does not have to be the default center of application state. Application databases do not need to be organized around tables, SQL strings, and separate systems for every new access pattern. A database can store facts, expose them through Datalog, keep transactions small and explicit, and still support key-value access, relational joins, graph traversal, documents, text search, vectors, operations, and persistent memory for intelligent systems.

That claim is not only about features. It is about a shape of software. The same body of facts can be updated transactionally, joined relationally, traversed as graph edges, searched as text, connected to vectors, recalled as memory, and audited as evidence. Datalevin's core choice is to keep the unit of information small and explicit: an entity, an attribute, and a value. Everything else in the guide builds from that choice.

The principles are meant to remain stable even as the implementation grows. Datalevin is current-state oriented by default, because most operational systems need to read and write the present quickly. It makes history, provenance, and memory explicit when the domain needs them, instead of treating every past database state as the universal answer. It keeps storage simple, using EAV and AVE indexes for the common Datalog path, while letting specialized indexes act as additional capabilities. It is embedded-first, because locality and simplicity matter, but it also has a server mode for teams, tools, and multi-language clients.

The next directions follow from the same model.

One direction is **semiring provenance**. Chapter 15 described this as a more general answer than transaction history alone. The chapters on agent memory develop the practical need: confidence, utility, source episodes, supersession links, and explanation trails should not be informal comments around the data. They should be queryable facts. A semiring-provenance layer can make this systematic: joins combine supporting annotations, alternative derivations merge competing explanations, and derived facts can carry the evidence that produced them. That opens a path to confidence propagation, rule-application tracing, RAG source attribution, min-cost explanations, and incremental maintenance of derived views.

Another direction is **fuller distribution**. Datalevin already uses Raft for high-availability control-plane work. A natural next step is to extend the distribution story from leader coordination toward data placement and data-plane replication, so Datalevin can grow beyond an embedded database with server and HA modes. That is a larger engineering project, but the fact model is a good fit for it. Datoms are small, typed, and independently addressable. Transactions have clear write sets. Index maintenance is explicit. Those properties make distributed boundaries easier to reason about than they would be in a system where the primary unit is an opaque document or a table row with many unrelated columns.

Distribution also leads naturally to **entity-oriented sharding**. If facts are partitioned by entity id, stable entity identity, or another entity-aware placement key, each shard can run the same Datalog query plan over its local facts. For many query shapes, the coordinator's work is then to merge compatible result sets, apply limits, combine aggregates, or perform a small amount of cross-shard joining. Not every query becomes trivial, and cross-entity joins still need careful planning, but a fact-oriented model gives the system a clean unit of placement. The same logical query can be sent to many nodes without rewriting the user's application model around the physical layout.

A third direction is **richer Datalog for analytics**. This guide used Datalog mostly for operational queries, rules, recursion, and memory assembly. But the same language can express more analytical computation if the syntax grows in the right way. One important extension is aggregation in rule heads. That would allow derived relations to summarize data as they are defined, instead of forcing every aggregation to live only at the outer query boundary. With recursive rules, indexes, and controlled aggregation, Datalevin could support more analytical work inside the database, including graph analytics and selected iterative algorithms over facts.

These directions should not be read as promises about release dates or compatibility details. They are design directions. The important point is that they do not require abandoning the model developed in this guide. Semiring provenance adds annotations to facts and derivations. Distribution places facts and indexes across nodes. Analytical Datalog extends how derived facts are computed. In each case, the center remains the same: small facts, explicit transactions, Datalog query forms as data, and application policy grounded in queryable state.

That is the larger bet behind Datalevin. Replacing SQL's default role is not only about storing bytes durably. A database for modern applications should help a system remember what it knows, explain why it knows it, revise that knowledge when the world changes, and carry the same model from a local process to a distributed service. The work ahead is to make more of that power direct, reliable, and easy to use without losing the simplicity that makes the current system practical.

PART VII

Appendices

APPENDIX A

Installation and Runtime

This appendix collects setup material that would otherwise interrupt the main chapters. Use Chapter 2 for the first embedded session. Use this appendix when you need platform requirements, package coordinates, server startup commands, `dtlv`, Docker, Babashka, MCP, or troubleshooting notes.

1. Choose a Mode

Datalevin can be used as an embedded database, a standalone server, a server with read-only replicas, a high-availability server cluster, a command-line tool, a Babashka pod, or a local MCP tool server. The same core concepts appear in each mode: a path or URI names the database, schema describes attribute behavior, transactions add or retract facts, and Datalog queries read through a current database state.

Most readers should start with embedded mode. It has the fewest moving parts and uses the same transactions, schema, and Datalog queries as the other modes.

Mode	Start here when	What you need first
Embedded library	One application process owns the database path.	Java 21+ and the language package.
Standalone server	Multiple services or machines need a shared database endpoint.	`dtlv` or the standalone JVM jar.
`dtlv` CLI	You want a REPL, shell scripts, import/export, backup, or maintenance commands.	The `dtlv` executable.
Babashka pod	You write fast Clojure scripts with bb.	Babashka and either the pod release or local `dtlv`.
MCP server	An AI client or agent needs local database tools.	`dtlv` and an MCP-compliant client.
Replication or HA	You need read scaling or automatic failover.	Server mode, WAL, and Chapter 22.

2. Supported Platforms

For local embedded storage, the standalone JVM jar, and pre-built `dtlv` command-line/server binaries, Datalevin supports the following host platforms:

Operating system	Architecture
Windows	x86_64 / AMD64
macOS	ARM64 / AArch64

Operating system	Architecture
Linux	x86_64 / AMD64
Linux	ARM64 / AArch64

The language libraries bundle native Datalevin components for these platforms. If your application must run on a different local platform, prefer server mode with Datalevin running on a supported host, or build and package the native components for that platform yourself.

Core Datalog and KV support is not always the same as support for every optional native feature. Vector and embedding search depend on vector-index native code; they are supported on Linux x86_64, Linux ARM64, and macOS ARM64, while Windows support is experimental.

3. Runtime Requirements

All language libraries load a JVM Datalevin runtime and require Java 21 or newer. Some language bindings have their own host-runtime requirements:

- Python: Python 3.10+ and Java 21+
- Node.js: Node.js 20+ and Java 21+

For Clojure and Java applications that load Datalevin in-process, pass these JVM flags to the JVM:

```
--add-opens=java.base/java.nio=ALL-UNNAMED
--add-opens=java.base/sun.nio.ch=ALL-UNNAMED
--enable-native-access=ALL-UNNAMED
```

Python and Node.js bindings add the required JVM flags automatically. The `--enable-native-access=ALL-UNNAMED` flag suppresses native-access warnings on newer Java releases.

For Clojure CLI, put the flags in an alias:

```
{:aliases
 {:datalevin/jvm
  {:jvm-opts ["--add-opens=java.base/java.nio=ALL-UNNAMED"
              "--add-opens=java.base/sun.nio.ch=ALL-UNNAMED"
              "--enable-native-access=ALL-UNNAMED"]}}}
```

Then run your program or REPL with that alias:

```
clojure -M:datalevin/jvm
```

For Leiningen, add top-level `:jvm-opts` in `project.clj`:

```
:jvm-opts ["--add-opens=java.base/java.nio=ALL-UNNAMED"
           "--add-opens=java.base/sun.nio.ch=ALL-UNNAMED"
           "--enable-native-access=ALL-UNNAMED"]
```

For Java launch commands, pass the same options before `-cp` or `-jar`:

```
java --add-opens=java.base/java.nio=ALL-UNNAMED \
     --add-opens=java.base/sun.nio.ch=ALL-UNNAMED \
     --enable-native-access=ALL-UNNAMED \
     -cp target/classes:your-dependencies.jar your.main.Class
```

4. Native Dependencies

Datalevin bundles its non-JVM native libraries for the supported platforms.

If loading fails with `java.lang.UnsatisfiedLinkError`, first confirm that the system `libc` for your OS is present. Some Linux distributions may not include it by default.

If bundled OpenMP/vector math libraries do not work on your machine, install the platform packages directly. On Linux, Datalevin needs OpenMP and vectorized math support from GCC; on Debian or Ubuntu, `libgomp1` is usually enough, while `g++-12 gcc-12` is a useful fallback on systems with older or incomplete GCC runtime libraries:

```
# Debian/Ubuntu
sudo apt-get install libgomp1

# Debian/Ubuntu fallback
sudo apt-get install g++-12 gcc-12

# macOS
brew install libomp llvm
```

5. Embedded Language Packages

Use the language-specific documentation and package pages below for API details, current package metadata, and additional examples:

- **Clojure:** Clojure API documentation on cljdoc (https://cljdoc.org/d/datalevin/datalevin).
- **Java:** JavaDoc (https://javadoc.io/doc/org.datalevin/datalevin-java/latest/datalevin/package-summary.html), plus Java server JavaDoc (https://javadoc.io/doc/org.datalevin/datalevin-java-server/latest/) for the optional in-process server artifact.
- **Python:** PyPI package page for `datalevin` (https://pypi.org/project/datalevin/).
- **Node.js:** npm package page for `datalevin-node` (https://www.npmjs.com/package/datalevin-node).

The default Java artifact, `org.datalevin:datalevin-java`, is for embedded database use and `dtlv://` client connections. Java applications that need to host a Datalevin server in-process should add the separate `org.datalevin:datalevin-java-server` artifact. Python and Node.js packages are embedded/client libraries: they can open local databases and connect to `dtlv://` servers, but they do not host a Datalevin server process themselves. Run `dtlv`, Docker, the standalone JVM jar, or the Java server artifact when an application needs to provide the server process.

For the current cross-language API surface, use Datalevin's language compatibility matrix (https://github.com/datalevin/datalevin/blob/master/doc/language-compatibility.md) as the source of truth. This guide does not reproduce that matrix.

API parity is broad, but not perfectly identical. Clojure, Java, Python, and JavaScript all expose the main Datalog connection, query, pull, explain, transaction, async transaction, listener, simulated transaction, datom/index read, bulk load, Datalog-backed KV, re-index, full-text, vector, idoc, search writer, UDF, snapshot, WAL, remote client, replica, and HA administration surfaces.

The examples in the main chapters therefore use the same conceptual API across languages: transaction data goes in, queries and pull patterns come out, and operational APIs are available without dropping to a raw JSON/exec escape hatch.

For quick orientation, the main areas look like this:

Area	Cross-language status
Datalog basics	Clojure, Java, Python, and JavaScript all support open, query, pull, explain, transactions, async transactions, listeners, simulated transactions, datoms, index reads, bulk load, Datalog-backed KV, and re-index.
Host-language composition	Python and JavaScript provide pure, immutable `q` query forms, typed pull selectors, and `tx` transaction forms. Java provides `DatalogQuery`, `QueryClause`, and transaction builders; Clojure uses native data.
KV APIs	All four support ordinary KV operations, list DB operations, range helpers, raw-buffer scans, stats, copy, sync, environment flag updates, KV transaction callbacks, and explicit transaction timeouts.
Search/vector/idoc	All four support Datalog full-text, custom analyzers through UDFs, vector search, idoc schema/options, standalone search/vector indexes, search index writers, and search re-indexing.
Operations	All four support backup/copy, sync, snapshots, tx log watermarks and GC, tx log inspection, remote clients, replica status, and HA membership administration.
UDFs and data helpers	All four support UDF registries, query/predicate/transaction UDFs, analyzer UDFs, EDN helpers, and schema/transaction construction helpers where the host language needs them.

The remaining notable differences to keep in mind while reading are:

- Clojure, Java, and Python expose the Datalog transaction callback API (`with-transaction`, `withTransaction`, and `with_transaction`), including per-call timeout options. JavaScript does not expose that Datalog callback; use a single `conn.transact(...)`, `conn.transactAsync(...)`, KV transaction APIs, or move the callback logic into a transaction function or application command.
- Existing entity objects are transactable only in Clojure. Java, Python, and JavaScript support lazy entity reads, but staged entity-object mutation should be written as transaction maps, datom forms, or binding transaction builders.

5.1 Idiomatic Python and JavaScript Forms

Python and JavaScript applications should normally compose queries and transactions with the `q` and `tx` namespaces. These forms are synchronous, pure host-language values: building one does not start the JVM or contact a database. The binding lowers them only at an API boundary such as `query`, `pull`, `explain`, `transact`, or simulated transaction reporting.

The lookup-ref pull assumes that `:user/name` is declared unique in the schema. Use an entity id instead when the selected attribute is not unique.

Relation queries preserve Datalog's set semantics while crossing the language bridge as ordinary sequences: Python returns a list of row lists, and JavaScript returns an Array of row Arrays. Row order is unspecified unless the query has `order_by/orderBy`.

In typed forms, grammar positions are explicit. `q.var("name")` is the Datalog variable `?name`; `q.kw("user.status/active")` is a keyword value; and ordinary strings—including strings beginning with `?` or `:`—remain strings. Attribute arguments to `q.datom`, pull-selector helpers, and `tx` operations are converted to attribute keywords automatically. Schema maps still use colon-prefixed keys and values because they are the legacy native-map API.

Where the grammar does not prescribe a token type—such as runtime input maps, transaction metadata, and pull `:as` values—use `q.kw` explicitly when a key or value must be a Datalevin keyword.

The search relations have position-aware builders. Use `q.fulltext` with `q.fulltext_options` in Python (`q.fulltextOptions` in JavaScript), `q.vec_neighbors` with `q.vector_search_options` (`q.vecNeighbors` and `q.vectorSearchOptions`), and `q.idoc_match` with `q.idoc_match_options` (`q.idocMatch` and `q.idocMatchOptions`). These helpers type attributes and option keys and validate a statically known `:display` mode against the width of its result binding.

Python lists and JavaScript Arrays represent EDN vectors. When syntax requires an EDN list, use `q.edn_list(q.sym(">="), 30)` or `q.ednList(q.sym(">="), 30)`; use `q.quote(value)` for the common `(quote value)` form. Both are pure values and avoid parsing an EDN string just to preserve list syntax.

Python uses its standard `uuid.UUID` values. JavaScript provides a synchronous, pure `uuid("...")` helper, and its `tx.entity(attrs)` overload lets Datalevin allocate an entity id; the older `tx.entity(null, attrs)` form remains compatible. Python returns UUIDs as `uuid.UUID`; JavaScript returns canonical UUID strings, which can be passed through `uuid(...)` before reuse in a typed input. An anonymous in-memory connection may likewise receive schema and options through `connect(None, schema=..., opts=...)` or `connect(null, { schema, opts })`.

For portable UDFs, construct an immutable `UdfDescriptor` and reuse it with the registry and typed forms. Python provides `q.bind_udf`, `q.udf_predicate`, `tx.call_udf`, and `tx.install_udf`; JavaScript provides the corresponding `q.bindUdf`, `q.udfPredicate`, `tx.callUdf`, and `tx.installUdf` helpers. The older `udf_descriptor/udfDescriptor` functions retain their native map/object shape for compatibility code.

Registry conveniences such as `query_udf`, `predicate_udf`, and `tx_udf` in Python, and `queryUdf`, `predicateUdf`, and `txUdf` in JavaScript, default to the callback's host language. Use an explicit descriptor when the same value must also appear in a query, transaction, schema, or search configuration. Database-valued UDF arguments arrive as the binding's ordinary `Database` wrapper, so callbacks can use methods such as `entity_map/entityMap` directly.

Use normal Python or JavaScript control flow to assemble clause and transaction lists. The builders recursively snapshot recognized containers, so mutating an input list or object later cannot alter an existing form. `q.raw` is the typed escape hatch for syntax that has no dedicated helper; tokens inside it must use `q.kw` and `q.sym` explicitly. EDN strings and native list/array forms remain supported for static queries, interoperability, and chapters that display Datalog syntax directly.

This guide targets the Datalevin 1.1.0 release, the same target shown in Chapter 22. The dependency examples below use `1.1.0` where a package manager, Babashka pod, or standalone-jar filename requires a released version string.

Some examples below use the shortest install form, such as `pip install datalevin`, `npm install datalevin-node`, `brew install`, `scoop install`, or `docker pull huahaiy/datalevin`. Treat those as convenience commands. For production, follow Chapter 22's version-pinning rule: pin the Datalevin version for the application, server, CLI, Docker image, Babashka pod, and language binding. Use package-manager version syntax or lockfiles for Python and Node.js, a fixed image tag for Docker, and a downloaded release artifact, standalone jar, or otherwise pinned `dtlv` package for CLI/server deployments. Do not let production hosts float to whatever Homebrew, Scoop, or the `latest` Docker tag currently resolves to.

For Clojure, `datalevin/datalevin` is the full library. Use it when the same process needs server, CLI, or Babashka pod runtime code, including the `datalevin.server` namespace shown later in this appendix. Use `org.datalevin/datalevin-embedded` when an application only needs the local embedded APIs and `datalevin.client`.

```
;; Full library with Clojure CLI
{:deps {datalevin/datalevin {:mvn/version "1.1.0"}}}

;; Full library with Leiningen
:dependencies [[datalevin "1.1.0"]]

;; Embedded-only artifact with Clojure CLI
{:deps {org.datalevin/datalevin-embedded {:mvn/version "1.1.0"}}}

;; Embedded-only artifact with Leiningen
:dependencies [[org.datalevin/datalevin-embedded "1.1.0"]]
```

6. Install `dtlv`

Datalevin's native image is a command-line program called `dtlv`. It can query, transact, import, export, back up, compact, and administer Datalevin databases. The same executable can also run a server or an interactive Datalevin REPL.

Pre-built `dtlv` distributions are available for Windows AMD64, macOS ARM64, Linux AMD64, and Linux ARM64.

On macOS or Linux with Homebrew:

```
brew install huahaiy/brew/datalevin
```

On Windows with Scoop:

```
scoop bucket add scoop-clojure https://github.com/littleli/scoop-clojure
scoop bucket add extras
scoop install datalevin
```

Direct release downloads are available from GitHub. Download the archive for your platform, unzip it, and put the `dtlv` executable on your `PATH`:

- macOS ARM64: `https://github.com/datalevin/datalevin/releases/download/1.1.0/dtlv-1.1.0-macos-14-aarch64.zip`
- Linux AMD64: `https://github.com/datalevin/datalevin/releases/download/1.1.0/dtlv-1.1.0-ubuntu-22.04-amd64.zip`
- Linux ARM64: `https://github.com/datalevin/datalevin/releases/download/1.1.0/dtlv-1.1.0-ubuntu-24.04-arm-aarch64.zip`
- Windows AMD64: `https://github.com/datalevin/datalevin/releases/download/1.1.0/dtlv-1.1.0-windows-latest-amd64.zip`

With Docker:

```
docker pull huahaiy/datalevin
```

The Docker image exposes the same `dtlv` command surface. To start a local server from Docker:

```
docker run --rm \
  -p 8898:8898 \
  -e DATALEVIN_DEFAULT_PASSWORD=secret \
  -v "$PWD/dtlv-data:/data" \
  huahaiy/datalevin --host 0.0.0.0 -r /data -p 8898 serv
```

The standalone JVM jar is another downloadable command-line and server runtime. It is the better choice when you want HotSpot's JIT and garbage collectors for a long-running server, or when no native `dtlv` build exists for your platform. Download it from:

```
https://github.com/datalevin/datalevin/releases/download/1.1.0/datalevin-1.1.0-
↪ standalone.jar
```

Then run it with the JVM flags from Section 3:

```
java --add-opens=java.base/java.nio=ALL-UNNAMED \
     --add-opens=java.base/sun.nio.ch=ALL-UNNAMED \
     --enable-native-access=ALL-UNNAMED \
     -jar datalevin-1.1.0-standalone.jar
```

7. Standalone Server Mode

Server mode is the right choice when multiple clients or services need a shared database endpoint.

You can run a Datalevin server either as a JVM standalone jar or a GraalVM native image. For production use, the JVM standalone jar is recommended: HotSpot's JIT and mature garbage collectors are well suited to long-running server workloads and can outperform a native image over time.

For getting started, start a local server with `dtlv`:

```
DATALEVIN_DEFAULT_PASSWORD=secret dtlv -r /path/to/data -p 8898 serv
```

By default the server listens on `127.0.0.1:8898`. Use `--host` to bind another address. Binding to a non-loopback address such as `0.0.0.0` requires `DATALEVIN_DEFAULT_PASSWORD`; exposing the server beyond localhost must not rely on an implicit or empty administrative credential.

Server authentication starts with a built-in administrative user named `datalevin`. Its default password is `datalevin`; setting `DATALEVIN_DEFAULT_PASSWORD` when the server starts changes that password for the built-in account, and is required when binding the server to a non-loopback address. Treat `datalevin` as an administration/bootstrap account. For normal applications, create separate users and roles, then connect with those credentials. In `dtlv://` URIs, the username and password are URL components, so percent-encode special characters such as `@`, `:`, `/`, `?`, `#`, and `%`.

You can start the same server from the standalone JVM jar:

```
DATALEVIN_DEFAULT_PASSWORD=secret \
java --add-opens=java.base/java.nio=ALL-UNNAMED \
     --add-opens=java.base/sun.nio.ch=ALL-UNNAMED \
     --enable-native-access=ALL-UNNAMED \
     -jar datalevin-1.1.0-standalone.jar \
     --host 0.0.0.0 -r /data/dtlv serv
```

In Clojure applications that depend on the full Datalevin library rather than the embedded-only artifacts, server mode can also be started in the host process with `datalevin.server/create`, `datalevin.server/start`, and `datalevin.server/stop`.

Java applications can do the same with the opt-in `org.datalevin:datalevin-java-server` artifact, which provides `datalevin.DatalevinServer`.

```
(require '[datalevin.server :as srv])

(def server
  (srv/create {:host "127.0.0.1"
               :port 8898
               :root "/data/dtlv"}))

(srv/start server)
;; Later, during application shutdown:
(srv/stop server)
```

For production deployment, `dtlv`, Docker, or the standalone JVM jar remains a cleaner operational boundary when the server should live independently from the application process.

8. Connect to a Server

Remote databases use `dtlv://` URIs:

```
(require '[datalevin.core :as d])

(def conn
  (d/get-conn "dtlv://datalevin:secret@localhost:8898/mydb"))
```

The URI shape is:

```
dtlv://<user>:<pass>@<host>:<port>/<db-name>?store=<datalog-or-kv>
```

The examples above use `dtlv://datalevin:secret@...` because the server commands in this appendix set `DATALEVIN_DEFAULT_PASSWORD=secret`. If you omit that variable for a localhost-only development server, the built-in account's default password is `datalevin`.

The `store` parameter defaults to `datalog`; use `store=kv` for the direct key-value store. Database names must be unique within a server.

Once connected to a server, the same transaction and query APIs used in embedded mode can be used with the remote connection.

9. Smoke Test the Server

A quick smoke test with `dtlv exec` is:

```
dtlv exec <<'EOF'
(require '[datalevin.core :as d])
(def conn (d/get-conn "dtlv://datalevin:secret@localhost:8898/getting-started"))
(d/transact! conn [{:user/name "Server Alice"}])
(d/q '[:find [?name ...] :where [_ :user/name ?name]] (d/db conn))
(d/close conn)
EOF
```

The query should return `["Server Alice"]`, possibly with set/vector rendering depending on the client surface. In Clojure, prefer `(d/db conn)` when a read API needs the current Datalog DB object.

10. Replication and High Availability

Datalevin has two distinct server-side replication modes.

For simple read scaling without automatic failover, configure a non-HA async read-only replica with `:replica/read-only? true` and `:replica/source`. The primary database must have WAL enabled. The replica bootstraps from the primary copy interface, tails durable WAL records, serves normal reads, and rejects user writes.

For failover, use consensus-lease HA. Think of it as one writable server at a time, with followers keeping copies and serving reads. If the writer fails, a follower can be promoted only after coordination confirms that it is safe to accept writes. HA databases require WAL, use the `:strict` WAL durability profile by default, and treat membership changes as explicit operator actions through `datalevin.client/ha-update-membership!`.

Chapter 22 covers the details: server behavior, client tuning, async read replicas, the Raft-backed control plane, bounded leases, replica lag checks, fencing hooks that stop stale leaders from accepting writes, and HA operations.

11. Command Line and Scripts

`dtlv` can be used in shell scripts directly. It supports common patterns such as pipes and heredocs for standard-I/O transactions and queries. The command line tool uses the same Clojure-oriented Datalevin API as embedded mode.

For a copy-paste local CLI example:

```
dtlv exec <<'EOF'
(require '[datalevin.core :as d])
(def conn (d/get-conn "/tmp/dtlv-cli"))
(d/transact! conn [{:msg "Hello from dtlv"}])
(d/q '[:find [?m ...] :where [_ :msg ?m]] (d/db conn))
(d/close conn)
EOF
```

This creates a local database under `/tmp/dtlv-cli`, writes one fact, queries it, and closes the connection.

12. Babashka Pod

For scripting, Babashka (`bb`) is a fast-starting Clojure runtime. When a database becomes necessary in a Babashka script, Datalevin can serve as a pod:

```
;; save as script.clj
(require '[babashka.pods :as pods])
(pods/load-pod 'huahaiy/datalevin "1.1.0")
(require '[pod.huahaiy.datalevin :as d])

(let [conn (d/get-conn "/tmp/bb-db")]
  (d/transact! conn [{:msg "Hello from Babashka"}])
  (println (d/q '[:find ?m :where [_ :msg ?m]] (d/db conn))))
```

Run it with:

```
bb script.clj
```

If the `dtlv` executable is already on your `PATH`, Babashka can load that local binary directly:

```
(require '[babashka.pods :as pods])
(pods/load-pod "dtlv")
(require '[pod.huahaiy.datalevin :as d])
```

Pod mode is intended to have parity with embedded Clojure for the public Datalevin APIs used in this guide. When testing examples, keep the pod and embedded library on the same Datalevin version; using `pods/load-pod "dtlv"` is a good way to test the exact local binary you downloaded. If a snippet works in embedded Clojure but fails in the pod at the same version, treat that as a bug or packaging issue rather than a semantic difference. Datalevin does not store `nil` values; pod mode should report the same validation failure as embedded mode, not crash.

The pod also provides `defpodfn`, which lets a script define a custom function and call it from a query:

```
(d/defpodfn greeting [name] (str "hello " name))

(d/q '[:find ?message
       :where [(greeting "world") ?message]])
;; => #{["hello world"]}
```

13. MCP Server

For local AI agent applications and MCP-compliant clients, `dtlv mcp` runs a Datalevin MCP server over `stdio`:

```
dtlv mcp
dtlv --allow-writes mcp
```

Read-only mode is the default. Write tools must be enabled explicitly. The MCP process can open local databases directly or use `dtlv://` URIs behind the local `stdio` process.

13.1 Codex

Codex can launch MCP servers from `~/.codex/config.toml`. To make the Datalevin tools available in read-only mode, add a server definition:

Restart Codex after changing the config. You can then ask Codex to inspect a local Datalevin database path or a remote `dtlv://` database through the MCP tools:

```
Use the Datalevin MCP server to open /path/to/mydb and list the schema
↪ attributes.
```

Keep the default server read-only for exploration. If you want Codex to run transactions, configure a separate, explicitly named writable server:

13.2 Claude Desktop and Claude Code

Claude Desktop uses a JSON `mcpServers` configuration. In the Developer settings, edit the Claude Desktop config file and add a server:

Claude Code can add the same local stdio server from the command line:

```
claude mcp add --transport stdio datalevin -- dtlv mcp
```

Restart the client, then ask it to inspect a local Datalevin database path or a remote `dtlv://` database URI.

14. Troubleshooting and Cleanup

If the code examples do not run, check these common issues before moving deeper into the guide:

- Run `java -version` and confirm Java 21 or newer.
- Make sure the Datalevin dependency version exists in your package repository. Chapter 2 uses the release line named in its dependency snippets.
- For Clojure and Java, make sure the JVM flags in Section 3 are actually passed to the process that starts Datalevin.
- Python and Node.js packages normally add the required JVM options for you.
- If you see `java.lang.UnsatisfiedLinkError`, revisit the native dependency notes in Sections 2 and 4.
- Do not repeatedly open the same local database path in the same process; use the connection lifecycle pattern from Chapter 22.

- If examples use `/tmp/datalevin-guide-getting-started`, `/tmp/mydb`, `/tmp/dtlv-cli`, `/tmp/bb-db`, or `$PWD/dtlv-data`, remove those directories before rerunning from a clean slate.
- For server examples, confirm the server process is still running, the port is `8898`, and the password in the URI matches `DATALEVIN_DEFAULT_PASSWORD`.
- Keep MCP servers read-only until you deliberately need write tools.

Summary

Embedded mode is the easiest starting point, but the same database model carries across the server, CLI, Babashka, Docker, and MCP surfaces. Chapter 2 gives the minimal embedded path; Chapter 22 explains production lifecycle and operations.

APPENDIX B

EDN Format

EDN, short for Extensible Data Notation, is the data notation used throughout this guide. Datalevin is written in Clojure, so its native examples use EDN directly: schemas are maps, attributes are keywords, transactions are maps or vectors, Datalog queries are vectors, rules are vectors of lists and vectors, and many examples use tagged literals such as `#inst` and `#uuid`.

Readers coming from JSON, SQL, Java, Python, or JavaScript do not need to learn Clojure as a programming language in order to read the examples. EDN is data, not code. It is closer to JSON than to a programming syntax, but with a more regular syntax, richer scalar types, and symbolic names. This appendix gives the EDN subset and Datalevin conventions needed to read and write the examples in the rest of this guide. The official EDN specification is maintained by the edn-format project [1].

1. EDN at a Glance

An EDN value is one readable value. It may be a scalar:

```
nil
true
false
42
3.14
"hello"
:user/email
?name
#inst "2026-06-09T12:00:00Z"
#uuid "00000000-0000-0000-0000-000000000001"
```

Or it may be a collection:

```
[:find ?name :where [?e :user/name ?name]]

{:user/email "ada@example.com"
 :user/name  "Ada"}

#{:admin :editor}

[(parent ?child ?parent)
 [?child :person/parent ?parent]]
```

Commas are optional whitespace in EDN. The following maps are equivalent:

```
{:user/name "Ada" :user/age 37}

{:user/name "Ada", :user/age 37}
```

Datalevin examples usually omit commas because aligned whitespace makes larger schemas and transactions easier to scan.

2. Common Literal Forms

EDN form	Meaning	Datalevin use
`nil`	Absence of a value	Query results, optional arguments, API options.
`true`, `false`	Booleans	Schema flags such as `:db/fulltext true`.
`42`, `-7`	Integers	Entity ids, counters, ranks, limits.
`3.14`	Floating-point number	Scores, measurements, vector coordinates.
`"Ada"`	String	Text fields and full-text content.
`:user/email`	Keyword	Attribute names, enum values, options.
`?name`	Symbol	Datalog variables.
`[a b c]`	Vector	Queries, transaction datoms, lookup refs, paths.
`(f x y)`	List	Rule heads and predicate/function clauses in queries.
`{:a 1}`	Map	Schemas, entity maps, idoc documents, options.
`#{:a :b}`	Set	Unordered collections and set-valued data.
`#inst "..."`	Instant	Time values.
`#uuid "..."`	UUID	Stable external identifiers.

Datalevin-specific caveat: normal datom transactions do not store `nil` as an attribute value. In transaction maps and `[:db/add ...]` forms, omit an unknown or inapplicable attribute, or retract the existing datom when a value should disappear. `nil` is still fine for optional API arguments, missing query results, and ordinary EDN data. `idoc` documents are the exception: inside a `:db.type/idoc` value, `nil` is normalized to `:json/null`.

The table above shows common EDN literals, not the full Datalevin schema value type reference. Appendix C lists every `:db/valueType`. A few Datalevin value types deserve special attention when writing portable data:

Datalevin value type	EDN / portable representation
`:db.type/bigint`	Clojure EDN can carry arbitrary-precision integers, such as `42N`, but non-Clojure clients should use the binding's helper or native big-integer type where available.
`:db.type/bigdec`	Clojure EDN can carry arbitrary-precision decimals, such as `3.14M`, but portable clients should use binding helpers or native decimal types where available.
`:db.type/bytes`	EDN has no simple portable byte-array literal; use the host binding's bytes value.
`:db.type/tuple`	Use vectors as the value shape, with `:db/tupleType` or `:db/tupleTypes` in schema. Composite tuple attributes with `:db/tupleAttrs` are maintained from component attributes; application code does not transact those tuple values directly.
`:db.type/vec`	Use a numeric vector/array with the client helper expected by the binding; configure dimensions and metric in vector options.

Datalevin value type	EDN / portable representation
`:db.type/idoc`	Use a nested map value, or a string payload for configured JSON, Markdown, or EDN idoc formats.

EDN has no object syntax and no assignment syntax. A map is just a collection of key-value pairs. A vector is just an ordered collection. A list is just another ordered collection, though Datalevin often gives lists a special meaning inside query and rule forms.

3. Keywords

Keywords are symbolic values that begin with `:`. In Datalevin, they are used for attribute names, schema properties, enum-like values, index names, options, and many API flags:

```
:user/email
:db/valueType
:db.type/string
:order.status/paid
:eav
:closed
:string
:long
```

A keyword may be unqualified, such as `:eav`, or qualified with a slash, such as `:user/email`. The part before the slash is a namespace-like prefix. It does not have to name a Clojure namespace, file, table, or class. It is simply a naming convention that keeps related names together and avoids collisions.

Use qualified keywords for application attributes:

```
:account/email
:account/created-at
:invoice/total
:invoice/customer
```

Use unqualified keywords when the surrounding API defines them as local operators, modes, flags, or type descriptors. Datalevin's key-value API uses unqualified keywords for KV encoding types and range operators in transactions and queries:

```
[:find ?e :where [?e :user/email]]
[:closed "a" "m"]
[:put "users" "u1" {:name "Ada"} :string :data]
[:closed 10 20]
[:string :instant]
```

Here `:string`, `:data`, and `:instant` are KV type descriptors, while `:closed` is a range operator. These are intentionally separate from Datalog schema value types such as `:db.type/string`, which appear under `:db/valueType`.

Do not quote keywords as strings in Clojure examples. These are different values:

```
:user/email     ;; keyword
"user/email"    ;; string
```

Datalevin's JSON and non-Clojure APIs sometimes encode keywords as strings at the language boundary, but the logical value is still a keyword.

4. Symbols and Datalog Variables

Symbols are names without a leading colon:

```
?e
?name
parent
fulltext
```

In Datalog queries, symbols beginning with ? are variables. Datalevin binds them by matching clauses:

```
[:find ?name
 :where [?e :user/email "ada@example.com"]
        [?e :user/name ?name]]
```

The same symbol means the same logical variable within a query. In the example above, both clauses use ?e, so they must match the same entity.

Symbols without ? often name rules or functions:

```
[(active? ?e)]
[(> ?age 18)]
[(fulltext $ "error budget") [[?e ?a ?v]]]
```

In Clojure source code, query forms are commonly quoted:

```
(d/q '[:find ?name
       :where [?e :user/name ?name]]
     db)
```

The leading quote tells Clojure to pass the vector as data instead of trying to evaluate symbols such as ?name. In Java, Python, JavaScript, and other client languages, the same query is commonly passed as an EDN string or constructed with an interop helper.

5. Vectors, Lists, Maps, and Sets

Datalevin uses EDN collection forms consistently.

5.1 Vectors

Vectors preserve order and are written with square brackets. Datalevin uses vectors for Datalog queries, datom transactions, lookup refs, idoc paths, tuple values, and range specifications:

```
[:find ?e :where [?e :user/email]]
```

```
[:db/add 42 :user/name "Ada"]
```

```
[:user/email "ada@example.com"]
```

```
[:profile :address :city]
```

```
["acct-42" #inst "2026-06-09T00:00:00Z"]
```

```
[:closed 10 20]
```

When a vector appears inside a Datalog query, position matters. In a data pattern such as `[?e :user/name ?name]`, the positions are entity, attribute, and value.

5.2 Lists

Lists are written with parentheses. In Datalevin query and rule syntax, lists usually represent rule heads, rule calls, or function/predicate calls:

```
(parent ?child ?parent)
```

```
[(parent ?child ?parent)
 [?child :person/parent ?parent]]
```

```
[(> ?age 18)]
```

Outside quoted data, a Clojure list means a function call. Inside quoted query or rule data, it is only EDN data until Datalevin interprets it.

5.3 Maps

Maps are written with curly braces and contain alternating keys and values. Datalevin uses maps for schema definitions, entity transactions, option maps, and idoc documents:

```
{:user/email {:db/valueType :db.type/string
              :db/unique    :db.unique/identity}
 :user/name  {:db/valueType :db.type/string}}
```

```
{:user/email "ada@example.com"
 :user/name  "Ada Lovelace"}
```

```
{:limit 10
 :offset 20}
```

Map order is not part of the meaning. Examples often align map values vertically to make related fields easier to compare.

5.4 Sets

Sets are written with `#{...}` and represent unordered unique values:

```
#{:admin :editor}
```

A plain `{...}` form is a map, not a set. Use `#{...}` when the value is a collection of unique elements without associated values.

6. Tagged Literals

EDN can attach a tag to the next value. Datalevin examples most often use the built-in `#inst` and `#uuid` tags.

Do not use custom tagged literals unless the reader or API path you are using has installed support for those tags. In particular, Datalevin reads EDN idoc string payloads with the standard EDN reader path; `#inst` and `#uuid` are the safe built-ins to rely on there.

`#inst` represents an instant in time:

```
#inst "2026-06-09T12:30:00Z"
```

`#uuid` represents a UUID:

```
#uuid "00000000-0000-0000-0000-000000000001"
```

These are not strings, even though their textual form contains quoted strings. The tag tells the EDN reader to produce a time or UUID value. This matters for typed schema attributes:

```
{:event/id          {:db/valueType :db.type/uuid
                     :db/unique    :db.unique/identity}
 :event/created-at {:db/valueType :db.type/instant}}
```

When using a non-Clojure client, prefer the client library's EDN helper or typed wrapper for keywords, UUIDs, instants, big integers, big decimals, and byte values. If you send raw JSON strings where the database expects typed values, Datalevin may store strings instead of UUID, instant, or keyword values.

7. EDN in Datalevin

The same EDN forms recur across Datalevin's API surface.

7.1 Schemas

A schema is a map from attribute keywords to property maps:

```
{:user/email {:db/valueType :db.type/string
              :db/unique    :db.unique/identity}
 :user/age   {:db/valueType :db.type/long}
 :user/tags  {:db/valueType   :db.type/keyword
              :db/cardinality :db.cardinality/many}}
```

The outer keys are application attributes. The inner keys are Datalevin schema properties. Most schema property values are also keywords.

7.2 Entity Map Transactions

Entity transactions are maps whose keys are attributes:

```
{:user/email "ada@example.com"
 :user/name  "Ada"
 :user/tags  #{:admin :researcher}}
```

With a unique identity attribute, the same map shape can insert a new entity or update the existing entity with that identity.

7.3 Datom Transactions

Datom transactions use vectors:

```
[:db/add 42 :user/name "Ada"]
[:db/retract 42 :user/old-email "ada@old.example"]
[:db/retractEntity 42]
```

The vector shape is compact and positional, so it is useful when you need to be explicit about the entity id and transaction operation.

7.4 Lookup Refs

A lookup ref is a two-element vector containing a unique attribute and a value:

```
[:user/email "ada@example.com"]
[:task/id #uuid "00000000-0000-0000-0000-000000000501"]
```

Lookup refs let transactions and queries refer to entities by domain identity instead of by numeric entity id.

7.5 Datalog Queries

A Datalog query is usually written as an EDN vector containing clauses and keywords such as `:find`, `:where`, `:in`, `:keys`, and `:limit`:

```
[:find ?e ?name
 :where [?e :user/email ?email]
        [?e :user/name ?name]
 :in $ ?email]
```

Datalevin also accepts the equivalent map form, which is convenient when code or configuration wants named query parts:

```
{:find  [?e ?name]
 :in    [$ ?email]
 :where [[?e :user/email ?email]
         [?e :user/name ?name]]
 :limit 10}
```

In Clojure source code, the same map is commonly written with quoted vectors so Clojure passes the query variables as data, while ordinary Clojure expressions such as a local `limit` value are evaluated before Datalevin receives the map.

```
{:find  '[?e ?name]
 :in    '[$ ?email]
 :where '[[?e :user/email ?email]
          [?e :user/name ?name]]
 :limit limit}
```

The query is data. Datalevin interprets that data as a logical query.

7.6 Rules

Rules are EDN data too. A rule set is a vector of rule definitions:

```
[[(active? ?e)
  [?e :user/status :user.status/active]]

 [(can-login? ?e)
  (active? ?e)
  [?e :user/email]]]
```

Because rule syntax contains lists and symbols, Clojure code usually quotes the rule set, while non-Clojure clients usually pass it through an EDN reader helper.

7.7 Indexed Documents

For `:db.type/idoc`, EDN can be the document format:

```
{:doc/id      "profile-1"
 :doc/content {:profile {:name "Ada"
                         :age  37
                         :tags [:math :computing]}}}
```

When `:db/idocFormat` is `:edn`, string payloads are read as EDN and must produce a map. Important document rules:

- The top-level idoc value must be a map.
- Keys may be keywords or strings, depending on the format and client language.
- Lists are not allowed; use vectors for arrays.
- Vectors are also the path collection shape used by idoc path operations.
- Sets are EDN values, but they are not array-like path collections; use vectors when the document shape should behave like an array.
- `nil` values are normalized to `:json/null`.
- The literal value `:json/null` is reserved and rejected on input.

Within those rules, EDN idocs can use keyword keys, string keys, vectors, numbers, strings, booleans, instants, UUIDs, sets, and nested maps.

7.8 Dumps and Configuration Files

Datalevin command-line dump/load workflows can use EDN files. EDN is useful for this because it preserves Datalevin-specific values such as keywords, UUIDs, instants, and sets without inventing JSON conventions for each type.

8. EDN and JSON

EDN and JSON are both textual data formats, but they make different trade-offs.

Feature	EDN	JSON
Map keys	Any EDN value, commonly keywords	Strings only
Symbolic names	Keywords and symbols	Usually strings

Feature	EDN	JSON
Sets	Built in with `#{...}`	Usually arrays by convention
Instants and UUIDs	Tagged literals	Usually strings by convention
Comments	Semicolon line comments	Not part of JSON
Commas	Optional whitespace	Required separators
Top-level value	Any single EDN value	Any JSON value

This is why this guide's Clojure examples can express Datalevin data directly:

```
{:task/id       #uuid "00000000-0000-0000-0000-000000000501"
 :task/status   :task.status/running
 :task/tags     #{:agent :memory}
 :task/started  #inst "2026-06-09T12:00:00Z"}
```

The equivalent JSON representation needs conventions for keyword, set, UUID, and instant values. Datalevin's non-Clojure clients provide helpers for those cases.

9. Reading and Printing EDN Safely

In Clojure, use `clojure.edn/read-string` to read EDN text:

```
(require '[clojure.edn :as edn])

(edn/read-string "{:user/name \"Ada\"}")
```

Do not use the general Clojure code reader for untrusted input. EDN is designed as a data reader; the general Clojure reader reads a larger language.

Use `pr-str` to print ordinary Clojure values in a readable form:

```
(pr-str {:user/name "Ada" :user/active? true})
;; => "{:user/name \"Ada\", :user/active? true}"
```

For Datalevin APIs, prefer passing native values when you are already in Clojure. Use EDN strings mainly at language boundaries, in configuration files, or when an API explicitly asks for an EDN form.

10. Common Mistakes

Do not write JSON object syntax when the example expects EDN:

```
;; EDN
{:user/email "ada@example.com"}

;; JSON, not EDN
{"user/email": "ada@example.com"}
```

Do not use strings when Datalevin expects keywords:

```
;; Keyword attribute
:user/email

;; String
"user/email"
```

Do not forget # when writing a set:

```
#{:read :write}  ;; set
{:read :write}   ;; map from :read to :write
```

Do not forget to quote Datalog queries in Clojure source code:

```
;; Correct: pass query data.
(d/q '[:find ?e :where [?e :user/email]] db)

;; Incorrect: Clojure tries to evaluate ?e and other symbols.
(d/q [:find ?e :where [?e :user/email]] db)
```

Do not use `null`; EDN uses `nil`.

For Datalevin attributes, do not transact `nil` as the value. Use omission for unknown values and retraction for values that should no longer be present. The exception is nested `idoc` content, where `nil` normalizes to `:json/null`.

Do not treat tagged literals as plain strings:

```
#uuid "00000000-0000-0000-0000-000000000001"  ;; UUID value
"00000000-0000-0000-0000-000000000001"        ;; string value
```

11. Quick Reference

Need	Use
Attribute name	`:user/email`
Enum value	`:order.status/paid`
Datalog variable	`?e, ?name, ?score`
Entity transaction	`{:user/email "ada@example.com"}`
Datom transaction	`[:db/add 42 :user/name "Ada"]`
Lookup ref	`[:user/email "ada@example.com"]`
Query	`[:find ?e :where [?e :user/email]]`
Rule set	`[[(rule ?x) [?x :attr/value]]]`
Set	`#{:a :b :c}`
Instant	`#inst "2026-06-09T12:00:00Z"`
UUID	`#uuid "00000000-0000-0000-0000-000000000001"`
Missing value	`nil`

EDN's main advantage in Datalevin is that the examples show the actual data structures the database consumes. Once the literal forms are familiar, schemas, transactions, queries, rules, and indexed documents all read as variations on the same small notation.

References

[1] Rich Hickey and contributors, "Extensible Data Notation," official edn-format specification. URL: https://github.com/edn-format/edn.

APPENDIX C

Datalog Schema Reference

This appendix is a compact reference for Datalevin Datalog schema maps. Schema maps are EDN data; projects using any Datalevin language binding can keep them in checked-in EDN files. For design guidance, see Chapter 5 and Chapter 11. For specialized index behavior, see the full-text, vector, embedding, and idoc chapters.

1. Schema Shape

A Datalevin Datalog schema is a map from attribute keywords to property maps:

```
(def schema
  {:user/email {:db/valueType :db.type/string
                :db/unique    :db.unique/identity
                :db/doc       "Primary account email."}
   :user/roles {:db/valueType   :db.type/ref
                :db/cardinality :db.cardinality/many}
   :post/body  {:db/valueType              :db.type/string
                :db/fulltext               true
                :db.fulltext/autoDomain true}})
```

Pass the schema when opening or creating a connection:

```
(def conn (d/get-conn "/data/app" schema))
```

Read the current effective schema with `d/schema`:

```
(d/schema conn)
```

The result includes application attributes, built-in attributes, and assigned internal attribute ids such as `:db/aid`. Use this for inspection, debugging, and migration checks. In Java examples, `Datalevin.schema()` is a builder for constructing schema input; it is not the same operation as reading the current schema from a database.

Attribute ids are encoded as 32-bit signed non-negative integers in Datalevin's Datalog indexes. This means one database can have at most about 2.1 billion distinct attribute ids, including built-in attributes. Normal application schemas will not approach this, but unbounded dynamic key spaces should be modeled as data rather than as newly minted attributes.

Use `update-schema` to change schema:

```
(d/update-schema conn schema-update)
(d/update-schema conn schema-update #{:old/attr})
(d/update-schema conn nil nil {:old/name :new/name})
```

2. Defaults and Built-Ins

- Schema is optional. Undefined attributes are created on write and store values as generic EDN data.
- `:db/cardinality` defaults to `:db.cardinality/one`.
- Attribute names must be keywords.
- Built-in schema includes `:db/ident`, which is unique and keyword-valued. It also includes internal/system attributes such as `:db/created-at`, `:db/updated-at`, `:db/fn`, and `:db/udf`.
- `:db/created-at` and `:db/updated-at` are maintained only when a database is opened with `{:auto-entity-time? true}`. They store epoch milliseconds as `:db.type/long` values and represent database mutation time, not domain event time.
- Use the store option `{:closed-schema? true}` to reject transactions that mention attributes not already defined in the schema.

3. Schema-Related Store Options

Some behavior that feels schema-related is controlled by connection or store options, not by entries inside the schema map:

Option	Default	Meaning
`:validate-data?`	`false`	When `true`, transaction values must already match declared `:db/valueType` runtime types. When `false`, Datalevin still uses declared types to coerce or canonicalize values where possible.
`:closed-schema?`	`false`	When `true`, transactions may mention only attributes already present in the schema. When `false`, Datalevin allows new attributes to be added as data is written.
`:auto-entity-time?`	`false`	When `true`, Datalevin maintains `:db/created-at` and `:db/updated-at` epoch-millisecond values for entities.

Pass these options when opening or creating a database:

```
(d/get-conn "/data/app"
            schema
            {:validate-data? true
             :closed-schema? true})
```

4. Attribute Properties

The public schema properties that Datalevin interprets are the following:

Property	Values	Meaning
`:db/valueType`	See value type table below	Encoded value type. Required for refs, range-friendly ordering, tuples, vectors, idocs, and typed validation.
`:db/cardinality`	`:db.cardinality/one`, `:db.cardinality/many`	One value or a set of values per entity. Defaults to one.
`:db/unique`	`:db.unique/value`, `:db.unique/identity`	Enforces uniqueness. Both modes support lookup refs; identity also enables upsert.
`:db/isComponent`	`true`, `false`	Owned child relationship. Must be used with `:db/valueType :db.type/ref`.
`:db/doc`	string	Human-readable attribute documentation.
`:db/fulltext`	`true`, `false`	Maintains a full-text secondary index. Values are converted with `str` before indexing.
`:db.fulltext/domains`	sequence of strings	Adds a full-text attribute to explicit search domains.
`:db.fulltext/autoDomain`	`true`, `false`	Adds an attribute-specific full-text domain named by the attribute without the leading colon.
`:db/embedding`	`true`, `false`	Maintains an embedding index for string values. Requires `:db/valueType :db.type/string`.
`:db.embedding/domains`	non-empty sequence of non-blank strings	Adds an embedding attribute to explicit embedding domains.
`:db.embedding/autoDomain`	`true`, `false`	Adds an attribute-specific embedding domain using vector-domain naming.
`:db.vec/domains`	sequence of strings	Adds a `:db.type/vec` attribute to shared vector domains in addition to its attribute domain.
`:db/idocFormat`	`:edn`, `:json`, `:markdown`	Format hint for `:db.type/idoc`; defaults to `:edn`.
`:db/domain`	string	Explicit idoc domain for a `:db.type/idoc` attribute. If omitted, Datalevin derives the idoc domain from the attribute name.
`:db.idoc/indexedPaths`	sequence of top-level keywords, top-level strings, or path vectors	Include-list of idoc path prefixes to index for a `:db.type/idoc` attribute. If absent, all paths are included.

Property	Values	Meaning
`:db.idoc/excludedPaths`	sequence of top-level keywords, top-level strings, or path vectors	Omit-list of idoc path prefixes for a `:db.type/idoc` attribute. Exclusions win over included paths.
`:db.attr/preds`	qualified symbol, predicate UDF descriptor map, registered UDF keyword id, or non-empty sequence of those	Additional checks for values written to this attribute. Each predicate receives the normalized attribute value and must return `true`.
`:db/tupleAttrs`	non-empty sequence of attribute keywords	Derived composite index entry maintained from other attributes.
`:db/tupleType`	single supported tuple element type	Homogeneous tuple element type.
`:db/tupleTypes`	sequence of more than one supported tuple element type	Heterogeneous tuple element types.

Use `:db/ensure` in transaction data for entity invariants over the would-be transaction result:

```
[:db/ensure 'my.app.payments/bank-account? entity]
```

The predicate may be a function, var, or qualified symbol. It receives the would-be database after applying the transaction followed by the resolved ensure arguments, so it can check required attributes, entity-shape rules, and cross-attribute invariants after all transaction data has been expanded to datoms. `:db/ensure` forms do not become datoms in `:tx-data`; they abort the transaction before commit if a predicate returns falsey or throws.

5. Value Types

Value type	Expected values	Notes
`:db.type/keyword`	Clojure keywords	Common for enums and small symbolic values.
`:db.type/symbol`	Clojure symbols	Symbolic values.
`:db.type/string`	strings	Required for `:db/embedding`; typical for full-text.
`:db.type/boolean`	booleans	`true` or `false`.
`:db.type/long`	64-bit integers	Use for integer range queries and timestamps stored as epoch values.
`:db.type/double`	double-precision numbers	Numeric range queries.
`:db.type/float`	single-precision numbers	Numeric range queries with float storage.
`:db.type/ref`	entity ids, idents, or lookup refs in transactions	Enables joins, reverse pull navigation, and component ownership.

Value type	Expected values	Notes
:db.type/bigint	arbitrary precision integers	Larger integer values.
:db.type/bigdec	arbitrary precision decimals	Decimal values.
:db.type/instant	java.util.Date / instants	Time values.
:db.type/uuid	UUID values	Stable external ids.
:db.type/bytes	byte arrays	Binary values.
:db.type/tuple	vectors	Stored homogeneous or heterogeneous tuple values. Composite indexes use :db/tupleAttrs and do not require this value type.
:db.type/vec	numeric vectors	Maintains vector similarity indexes. Configure dimensions and metric in store options.
:db.type/idoc	map values or supported document string payloads that normalize to maps	Stored as one datom and indexed by path.

There is no `:db.type/uri`; store URI values as strings.

If `:db/valueType` is omitted, Datalevin stores values as serialized EDN data, which is flexible but not as useful for ordered range access.

6. Identity and Uniqueness

`:db.unique/value` means no two entities may have the same value for that attribute. Duplicates will be rejected with a transaction error.

Any `:db/unique` attribute can be used in a lookup ref. `:db.unique/identity` also enables upsert:

```
{:user/email {:db/valueType :db.type/string
              :db/unique    :db.unique/identity}}

(d/pull db '[*] [:user/email "a@example.com"])

(d/transact! conn
  [{:user/email "a@example.com"
    :user/name  "Ada"}])
```

If an entity with that email already exists, the map transaction updates that entity instead of creating a duplicate.

`:db/ident` is the built-in identity attribute for globally named entities:

```
(d/transact! conn [{:db/ident :order.status/shipped}])
```

7. References and Components

Reference attributes must declare `:db/valueType :db.type/ref`:

```
{:order/customer {:db/valueType :db.type/ref}
 :order/items    {:db/valueType   :db.type/ref
                  :db/cardinality :db.cardinality/many
                  :db/isComponent true}}
```

Use `:db/isComponent true` only for owned children. Component entities are pulled recursively by wildcard pull and are retracted with the parent when using `:db/retractEntity`.

`:db/isComponent` does not enforce unique ownership. Datalevin will allow more than one parent to refer to the same component child, but that shape is not a well-formed component model. Parent retraction with `:db/retractEntity`, or attribute retraction with `:db.fn/retractAttribute`, treats the child as owned data and retracts it. Retracting the child also retracts declared reference datoms that point to it, so any other parent loses its reference. A value-specific `[:db/retract parent component-attr child]` removes only that one reference and does not cascade.

`:db/isComponent true` without `:db/valueType :db.type/ref` is invalid.

8. Tuples

Datalevin has two related tuple features that should not be confused:

- **Composite indexes** use `:db/tupleAttrs` to derive an index entry from existing attributes. Application code does not transact the composite tuple value directly; Datalevin maintains it from the component attributes in the normal Datalog indexes.
- **Stored tuple values** use `:db.type/tuple` with `:db/tupleType` or `:db/tupleTypes` when the tuple itself is application data and its elements usually do not need to stand alone as separate facts.

Composite index derived from other attributes:

```
{:order/customer {:db/valueType :db.type/ref}
 :order/date     {:db/valueType :db.type/instant}
 :order/customer+date
 {:db/tupleAttrs [:order/customer :order/date]}}
```

Rules for `:db/tupleAttrs`:

- It must be a non-empty sequential collection.
- The tuple attribute must be cardinality one.
- It cannot depend on another tuple attribute.
- It cannot depend on a cardinality-many attribute.
- It does not require `:db/unique`. Add uniqueness only when the component combination is a real identity or uniqueness constraint; a non-unique tuple can be used as a repeated-query access path.
- The derived tuple uses the same tuple value encoding as a transacted typed tuple, so tuple element limitations still apply.

Datalevin stores the derived tuple in the normal EAV and AVE indexes and updates it whenever a component changes. A query must reference the tuple attribute to use that composite path. Its read advantage comes from scanning the qualifying joint combinations instead of emitting a potentially much larger intermediate relation from separate component EAV datoms. It increases write work and index storage; Chapter 11, Section 1.2 gives the IC5 cardinalities and the selection criteria.

Stored homogeneous tuple value:

```
{:point/xy {:db/valueType :db.type/tuple
            :db/tupleType :db.type/double}}
```

Stored heterogeneous tuple value:

```
{:image/bounds {:db/valueType  :db.type/tuple
                :db/tupleTypes [:db.type/double
                                :db.type/double
                                :db.type/long
                                :db.type/long]}}
```

Rules for stored typed tuples:

- Stored tuple values are transacted as non-empty vectors. An empty vector is not a valid tuple value.
- `:db.type/tuple` must include `:db/tupleType` or `:db/tupleTypes` when the tuple value is stored directly.
- Tuple element declarations support these scalar Datalevin value types: `:db.type/string`, `:db.type/bigdec`, `:db.type/bigint`, `:db.type/boolean`, `:db.type/double`, `:db.type/float`, `:db.type/long`, `:db.type/ref`, `:db.type/instant`, `:db.type/uuid`, `:db.type/keyword`, and `:db.type/symbol`.
- `:db/tupleType` must be one supported tuple element type. It describes a homogeneous tuple: one or more elements, all of the same type.
- `:db/tupleTypes` must contain more than one supported tuple element type. It describes a heterogeneous tuple: the vector arity must match the declaration.
- Tuple elements cannot themselves be `:db.type/tuple`, `:db.type/vec`, or `:db.type/idoc`.
- Variable-length tuple elements, such as strings, must fit within Datalevin's tuple element encoding limit of 255 bytes. Keep stored tuples small; use ordinary attributes or document values when the structure is open-ended.

For example, this homogeneous tuple schema is valid:

But transacting an empty tuple fails:

Datalevin rejects this data with `Cannot store an empty homogeneous tuple for attribute :sf/emails`. If the entity has no tuple value, omit `:sf/emails` from the transaction rather than transacting an empty vector.

9. Full-Text Schema Keys

Use `:db/fulltext true` for attributes that should participate in full-text search:

```
{:post/body {:db/valueType              :db.type/string
             :db/fulltext               true
             :db.fulltext/domains       ["posts"]
             :db.fulltext/autoDomain true}}
```

If no explicit domain is given, full-text attributes participate in the default `"datalevin"` domain. `:db.fulltext/autoDomain true` also creates an attribute-specific domain named by the attribute without the leading colon: `:title` becomes `"title"` and `:post/body` becomes `"post/body"`. Unlike vector and embedding domains, full-text auto domains keep / in namespaced attribute names.

Store-level `:search-opts` and `:search-domains` configure search domain options such as `:indexing-mode :async`, `:index-position? true`, and `:include-text? true`. If a custom analyzer uses ngrams, prefer gram sizes of 3 characters or longer; shorter grams produce many low-selectivity tokens and can make the index much larger. Full-text indexes store document references by default. In Datalog, the document reference is the indexed datom itself; the index does not store a duplicate raw-text copy unless `:include-text? true` opts a domain into storing text for `:display :texts`, `:display :texts+offsets`, and re-indexing.

10. Vector and Embedding Schema Keys

Use `:db.type/vec` when your application supplies vectors:

```
{:item/embedding {:db/valueType   :db.type/vec
                  :db.vec/domains ["items"]}}
```

Every vector attribute has an attribute-specific domain derived from the attribute name with / replaced by _: `:embedding` becomes `"embedding"` and `:item/embedding` becomes `"item_embedding"`. `:db.vec/domains` adds shared domains in addition to this attribute domain. Configure dimensions, metric, and indexing mode in `:vector-opts` or `:vector-domains`, and keep every domain that receives the same vector values compatible.

Use `:db/embedding true` when Datalevin should embed string datoms:

```
{:doc/text {:db/valueType              :db.type/string
            :db/embedding              true
            :db.embedding/domains      ["docs"]
            :db.embedding/autoDomain true}}
```

Embedding rules:

- The attribute must have `:db/valueType :db.type/string`.
- `:db.embedding/domains` must be a non-empty sequence of non-blank strings.
- If no explicit domain is given, the attribute participates in the default `"datalevin"` embedding domain.
- If `:db.embedding/domains` is present, the attribute participates in those domains instead of the default unless `"datalevin"` is listed.
- `:db.embedding/autoDomain true` adds an attribute-specific embedding domain using the same slash-to-underscore naming rule as vector domains.
- Changing `:db/embedding`, `:db.embedding/domains`, or `:db.embedding/autoDomain` on a populated attribute is rejected; rebuild explicitly instead.

Store-level `:embedding-opts`, `:embedding-domains`, and `:embedding-providers` configure providers, dimensions, metric, and indexing mode.

11. Idoc Schema Keys

Use `:db.type/idoc` for nested document values that should be indexed by path:

```
{:doc/json {:db/valueType  :db.type/idoc
            :db/idocFormat :json
            :db/domain     "profiles"}
 :doc/edn  {:db/valueType :db.type/idoc}
 :doc/md   {:db/valueType  :db.type/idoc
            :db/idocFormat :markdown}}
```

Allowed `:db/idocFormat` values are `:edn`, `:json`, and `:markdown`. The default is `:edn`.

Idoc input must normalize to a map at the root. Map keys must be keywords or strings. Use vectors for arrays; lists are rejected. `nil` values normalize to `:json/null`, and the literal value `:json/null` is reserved and rejected on input.

`:db/domain` names the idoc path index used for that attribute. If it is omitted, Datalevin derives the domain from the attribute name without the leading colon, so `:doc/edn` becomes `"doc/edn"`. Several idoc attributes may share a domain intentionally, but keep formats coherent: if attributes with different `:db/idocFormat` values share a domain, Datalevin treats the domain as mixed-format.

Use `:db.idoc/indexedPaths` and `:db.idoc/excludedPaths` to control which path prefixes enter the idoc path index. Changing those keys on a populated attribute is rejected; rebuild explicitly instead. Chapter 14 covers path selectors and store-level idoc domain options.

Python and JavaScript expose convenience builders for the same maps: `idoc_attr` / `idocAttr` build `:db.type/idoc` schema attributes, and `idoc_domain` / `idocDomain` build idoc option maps for `:idoc-opts` or entries under `:idoc-domains`. Clojure can write the EDN maps directly; Java can use the schema builder methods shown in Chapter 14.

12. Schema Evolution

Use `d/update-schema` to add, change, delete, or rename schema attributes on an open connection:

```
(d/update-schema conn
  {:user/age {:db/valueType :db.type/long}})

(d/update-schema conn
  {:user/age {:db/doc "Age in whole years"}})

(d/update-schema conn
  {:user/age {:db/doc :db/retract}})

(d/update-schema conn
  {:new/name {:db/valueType :db.type/string}}
```

```
  #{:old/name})

(d/update-schema conn nil nil
  {:old/name :new/name})
```

The `schema-update` map and each nested property map are patches:

Operation	Effect
Omit an attribute or property	Retain its stored definition.
Supply a property value	Replace that property after validating the resulting complete definition.
Set a property to `:db/retract`	Remove that property.
Add `:db/valueType` to a populated untyped attribute	Atomically re-encode its existing EDN values, or fail without changing data.
Change a populated typed attribute's `:db/valueType`	Reject the change.
Change populated `:db/cardinality` from many to one	Reject the change.
Add `:db/unique`	Accept only when existing values satisfy the requested uniqueness.
Delete an attribute	Accept only when no datoms use it.
Rename an attribute	Keep its internal attribute id and make existing datoms readable under the new name.
Change secondary-index metadata	Plan any required re-index or rebuild; populated embedding and idoc path-index changes require an explicit rebuild.

Existing attributes retain their internal `:db/aid`, and supplied ids are ignored. The complete request is validated before writing, committed atomically, and safe to retry unchanged. Conflicting renames, rename chains or cycles, duplicate targets, and conflicting patch/delete or delete/rename operations are rejected before writing.

A schema supplied while opening an existing database uses the same patch semantics and persists its changes; it is not a read-only drift check. For production evolution, open from the stored schema, apply deliberate changes through `update-schema`, and verify with `d/schema`. Chapter 11, Section 6.6 explains declarative convergence and ordered migrations.

13. Datomic and DataScript Differences

Datalevin's Datalog model is familiar to Datomic and DataScript users, but the schema mechanics differ:

- Schema is a map of maps, not a vector of transaction maps.
- Schema is passed to `get-conn`, `create-conn`, `empty-db`, or `init-db`, and later changed with `update-schema`.
- Undefined attributes are allowed unless `:closed-schema? true` is set.
- Unspecified value types are stored as EDN data.
- AVE indexing is automatic for every attribute.
- Use `:db/ident` for named enum/system entities.

- Use `:db.attr/preds` for attribute-local value predicates, and use `:db/ensure` for explicit entity invariants over the would-be transaction result.

14. Minimal Production Template

```
(def schema
  {:user/id     {:db/valueType :db.type/uuid
                 :db/unique    :db.unique/identity
                 :db/doc       "Stable external user id."}
   :user/email {:db/valueType :db.type/string
                 :db/unique    :db.unique/identity
                 :db/doc       "Login and notification email."}
   :user/name   {:db/valueType :db.type/string}

   :post/id     {:db/valueType :db.type/uuid
                 :db/unique    :db.unique/identity}
   :post/author {:db/valueType :db.type/ref}
   :post/body   {:db/valueType           :db.type/string
                 :db/fulltext            true
                 :db.fulltext/autoDomain true}
   :post/tags   {:db/valueType   :db.type/keyword
                 :db/cardinality :db.cardinality/many}})

(def conn
  (d/get-conn "/data/app" schema
              {:closed-schema? true}))
```

APPENDIX D

Datalog Built-Ins

This appendix lists the functions and predicates that Datalevin resolves without namespace qualification inside Datalog query clauses. It also lists the built-in aggregate functions available in `:find` and `:having`, plus the special forms that appear in `:find` and `:where`.

Most of these functions mirror `clojure.core`. Datalevin-specific functions are called out separately because they interact with the database, full-text search, vector indexes, idoc indexes, or runtime UDFs.

1. Where Built-Ins Can Appear

Built-ins can be used in predicate clauses:

```
[(> ?age 18)]
[(like ?name "A%")]
```

They can also be used in binding clauses:

```
[(+ ?subtotal ?tax) ?total]
[(get-some $ ?e :profile/email :user/email) [?attr ?email]]
```

Aggregate functions appear in `:find` and `:having`:

```
[:find ?city (count ?e)
 :where [?e :user/city ?city]
 :having [(> (count ?e) 10)]]
```

Fully qualified Clojure functions can also be used when they are available to the runtime, but they are not part of this built-in table.

On the JVM, a symbol whose name starts with `.` is treated as a reflective instance-method call on its first argument:

```
[(.toUpperCase ?name) ?upper]
```

This is useful for local Clojure/JVM queries, but it is not a portable cross-language pattern. Prefer documented built-ins or runtime UDFs when the query will cross process or language boundaries.

Find-only forms appear in `:find`, not as ordinary `:where` clauses:

Form	Purpose
`(pull ?e pattern)`	Pull a shaped map for each entity found by the query.
`(pull $source ?e pattern)`	Pull from an explicit query source.
`(sum ?x)`, `(count ?x)`, etc.	Built-in aggregate calls.
`(+ (sum ?x) (sum ?y))`	Arithmetic expression over aggregate results.
`(aggregate ?f ?x)`	Custom aggregate form, where `?f` is an aggregate function input.

`:where` also has special clause forms. These are parsed as Datalog clauses, not resolved as ordinary functions:

Form	Purpose
`(or clause...)`	Match one of several alternative branches.
`(or-join [?var ...] branch...)`	Match alternatives with explicit branch output variables.
`(not clause...)`	Exclude rows for which the nested clauses match.
`(not-join [?var ...] clause...)`	Exclude rows with explicit variables connecting the nested clauses to the surrounding query.
`(and clause...)`	Group multiple clauses inside an `or` or `or-join` branch.

Do not confuse these clause forms with predicate calls such as `[(not ?flag)]` or `[(or ?a ?b)]`. Bracketed predicate clauses call functions; the forms above contain nested Datalog clauses. Chapter 8 covers their evaluation rules and examples.

2. Predicates and Comparisons

Function	Purpose
`=`	Equality predicate.
`==`	Numeric equality predicate.
`not=`	Inequality predicate.
`!=`	Alias for `not=`.
`<`	Datalevin typed less-than comparison.
`<=`	Datalevin typed less-than-or-equal comparison.
`>`	Datalevin typed greater-than comparison.
`>=`	Datalevin typed greater-than-or-equal comparison.
`compare`	Clojure-style comparison function.
`true?`	True only for the boolean value `true`.
`false?`	True only for the boolean value `false`.
`nil?`	True for `nil`.
`some?`	True for non-`nil` values.
`not`	Logical negation.
`and`	Logical conjunction over arguments.
`or`	Logical disjunction over arguments.
`complement`	Return a function that negates another predicate.
`identical?`	Reference identity predicate.
`zero?`	Numeric zero predicate.
`pos?`	Positive-number predicate.
`neg?`	Negative-number predicate.
`even?`	Even integer predicate.
`odd?`	Odd integer predicate.
`in`	Membership predicate, e.g. `[(in ?status [:open :ready])]`.
`not-in`	Negated membership predicate.
`like`	SQL `LIKE`-style string predicate using `%` and `_` wildcards; accepts `{:escape ?esc}` as an optional third argument.
`not-like`	Negated `like` predicate; also accepts `{:escape ?esc}`.
`re-find`	Regex search.

Function	Purpose
`re-matches`	Regex full-match test.

3. Numeric Functions

Function	Purpose
`+`	Addition.
`-`	Subtraction or numeric negation.
`*`	Multiplication.
`/`	Division.
`quot`	Integer quotient.
`rem`	Remainder.
`mod`	Modulus.
`inc`	Increment.
`dec`	Decrement.
`min`	Minimum value using Datalevin typed comparison.
`max`	Maximum value using Datalevin typed comparison.
`rand`	Random floating-point number.
`rand-int`	Random integer below a bound.

4. General Value and Collection Functions

Function	Purpose
`identity`	Return its argument unchanged.
`ground`	Return a literal or otherwise-bound argument unchanged; an unbound argument is invalid.
`quote`	Return its argument unchanged; useful for nested query literals in EDN strings.
`keyword`	Create a keyword.
`meta`	Return metadata.
`name`	Return the name part of a keyword or symbol.
`namespace`	Return the namespace part of a keyword or symbol.
`type`	Return the Java/Clojure runtime type.
`vector`	Construct a vector.
`list`	Construct a list.
`set`	Construct a set.
`hash-map`	Construct a hash map.
`array-map`	Construct an array map.
`count`	Count items in a collection or characters in a string.
`range`	Produce a numeric range.
`not-empty`	Return a collection if non-empty, otherwise `nil`.
`empty?`	True for empty collections.
`contains?`	True when a key/index is present.
`get`	Look up a value in a collection or map.
`apply`	Apply a built-in or fully qualified function to a sequence of arguments.

Function	Purpose
`tuple`	Construct one vector from its arguments; equivalent to `vector` in query context.
`untuple`	Return its argument unchanged so a tuple binding can destructure it.

Examples:

```
;; Bind each vowel from a literal collection.
(d/q '[:find ?vowel
       :where [(ground [:a :e :i :o :u]) [?vowel ...]]])

;; Construct one tuple, then bind its positions.
(d/q '[:find ?key ?sku2 ?region2
       :where [(ground "book") ?sku]
              [(ground "us") ?region]
              [(tuple ?sku ?region) ?key]
              [(untuple ?key) [?sku2 ?region2]]])
;; => #{[["book" "us"] "book" "us"]}

;; Build transaction data in a query.
(d/q '[:find [?tx-data ...]
       :where [(ground :db/add) ?op]
              [(vector ?op -1 :user/name "Alice") ?tx-data]])

;; Sum a tuple stored as a value.
(d/q '[:find ?e ?sum
       :where [?e :nums ?nums]
              [(apply + ?nums) ?sum]]
     db)
```

`ground` and `untuple` are identity functions at runtime: neither expands nor decomposes its argument. `ground` accepts a literal or a variable bound by another clause or query input; an unbound variable produces an insufficient- binding error. The binding shape on the right performs the expansion or destructuring. For example, `[?vowel ...]` expands a collection, while `[?sku2 ?region2]` binds positions from one tuple. `tuple` constructs the vector that represents one query-time tuple value. These helpers are distinct from schema tuple features such as `:db.type/tuple` and `:db/tupleAttrs`.

5. String and Regex Functions

Function	Purpose
`str`	Concatenate values as strings.
`pr-str`	Print values readably to a string.
`print-str`	Print values to a string using `print` semantics.
`println-str`	Print values with a trailing newline.
`prn-str`	Print values readably with a trailing newline.
`subs`	Return a substring.
`re-pattern`	Compile a regex pattern.
`re-find`	Find a regex match.

Function	Purpose
`re-matches`	Match an entire string against a regex.
`re-seq`	Return a sequence of regex matches.

6. Database-Aware Functions

These functions take a Datalevin DB object, usually $, or interact with Datalevin indexes and runtime facilities.

Function	Purpose
`get-else`	Return an entity attribute value or a non-`nil` default.
`get-some`	Return the first present attribute and value from a list of attributes.
`get-some-else`	Return the first present attribute and value, or `[nil fallback]`.
`missing?`	True when an entity has no value for an attribute.
`q`	Run a nested Datalog query; useful for factoring an aggregate or small result relation into a later join.
`fulltext`	Search full-text indexes and bind matching `[e a v]` tuples.
`vec-neighbors`	Search vector indexes with a vector query and bind matching tuples.
`embedding-neighbors`	Embed a text query and search embedding indexes.
`idoc-match`	Search indexed document attributes and bind matching tuples.
`idoc-get`	Extract a nested value from a bound idoc document by path.
`udf`	Invoke a registered runtime UDF descriptor in query context.

6.1 Attribute Fallbacks

```
;; Bind either the user's height or a default.
(d/q '[:find ?e ?height
       :where [?e :user/name]
              [(get-else $ ?e :user/height 0) ?height]]
     db)

;; Bind the first present contact attribute.
(d/q '[:find ?e ?attr ?value
       :where [?e :user/name]
              [(get-some $ ?e :user/email :user/phone) [?attr ?value]]]
     db)

;; Keep the entity when neither contact attribute is present.
(d/q '[:find ?e ?attr ?value
       :where [?e :user/name]
              [(get-some-else $ ?e "unavailable"
                               :user/email :user/phone)
               [?attr ?value]]]
     db)

;; Filter entities that do not have an attribute.
(d/q '[:find ?e
```

```
        :where [?e :user/name]
               [(missing? $ ?e :user/deleted-at)]]
     db)
```

`get-else` does not accept `nil` as its default value. Use a concrete sentinel or handle the absence with `missing?`.

`get-some` and `get-some-else` try the listed cardinality-one attributes in order. Both return `[attribute value]` for the first one present. When none is present, `get-some` returns no value, so its binding clause removes that input row. `get-some-else` instead returns `[nil fallback]`, preserving one result tuple for the input row. Its fallback appears before the attribute list and may itself be `nil`:

```
[(get-some-else $ ?e nil :user/email :user/phone) [?attr ?value]]
```

6.2 Nested Queries

```
(d/q '[:find ?e ?age
       :where [(q (quote [:find (min ?age)
                          :where [_ :user/age ?age]])
                  $) [[?age]]]
              [?e :user/age ?age]]
     db)
```

Nested `q` is available, but it can be expensive when executed per candidate tuple. It is useful when the query genuinely has separate stages, as above: the uncorrelated inner query computes the global minimum, and the outer query joins that aggregate result back to every matching entity. Inner variables are local to the nested query; its result binding is the explicit interface to the outer query. Prefer direct joins, `not-join`, rules, or aggregates when they express the same operation without a per-candidate subquery.

6.3 Full-Text Search

```
;; Search all full-text domains.
(d/q '[:find ?e ?a ?v
       :in $ ?query
       :where [(fulltext $ ?query) [[?e ?a ?v]]]]
     db "red fox")

;; Search one attribute-specific domain.
(d/q '[:find ?e ?a ?v
       :in $ ?query
       :where [(fulltext $ :article/body ?query) [[?e ?a ?v]]]]
     db "red fox")

;; Search explicit full-text domains.
(d/q '[:find ?e ?a ?v
       :in $ ?query
       :where [(fulltext $ ?query {:domains ["articles"]})
               [[?e ?a ?v]]]]
     db "red fox")
```

By default, `fulltext` returns tuples that destructure as `[e a v]`. With `{:display :refs+scores}`, bind `[e a v score]`; with `:texts`, bind `[e a v text]`; with `:offsets`, bind `[e a v offsets]`; and with `:texts+offsets`, bind `[e a v text offsets]`. Attribute-specific search requires `:db.fulltext/autoDomain true` on the attribute. Phrase search and offset display require the relevant search domain to be configured with `:index-position? true` before indexing. Text display requires `:include-text? true`; full-text indexes store document references by default, not a duplicate raw-text copy.

For a finite relation query, `fulltext` can be an optimizer-selected complete, resumable access path. A single-domain `:refs+scores` result also provides descending score order when the score is the leading `:order-by` term. The function's own `:top`, `:limit`, and `:offset` remain source-local; query-level bounds apply after the rest of the Datalog query.

6.4 Vector and Embedding Search

Vector and embedding search is always domain-scoped. Unlike `fulltext` and `idoc-match`, `vec-neighbors` and `embedding-neighbors` do not search all domains when no domain is supplied. Use an attribute-specific call or pass an options map with `:domains`.

```
;; Vector query against an attribute-specific vector domain.
(d/q '[:find ?e ?a ?v
       :in $ ?query-vector
       :where [(vec-neighbors $ :item/embedding ?query-vector)
               [[?e ?a ?v]]]]
     db query-vector)

;; Vector query against explicit vector domains.
(d/q '[:find ?e ?a ?v
       :in $ ?query-vector
       :where [(vec-neighbors $ ?query-vector {:domains ["items"]})
               [[?e ?a ?v]]]]
     db query-vector)

;; Text query against an attribute-specific embedding domain.
(d/q '[:find ?e ?a ?v
       :in $ ?text
       :where [(embedding-neighbors $ :article/body ?text)
               [[?e ?a ?v]]]]
     db "database internals")
```

With `{:display :refs+dists}`, vector and embedding searches can bind `[e a v distance]`. Attribute-specific embedding search requires `:db.embedding/autoDomain true` on the attribute. Explicit embedding-domain search uses the provider configured for each named domain.

For a finite relation query, either neighbor function can provide a complete, resumable access path over its existing approximate result. A single-domain `:refs+dists` result provides ascending distance order when distance is the leading `:order-by` term. The optimizer does not widen `:top` or turn the HNSW result into exact nearest-neighbor search.

6.5 Indexed Documents

```
;; Search idoc documents by query map.
(d/q '[:find ?e ?a ?doc
       :where [(idoc-match $ {:status "active"}) [[?e ?a ?doc]]]]
     db)

;; Extract a nested field from the matched document.
(d/q '[:find ?e ?email
       :where [(idoc-match $ :person/profile {:status "active"})
                [[?e ?a ?doc]]]
               [(idoc-get ?doc :contact :email) ?email]]
     db)

;; Search explicit idoc domains.
(d/q '[:find ?e ?a ?doc
       :where [(idoc-match $ {:status "active"} {:domains ["profiles"]})
                [[?e ?a ?doc]]]]
     db)
```

Attribute-specific `idoc-match` requires an attribute with `:db/valueType :db.type/idoc`. `idoc-get` accepts path segments as separate arguments or a single vector path.

An uncorrelated `idoc-match` can provide a complete, resumable access path. Finite unordered relation queries may consume it in batches until enough final tuples survive later joins and filters. Idoc matching advertises no ranked order, so a limited result without `:order-by` has no stable ordering guarantee.

6.6 Runtime UDFs

```
(d/q '[:find ?normalized
       :in $ ?email
       :where [(udf :normalize-email ?email) ?normalized]]
     db
     "ALICE@EXAMPLE.COM")
```

The descriptor must be installed in the database or available from the runtime UDF registry, and its kind must be usable as a query function or predicate. Clojure, Java, Python, and JavaScript use the same Datalog `udf` form; only the registry helper names differ by host language.

In server mode, query resolution is server-safe: remote queries can call Datalevin built-ins and descriptor-backed UDFs registered on the server, but cannot resolve arbitrary host-language functions supplied by the client. Use `udf` for custom query logic that must run in a server process.

7. Aggregate Functions

Aggregate functions are used in `:find`, aggregate expressions, and `:having`.

Function	Forms	Purpose
`sum`	`(sum ?x)`	Sum numeric values.
`avg`	`(avg ?x)`	Arithmetic mean.

Function	Forms	Purpose
`median`	`(median ?x)`	Median value.
`variance`	`(variance ?x)`	Population variance.
`stddev`	`(stddev ?x)`	Population standard deviation.
`count`	`(count ?x)`	Count values.
`count-distinct`	`(count-distinct ?x)`	Count distinct values.
`distinct`	`(distinct ?x)`	Return a set of distinct values.
`vec`	`(vec ?x)`	Return a vector of grouped values.
`min`	`(min ?x)`, `(min n ?x)`	Smallest value, or vector of the `n` smallest values.
`max`	`(max ?x)`, `(max n ?x)`	Largest value, or vector of the `n` largest values.
`rand`	`(rand ?x)`, `(rand n ?x)`	Random value, or vector of `n` random draws.
`sample`	`(sample n ?x)`	Vector of up to `n` sampled values.

The table above lists built-in aggregates that are available without resolving host functions. A query can also use custom aggregates when the runtime permits them. The custom aggregate form `(aggregate ?f ?x)` receives the collector function as an input. Host-resolved aggregate symbols beyond the built-ins above depend on resolver mode and are not portable to server-safe query execution.

Aggregate expressions can combine aggregate results:

```
(d/q '[:find ?name (+ (sum ?debits) (sum ?credits))
       :in [[?name ?debits ?credits]]]
     rows)
```

Aggregate expressions are evaluated after grouping. The supported expression operators are +, -, *, /, `mod`, `rem`, and `quot`; arguments may be aggregate calls, constants, or nested aggregate expressions.

Custom aggregate functions are also supported with the special aggregate form:

```
(d/q '[:find ?group (aggregate ?agg ?x)
       :in [[?group ?x]] ?agg]
     rows
     my-aggregate-fn)
```

8. Public Name Index

This index lists the public query names documented for application use. Internal resolver entries, such as implementation helper predicates with leading hyphens, are intentionally omitted.

Query functions and predicates:

`!=`, `*`, `+`, `-`, `/`, `<`, `<=`, `=`, `==`, `>`, `>=`, `and`, `apply`, `array-map`, `compare`, `complement`, `contains?`, `count`, `dec`, `embedding-neighbors`, `empty?`, `even?`, `false?`, `fulltext`, `get`, `get-else`, `get-some`, `get-some-else`, `ground`, `hash-map`, `identical?`, `identity`, `idoc-get`, `idoc-match`, `in`, `inc`, `keyword`, `like`, `list`, `max`, `meta`, `min`, `missing?`, `mod`, `name`, `namespace`, `neg?`, `nil?`, `not`, `not-empty`, `not-in`, `not-like`, `not=`, `odd?`, `or`, `pos?`, `pr-str`, `print-str`, `println-str`, `prn-str`, `q`, `quot`, `quote`, `rand`, `rand-int`, `range`, `re-find`, `re-matches`, `re-pattern`, `re-seq`, `rem`, `set`, `some?`, `str`, `subs`, `true?`, `tuple`, `type`, `udf`, `untuple`, `vec-neighbors`, `vector`, `zero?`.

Find-only forms:

`pull`, `aggregate`, arithmetic aggregate expressions using `+`, `-`, `*`, `/`, `mod`, `rem`, and `quot`.

JVM method-call forms:

Symbols beginning with `.`, such as `.toUpperCase`.

Aggregate functions:

`avg`, `count`, `count-distinct`, `distinct`, `max`, `median`, `min`, `rand`, `sample`, `stddev`, `sum`, `variance`, `vec`.

Other host-resolved aggregate functions are allowed only when the resolver mode permits host function resolution, or when supplied through an explicit custom aggregate mechanism.

APPENDIX E

Core API Helpers

This appendix is a lookup reference for public Datalevin helpers that sit around the main connection, transaction, query, pull, indexing, and provider APIs. For examples and rationale, use the narrative chapters referenced below. Host-language bindings follow the naming conventions and coverage notes in Appendix A.

1. Connection and DB Object Helpers

Function	Use
`conn?`	Returns true for an open Datalog connection.
`closed?`	Returns true when a connection is closed, nil, or points at a nil DB.
`db`	Returns the current Datalog DB object behind a connection. Call it again after writes.
`opts`	Returns the option map of the Datalog DB.
`schema`	Returns the current effective schema.
`empty-db`	Creates a raw Datalog DB object without wrapping it in an application connection.
`datalog-kv`	Returns the KV handle backing a Datalog connection or DB.
`datalog-index-cache-limit`	Gets or sets the Datalog index cache limit.
`close`	Closes a Datalog connection.
`close-db`	Closes a raw Datalog DB object.
`with-conn`	Clojure macro that opens or reuses a connection for a body and closes it afterward.
`conn-from-db`	Wraps an existing DB object in a mutable connection.
`conn-from-datoms`	Creates a connection from trusted datoms, with optional dir, schema, and opts.
`reset-conn!`	Replaces the contents of a connection with a supplied DB and returns a tx report.

Use `get-conn` or `create-conn` for normal application opening. `d/datalog-kv` returns a borrowed handle that is safe to retain and reuse for as long as its owning Datalog connection or raw DB remains open. This differs from the Datalog DB object returned by `d/db`, which should be obtained again for current reads after writes. Close the owning connection or raw DB, not the borrowed KV handle.

2. Pull Pattern Syntax Reference

Selector form	Meaning
`[:user/name :user/email]`	Pull named forward attributes.
`[*]`, `[:*]`, or `["*"]`	Pull all forward attributes. Datalevin includes `:db/id` if the pattern does not name it.
`[[:user/name :as :name]]`	Rename a pulled value in the returned map.
`[(:user/name :as :name)]`	List-form attribute expressions are accepted for compatibility. Prefer vector expressions in new code.
`[[:user/name :as :name :default "unknown"]]`	Combine attribute options in one expression.
`[[:user/friends :limit 5]]`	Limit a cardinality-many attribute. Without an explicit limit, Datalevin returns at most 1000 values for a cardinality-many attribute. `nil` means no limit.
`[[:user/active? :default true]]`	Return a default value when the attribute is missing.
`[[:user/name :xform clojure.string/upper-case]]`	Transform the pulled value before returning it. Prefer resolvable symbols across process or language boundaries.
`[{:order/customer [:user/name]}]`	Follow a forward reference and pull a nested shape.
`[{[:user/friends :limit 5] [:user/name]}]`	Use an attribute expression as a nested map key.
`[:order/_customer]`	Pull reverse references for entities that point to the current entity.
`[{:order/_customer [:order/id]}]`	Pull a nested shape through a reverse reference.
`[{:category/children ...}]`	Recursively pull through a reference until no new entities remain.
`[{:category/children "..."}]`	String recursion marker; equivalent to `...`.
`[{:category/children 2}]`	Recursively pull through a reference with a positive depth limit.
`[{:category/_children 2}]`	Recursively pull through a reverse reference with a positive depth limit.
`[(limit :user/friends 5)]`	Legacy limit form. Prefer `[[:user/friends :limit 5]]` in new code.
`[(default :user/nickname "unknown")]`	Legacy default form. Prefer `[[:user/nickname :default "unknown"]]` in new code.
`[(default (limit :user/friends 5) [])]`	Legacy option forms may be nested or wrapped around attribute expressions; keep this shape for compatibility code only.

Pull patterns are EDN data. `:limit` applies only to cardinality-many attributes. Recursive pull tracks visited entities and represents cycles with `{:db/id ...}` rather than expanding forever. For multi-source Datalog queries, qualify pull with the source that owns the entity ids. Chapter 7 gives the worked examples.

3. Transaction Observation and Simulation

Function	Use
`transact-async`	Submits a Datalog transaction and returns a future/promise immediately.
`transact`	Blocks for the same asynchronous transaction machinery and returns an already realized future.
`with-transaction`	Groups multiple Datalog writes, and optionally borrowed KV writes, into one explicit transaction; accepts `{:timeout-ms n}` in the binding vector.
`abort-transact`	Aborts the open explicit Datalog transaction from inside `with-transaction`; also available through host-language bindings.
`explicit-transaction-timeout`	Gets or sets the default timeout, in milliseconds, for explicit Datalog and KV transactions.
`set-explicit-transaction-timeout!`	Sets or clears the default explicit transaction timeout; pass `nil` to disable it.
`listen!`, `unlisten!`	Register or remove transaction-report callbacks on a connection.
`tx-data->simulated-report`	Builds a transaction report from a DB object and tx data without mutating the connection.
`resolve-tempid`	Helper for looking up tempids in a transaction report.
`update-schema`	Atomically applies schema property patches, whole-attribute deletions, or renames to an open connection. The operation is idempotent on an unchanged retry. Omitted properties remain; `:db/retract` removes a property.

`listen!` replaces an existing callback when the same key is registered again. A simulated report has the normal report shape, including `:new-attributes` when the tx data introduce attributes not already present in the schema. Chapter 6 covers transaction semantics and callbacks.

4. Datom, Entity, and Type Helpers

Function	Use
`db?`	Tests whether a value is a Datalog DB object and refreshes a stale DB cache when needed.
`datom`	Creates a raw datom for trusted low-level import APIs such as `init-db` and `fill-db`.
`datom?`	Tests whether a value is a Datalevin datom.
`datom-e`, `datom-a`, `datom-v`	Return the entity, attribute, or value component of a datom.
`entity`	Returns a lazy, map-like entity for an eid or lookup ref.
`entid`	Resolves a lookup ref to a numeric eid, returns nil when it does not exist, and returns numeric eids unchanged.
`entity-db`	Returns the DB object from which an entity was created.
`touch`	Eagerly realizes an entity's forward attributes, mainly for REPL inspection and debugging.
`add`, `retract`	Clojure transactable-entity helpers for staging additions or retractions on an entity before transacting it.

Application transactions should normally use maps or transaction datom vectors. Raw datoms are for trusted bulk-loading paths. Lazy entity reads are available across Clojure, Java, Python, and JavaScript; staged entity mutation is Clojure-only.

5. Datalog Index, Statistics, and Bulk Helpers

Function	Use
datoms	Reads datoms from a named index, currently :eav or :ave, using supplied leading key components.
search-datoms	Describes an (e, a, v) pattern with nil wildcards and lets Datalevin choose an efficient index.
count-datoms	Counts datoms matching an (e, a, v) wildcard pattern without realizing the datoms.
cardinality	Counts distinct current values for one attribute. This is a statistic, not the schema property :db/cardinality.
max-eid	Returns the current maximum entity id, useful when a bulk import allocates numeric eids.
seek-datoms, rseek-datoms	Starts a forward or reverse index cursor near a supplied key prefix.
index-range	Reads the AVE range for one attribute between lower and upper bounds.
fulltext-datoms	Runs a full-text query and returns matching datoms rather than shaped search results; accepts search options such as :top, :limit, and :offset.
analyze	Refreshes attribute statistics used by the query planner.
init-db	Creates a database quickly from trusted prepared datoms.
fill-db	Adds trusted prepared datoms to an existing DB object.
re-index	Dumps, clears, reloads, and rebuilds a Datalog store, KV store, search engine, or vector index with supplied options.

init-db and fill-db bypass ordinary transaction processing and the WAL transaction log; use them only with trusted prepared datoms. Chapter 15 covers direct index reads, Chapter 20 covers bulk loading, and Chapter 21 covers analyze.

re-index closes, dumps, clears, reloads, and rebuilds the supplied store or index. Call it only when no other threads or programs are accessing the same database or index. Pass :backup? true to create a temporary backup named dtlv-re-index-<timestamp> in the system temporary directory. Search-engine re-indexing requires the search engine to have stored source text with :include-text? true.

6. Secondary Index Job Helpers

Function	Use
secondary-index-status	Returns async secondary-index job status for a local Datalog connection.

Function	Use
`process-secondary-index-jobs!`	Processes pending jobs, with options such as `:max-jobs` and `:retry-failed?`.
`wait-for-secondary-index`	Waits until matching secondary-index jobs are complete, with options such as `:tx`, `:type`, `:domain`, `:timeout-ms`, `:poll-ms`, `:process?`, `:max-jobs`, and `:retry-failed?`.

These helpers apply to domains configured for async full-text, vector, or embedding indexing. Synchronous domains update the secondary index before the transaction returns. Binding coverage for these helpers is listed in the language compatibility matrix linked from Appendix A.

7. Standalone Search and Vector Helpers

7.1 Full-Text Search

Function	Use
`new-search-engine`	Creates a full-text search engine backed by an `open-kv` handle.
`add-doc`, `remove-doc`, `clear-docs`	Add, remove, or clear indexed documents.
`doc-indexed?`, `doc-count`	Inspect document presence and document count.
`search`	Runs a full-text query and returns refs, scores, offsets, or stored texts depending on `:display`.
`search-index-writer`, `write`, `commit`	Bulk-writer API for faster standalone search indexing. Call `commit` after all `write` calls.

7.2 Vector Search

Function	Use
`new-vector-index`	Creates an HNSW vector index backed by an `open-kv` handle.
`add-vec`, `remove-vec`, `search-vec`	Add, remove, and search vectors by semantic refs.
`vector-index-info`	Returns size, memory, capacity, dimensions, metric, quantization, domain, and checkpoint metadata.
`force-vec-checkpoint!`, `vector-checkpoint-state`	Persist or inspect the vector checkpoint stored in LMDB.
`close-vector-index`, `clear-vector-index`	Close index resources, or close and delete all vectors.

Use Datalog full-text, vector, embedding, or idoc attributes when search results should participate in entity modeling and transactions. Use standalone engines for direct KV-backed search and vector indexes. Chapters 16 and 17 own the tutorial coverage.

8. Direct Provider Helpers

8.1 Embedding Providers

Function	Use
`new-embedding-provider`	Creates a local or OpenAI-compatible embedding provider.
`embedding-metadata`	Returns stable metadata describing the embedding space.
`embedding-dimensions`	Returns the vector dimension count.
`embed-text`, `embed-texts`	Embed one string or a batch of strings.
`token-count`, `token-counts`	Count embedding input tokens.
`truncate-item`, `truncate-text`	Truncate input to fit a token limit.
`close-embedding-provider`	Releases provider resources. Safe to call more than once.

8.2 LLM Providers

Function	Use
`new-llm-provider`	Creates a local llama.cpp generation provider.
`llm-metadata`	Returns metadata for the configured LLM provider.
`llm-context-size`	Returns the model context size.
`generate-text`	Generates completion text for a prompt.
`summarize-text`	Summarizes an input string.
`llm-token-count`	Counts tokens for one LLM input string.
`close-llm-provider`	Releases provider resources. Safe to call more than once.

Provider helpers are direct model calls. Persisted facts, episodes, embeddings, summaries, and provenance are ordinary application data.

9. Small Utility Helpers

Function	Use
`tempid`	Datomic-compatibility helper for negative tempids; `:db.part/tx` resolves to `:db/current-tx`. Prefer explicit negative ids when possible.
`squuid`, `squuid-time-millis`	Semi-sequential UUID helpers. Use application identity attributes for durable domain identity.
`read-csv`	Reads CSV data from a string or reader into lazy vectors of strings.
`write-csv`	Writes CSV rows to a writer.
`hexify-string`, `unhexify-string`	Encode a string as hex text and decode it back.

10. Python and JavaScript Schema Builders

These binding-level helpers build ordinary schema and option maps. They are most useful where the host language needs a typed helper instead of writing EDN by hand. Chapter 14 shows them in context.

Helper	Use
`idoc_attr` / `idocAttr`	Python and JavaScript helpers for building a `:db.type/idoc` schema attribute map. They accept idoc-specific options such as `format`, `domain`, `indexed_paths` / `indexedPaths`, and `excluded_paths` / `excludedPaths`, plus common schema attribute keys.
`idoc_domain` / `idocDomain`	Python and JavaScript helpers for building an idoc domain/default option map for `:idoc-opts` or an entry under `:idoc-domains`. They accept `indexed_paths` / `indexedPaths`, `excluded_paths` / `excludedPaths`, and `extra`.

11. Other Public API Namespaces

Namespace	Main use
`datalevin.interpret`	Interpreted functions and code execution for CLI, pods, remote calls, transaction functions, and KV visitor functions.
`datalevin.udf`	Runtime registry and descriptor handling for descriptor-backed user-defined functions.
`datalevin.constants`	Selected stable defaults, documented tuning vars, and system constants shared by the implementation.
`datalevin.main`	Programmatic support for the `dtlv` command-line tool.

11.1 `datalevin.interpret`

Var	Use
`load-edn`	Loads EDN forms from a file in Datalevin's interpreter, often for schema or setup data.
`exec-code`	Executes a string of Datalevin REPL-compatible code and prints results.
`inter-fn`	Creates a serializable interpreted function.
`definterfn`	Defines a named `inter-fn`.
`inter-fn?`	Tests whether a value is an interpreted function.
`inter-fn-from-reader`	Reads a printed interpreted function back into a callable value.

11.2 `datalevin.udf`

Descriptor key	Meaning
`:udf/lang`	Descriptor namespace used in the registry key and resolver lookup, often the implementation language, such as `:clojure`, `:java`, `:python`, `:javascript`, or an application-defined keyword.
`:udf/kind`	Function role, such as `:tx-fn`, `:query-fn`, or `:predicate`.
`:udf/id`	Stable function id keyword.
`:udf/version`	Optional keyword, string, or integer version.

Function	Use
`descriptor?`, `descriptor`	Test or validate descriptor maps.
`create-registry`	Creates a mutable UDF registry atom.
`generation`	Returns the registry generation, which changes after registration updates.
`register!`	Registers a descriptor to a callable value.
`register-resolver!`	Registers a language-level resolver that can materialize functions on demand.
`unregister!`, `registered?`, `registered-descriptor`	Manage or inspect registry entries.
`ensure-kind`, `descriptor-or-registered`	Validate that a descriptor/id is suitable for a query, predicate, or transaction context.
`materialize`, `resolve`	Resolve a descriptor to a callable function.

Use `datalevin.interpret/inter-fn` when embedded Clojure source should travel with a transaction or query. Use `datalevin.udf` when a host application should look up functions by descriptor.

11.3 Selected `datalevin.constants` Vars

The constants namespace contains more implementation constants than this table lists. This is a selected index of public defaults and tuning vars that readers are most likely to encounter in examples, diagnostics, or operational tuning.

Var	Use
`version`	Current Datalevin version string.
`default-env-flags`	Default LMDB environment flags for `open-kv`.
`default-dbi-flags`	Default DBI flags, including `:create`, `:counted`, and `:prefix-compression`.
`default-put-flags`	Default LMDB put flags.
`+max-key-size+`	Maximum LMDB key size in bytes.
`default-spill-threshold`, `default-spill-root`, `tmp-dbi`	Defaults for spill-to-disk temporary storage.
`default-port`, `default-idle-timeout`	Server defaults.
`default-connection-pool-size`, `default-connection-timeout`	Client connection defaults.

Var	Use
default-username, default-password	Built-in development defaults; production servers should override them.
default-domain, default-top	Search defaults.
default-metric-type, default-quantization, default-connectivity, default-expansion-add, default-expansion-search, default-multi?	Vector index defaults.
vec-max-buffer-bytes, *vec-chunk-bytes*	Vector blob buffering and LMDB chunk-size tuning.
wal?, *wal-durability-profile*, *datalog-wal?*, *datalog-wal-durability-profile*	Dynamic WAL defaults for new raw KV and Datalog stores. Both durability-profile vars default to :strict.
fill-db-batch-size	Batch size, in datoms, used by fill-db.
async-secondary-index-worker-max-jobs, *async-secondary-index-worker-lease-ms*	In-process async secondary-index worker limits.
async-secondary-index-retry-base-ms, *async-secondary-index-retry-max-ms*	Retry backoff bounds for failed async secondary-index jobs.
remote-db-last-modified-check-interval-ms	Remote freshness-check throttle for client DB checks.
wire-compression-threshold, *wire-compression-level*	Client/server wire compression tuning.
query-result-cache-size, query-plan-cache-size	Global Datalog query result and query plan cache sizes.
init-exec-size-threshold, plan-search-max	Query planner sampling threshold and plan-search cap.
idoc-pattern-cache-size, idoc-range-cache-size	Per-idoc-index wildcard-pattern and range-scan cache sizes.
query-pipe-timeout, query-pipe-capacity, query-pipe-batch-size	Query pipeline timeout, back-pressure, and batched scan tuning.
index-writer-batch-size, *index-writer-batch-size-pos*	Full-text search writer batch sizes without and with position indexing.

Prefer documented option maps and API functions over persisting implementation constants in application data. Consult the API docs or source when you need the complete current set of constants for a specific release.

11.4 datalevin.main

Function	Use
exec	Executes Datalevin code supplied as command-line style argument strings.

Function	Use
`copy`	Copies a database directory, optionally compacting while copying.
`drop`	Drops or clears named key-value DBIs in a database directory.
`dump`	Dumps Datalog or key-value database content.
`load`	Loads Datalog or key-value content into a database directory.
`-main`	Entry point for the `dtlv` command, dispatching `copy`, `drop`, `dump`, `exec`, `help`, `load`, `mcp`, `pods`, `repl`, `serv`, and `stat`.
`default-root-dir`	Default server root directory used by the CLI.

When neither `--datalog` nor explicit DBI names are supplied, `dump` and `load` use auto mode. Auto mode detects a Datalog store and writes a mixed text dump: the Datalog options, schema, and datoms, followed by user-created KV DBIs in the same LMDB environment. This is the right export/import path for a Datalog database that also keeps application state through `datalog-kv`. Datalevin skips its own internal and derived Datalog DBIs, so full-text, vector, embedding, and idoc indexes are treated as rebuildable indexes rather than separate user data.

```
dtlv -d /data/companydb -f companydb.edn dump
dtlv -d /data/companydb-restored -f companydb.edn load
```

Use `--datalog` when you intentionally want only the Datalog schema and datoms. Use `--all` or explicit DBI names when you want raw KV dumps. Mixed Datalog/KV auto dump is a text format; for Nippy binary dumps, choose an explicit Datalog or KV mode.

APPENDIX F

Key-Value API

This appendix is a compact reference for public Datalevin KV functions exposed from `datalevin.core`. Chapter 10 is the tutorial introduction. Chapters 19 and 20 cover durability, snapshots, WAL, re-indexing, and ingestion strategy. Appendix A covers host-language binding conventions.

1. Store Lifecycle

Function	Common forms	Purpose
`open-kv`	`(d/open-kv dir)`, `(d/open-kv dir opts)`	Open a local KV store or a remote `dtlv://` KV database.
`close-kv`	`(d/close-kv db)`	Close a KV store.
`closed-kv?`	`(d/closed-kv? db)`	Return true if the KV store is closed.
`dir`	`(d/dir db)`	Return the path or URI string for the store.
`datalog-kv`	`(d/datalog-kv conn)`, `(d/datalog-kv db)`	Return the KV handle backing a Datalog connection or DB.
`copy`	`(d/copy db dest)`, `(d/copy db dest compact?)`	Copy a Datalog or KV database to another directory, optionally compacting.
`sync`	`(d/sync db)`	Force a synchronous flush to disk.
`re-index`	`(d/re-index db opts)`	Dump, clear, reload, and rebuild a KV store with new options.

This table lists common option groups, not every accepted key.

Common option group	Keys
Capacity and environment	`:mapsize`, `:max-dbs`, `:max-readers`, `:flags`, `:temp?`, `:inmemory?`
Store identity	`:db-name`, `:db-identity`
WAL commit path	`:wal?`, `:wal-durability-profile`, `:wal-group-commit`, `:wal-group-commit-ms`
LMDB checkpointing	`:lmdb-sync-interval-ms`, `:lmdb-sync-interval`, `:wal-checkpoint-stale-threshold-ms`
WAL segments and retention	`:wal-segment-max-bytes`, `:wal-segment-max-ms`, segment preallocation options, `:wal-retention-bytes`, `:wal-retention-ms`, retention backpressure options

Common option group	Keys
Snapshots	`:snapshot-scheduler?`, `:snapshot-dir`, `:snapshot-interval-ms`, `:snapshot-max-lsn-delta`, `:snapshot-max-log-bytes-delta`, `:snapshot-max-age-ms`, snapshot contention and defer options
Client	`:client-opts`
Spill to disk	`:spill-opts`
HA	`:ha-mode`, `:ha-control-plane`, lease timing, fencing, promotion, and clock-skew options

Chapter 19 covers WAL, snapshot, checkpoint, and retention tuning in detail. WAL is disabled by default; for a new store, when it is enabled and no durability profile is specified, the profile defaults to `:strict`. Existing stores retain their persisted profile as Chapter 19 describes. The default LMDB environment flags are `#{:nordahead :notls}`. `d/datalog-kv` returns a borrowed handle that is safe to retain and reuse for as long as its owning Datalog connection or raw DB remains open. Unlike the Datalog DB object returned by `d/db`, it does not need to be obtained again after writes. Close the owner, not the borrowed KV handle. `re-index` is an offline maintenance operation; use it only when other threads or programs are not accessing the same store.

For a Datalog database that also stores application KV DBIs through `datalog-kv`, the `dtlv dump`/`dtlv load` auto mode can round-trip the mixed store: Datalog schema and datoms plus user-created KV DBIs in the same LMDB environment. Appendix E lists the `datalevin.main` functions and shows the command-line form. Use `copy` when you want a physical database copy; use dump/load when you want a portable text export/import workflow.

2. DBI Management

Function	Common forms	Purpose
`open-dbi`	`(d/open-dbi db name)`, `(d/open-dbi db name opts)`	Open a regular named DBI.
`open-list-dbi`	`(d/open-list-dbi db name)`, `(d/open-list-dbi db name opts)`	Open a DUPSORT/list DBI.
`clear-dbi`	`(d/clear-dbi db name)`	Remove data from a DBI but leave the DBI open.
`drop-dbi`	`(d/drop-dbi db name)`	Clear and delete a DBI.
`list-dbis`	`(d/list-dbis db)`	Return names of sub-databases in the store.
`stat`	`(d/stat db)`, `(d/stat db name)`	Return LMDB statistics for the environment or a DBI.
`entries`	`(d/entries db name)`	Return the number of entries in a DBI.

Use regular DBIs for one value per key and list DBIs for sorted multi-values per key. `:validate-data?` checks keys and values against KV type descriptors; it is independent of Datalog schema validation.

Option or flag group	Values
DBI options	`:key-size`, `:val-size`, `:validate-data?`, `:closed-schema?`, `:flags`
Default DBI flags	`#{:create :counted :prefix-compression}`
Common LMDB flags	`:create`, `:reversekey`, `:dupsort`, `:integerkey`, `:dupfixed`, `:integerdup`, `:reversedup`, `:counted`, `:prefix-compression`

3. Writes and Transactions

Transaction functions:

Function	Common forms	Purpose
`transact-kv`	`(d/transact-kv db txs)`, `(d/transact-kv db name txs k-type v-type)`	Synchronously apply KV write operations.
`transact-kv-async`	`(d/transact-kv-async db txs)`, `(d/transact-kv-async db name txs k-type v-type callback)`	Batch asynchronous KV writes; returns a future.
`with-transaction-kv`	`(d/with-transaction-kv [tx db] body...)`, `(d/with-transaction-kv [tx db {:timeout-ms n}] body...)`	Run reads and writes in one explicit KV transaction, optionally with a timeout.
`abort-transact-kv`	`(d/abort-transact-kv tx)`	Roll back writes from inside `with-transaction-kv`.

Transaction data forms:

Transaction item	Shape
Self-contained put	`[:put dbi-name key value]`, `[:put dbi-name key value key-type value-type]`, or `[:put dbi-name key value key-type value-type flags]`
Self-contained delete	`[:del dbi-name key]` or `[:del dbi-name key key-type]`
Single-DBI put	`[:put key value]` or `[:put key value flags]`
Single-DBI delete	`[:del key]`
List operators	`:put-list` and `:del-list` use the same transaction shapes as puts, with a sequential value collection.

`key-type`, `value-type`, and `flags` are optional. Common write flags include `:nooverwrite`, `:nodupdata`, `:current`, `:reserve`, `:append`, `:appenddup`, and `:multiple`. Use `:append` only with input already sorted in the target DBI's key order.

KV keys and put values cannot be `nil`. For list DBI writes, prefer `put-list-items` and `del-list-items`; they wrap the lower-level `:put-list` and `:del-list` transaction operators.

4. Encoding Helpers

For raw key and value encode and decode:

Function	Common forms	Purpose
`put-buffer`	`(d/put-buffer bf x)`, `(d/put-buffer bf x x-type)`	Encode a value into a `ByteBuffer`. Mostly useful in raw visitors.
`read-buffer`	`(d/read-buffer bf)`, `(d/read-buffer bf v-type)`	Decode a value from a `ByteBuffer`. Mostly useful in raw visitors.
`k`	`(d/k kv)`	Return the raw key buffer from an `IKV` visitor object.
`v`	`(d/v kv)`	Return the raw value buffer from an `IKV` visitor object.

Data types:

Descriptor form	Meaning
`:data`	General serialized data.
`:string`, `:long`, `:float`, `:double`, `:bigint`, `:bigdec`	Scalar numeric or string encodings.
`:bytes`, `:keyword`, `:symbol`, `:boolean`, `:instant`, `:uuid`	Other scalar encodings.
`[:string :instant]`, `[:long]`	Tuple key or value encodings.

KV type descriptors use unqualified keywords, not Datalog value-type keywords such as `:db.type/string`. Prefer one consistent key type, or one tuple key type, per application DBI.

`:ignore` is not a `put-buffer` or `read-buffer` descriptor. It is a range-scan sentinel accepted by scan APIs when the caller wants to skip value decoding.

5. Point Reads and Order Statistics

Function	Common forms	Purpose
`get-value`	`(d/get-value db name k)`, `(d/get-value db name k k-type v-type ignore-key?)`	Get one key's value, or `[k v]` when `ignore-key?` is false.
`get-rank`	`(d/get-rank db name k)`, `(d/get-rank db name k k-type)`	Return a key's zero-based rank in sorted key order.
`get-by-rank`	`(d/get-by-rank db name rank)`, `(d/get-by-rank db name rank k-type v-type ignore-key?)`	Return the value or pair at a zero-based key rank.

`sample-kv`	`(d/sample-kv db name n)`, `(d/sample-kv db name n k-type v-type ignore-key?)`	Return `n` random samples from a DBI, or `nil` if `n` exceeds the entry count.

Rank and sampling functions use DLMDB order-statistics support.

6. Key Ranges

Function	Common forms	Purpose
`get-first`	`(d/get-first db name k-range)`, `(d/get-first db name k-range k-type v-type ignore-key?)`	Return the first value or pair in a key range.
`get-first-n`	`(d/get-first-n db name n k-range)`, `(d/get-first-n db name n k-range k-type v-type ignore-key?)`	Return the first `n` values or pairs in a key range.
`get-range`	`(d/get-range db name k-range)`, `(d/get-range db name k-range k-type v-type ignore-key?)`	Eagerly return values or pairs in a key range.
`range-seq`	`(d/range-seq db name k-range)`, `(d/range-seq db name k-range k-type v-type ignore-key? opts)`	Lazily stream a key range in batches; close it with `with-open`.
`key-range`	`(d/key-range db name k-range)`, `(d/key-range db name k-range k-type)`	Return only keys in a range.
`range-count`	`(d/range-count db name k-range)`, `(d/range-count db name k-range k-type)`	Count KV pairs in a key range without materializing them.
`key-range-count`	`(d/key-range-count db name k-range k-type)`, `(d/key-range-count db name k-range k-type cap)`	Count keys in a range, optionally stopping at cap.
`key-range-list-count`	`(d/key-range-list-count db name k-range k-type)`, `(d/key-range-list-count db name k-range k-type cap)`	Count list items across keys in a list DBI key range.

Range specs use the keywords in Section 11. Use `get-range` for bounded results and `range-seq` for large scans.

7. Range Predicates and Visitors

Function	Common forms	Purpose
`get-some`	`(d/get-some db name pred k-range ...)`	Return the first KV pair whose predicate is true.
`range-filter`	`(d/range-filter db name pred k-range ...)`	Eagerly return KV pairs whose predicate is true.
`range-keep`	`(d/range-keep db name pred k-range ...)`	Return non-nil predicate results over a key range.
`range-some`	`(d/range-some db name pred k-range ...)`	Return the first logical true predicate result.
`range-filter-count`	`(d/range-filter-count db name pred k-range ...)`	Count KV pairs whose predicate is true.
`visit-key-range`	`(d/visit-key-range db name visitor k-range ...)`	Visit only keys in a key range for side effects.
`visit`	`(d/visit db name visitor k-range ...)`	Visit KV pairs in a key range for side effects.

`raw-pred?` controls whether callbacks receive raw storage objects or decoded values, and the raw shape depends on the helper. `visit-key-range` receives a raw key `ByteBuffer`; `visit` and range/list-range predicates over key/value pairs receive an `IKV` object; `visit-list` for one key receives a raw value `ByteBuffer`. Raw callbacks expose short-lived buffers; decode or copy data inside the callback if it must outlive that call.

8. List DBI Operations

Write functions:

Function	Common forms	Purpose
`put-list-items`	`(d/put-list-items db name k vs k-type v-type)`	Add multiple values under one key.
`del-list-items`	`(d/del-list-items db name k k-type)`, `(d/del-list-items db name k vs k-type v-type)`	Delete a whole list or selected list items.

Read functions:

Function	Common forms	Purpose
`get-list`	`(d/get-list db name k k-type v-type)`	Return all values for one key.
`visit-list`	`(d/visit-list db name visitor k k-type v-type raw-pred?)`	Visit values for one key.
`list-count`	`(d/list-count db name k k-type)`	Count values associated with one key.
`in-list?`	`(d/in-list? db name k v k-type v-type)`	Test whether one value is present under a key.

Function	Common forms	Purpose
`list-range`	`(d/list-range db name k-range k-type v-range v-type)`	Return key/value pairs matching both ranges.
`list-range-count`	`(d/list-range-count db name k-range k-type)`	Approximate count across a key range; ignores value range boundaries.
`list-range-first`	`(d/list-range-first db name k-range k-type v-range v-type)`	Return the first matching list item.
`list-range-first-n`	`(d/list-range-first-n db name n k-range k-type v-range v-type)`	Return the first n matching list items.
`list-range-filter`	`(d/list-range-filter db name pred k-range k-type v-range v-type raw-pred?)`	Filter list range results with a predicate.
`list-range-keep`	`(d/list-range-keep db name pred k-range k-type v-range v-type raw-pred?)`	Return non-nil predicate results over list range matches.
`list-range-filter-count`	`(d/list-range-filter-count db name pred k-range k-type v-range v-type raw-pred?)`	Count list range matches whose predicate is true.
`list-range-some`	`(d/list-range-some db name pred k-range k-type v-range v-type raw-pred?)`	Return the first logical true predicate result.
`visit-list-range`	`(d/visit-list-range db name visitor k-range k-type v-range v-type raw-pred?)`	Visit list range matches for side effects.

List DBIs are opened with `open-list-dbi`. Each key maps to a sorted set of values, and list items must fit within the LMDB key-size limit.

9. WAL and Snapshots

Function	Common forms	Purpose
`txlog-watermarks`	`(d/txlog-watermarks db)`	Return WAL watermarks and runtime sync state, or `{:wal? false}` when WAL is disabled.
`open-tx-log`	`(d/open-tx-log db from-lsn)`, `(d/open-tx-log db from-lsn upto-lsn)`	Read committed WAL records from an LSN range.
`force-lmdb-sync!`	`(d/force-lmdb-sync! db)`	Force an LMDB environment sync for the store.

Function	Common forms	Purpose
`create-snapshot!`	`(d/create-snapshot! db)`	Create or rotate LMDB snapshots and update WAL snapshot bookkeeping.
`list-snapshots`	`(d/list-snapshots db)`	Return available snapshot metadata.
`gc-txlog-segments!`	`(d/gc-txlog-segments! db)`, `(d/gc-txlog-segments! db retain-floor-lsn)`	Garbage-collect WAL segments, optionally retaining records from a floor LSN.

10. Environment Flags

Functions:

Function	Common forms	Purpose
`set-env-flags`	`(d/set-env-flags db flags on?)`	Set or clear LMDB environment flags.
`get-env-flags`	`(d/get-env-flags db)`	Return currently active LMDB environment flags.

Flags:

Common flag group	Keywords
Read and locking	`:rdonly-env`, `:notls`, `:nolock`
Durability and write behavior	`:nosync`, `:nometasync`, `:writemap`, `:mapasync`
Filesystem and memory	`:nordahead`, `:nomeminit`, `:nosubdir`, `:fixedmap`

For a Datalog connection, call `datalog-kv` first and apply `set-env-flags`, `get-env-flags`, or `sync` to the backing KV handle.

11. Range Keywords

Range type	Meaning
`:all` / `:all-back`	Whole keyspace, forward or backward.
`:at-least` / `:at-most-back`	Inclusive lower bound, forward or backward.
`:at-most` / `:at-least-back`	Inclusive upper bound, forward or backward.
`:closed` / `:closed-back`	Inclusive lower and upper bounds.
`:closed-open` / `:closed-open-back`	Inclusive lower bound, exclusive upper bound.
`:open-closed` / `:open-closed-back`	Exclusive lower bound, inclusive upper bound.
`:open` / `:open-back`	Exclusive lower and upper bounds.
`:greater-than` / `:less-than-back`	Exclusive lower bound.

Range type	Meaning
`:less-than` / `:greater-than-back`	Exclusive upper bound.

The same range shapes are used for key ranges and list value ranges.

APPENDIX G

Client API

The `datalevin.client` namespace is Datalevin's public administrative client API for a remote Datalevin server. Use it when you need to manage server-side databases, users, roles, permissions, read replicas, HA membership, or connected clients.

This appendix is a reference. Chapter 22 explains the operational model: server mode, RBAC, replicas, HA, monitoring, and deployment choices.

Most application reads and writes should still use ordinary connections from `datalevin.core/get-conn` or `datalevin.core/create-conn`. Use `datalevin.client` for server administration and cluster operations.

The appendix uses the Clojure names. Java, Python, and JavaScript expose the same administrative families with their normal naming conventions: for example `Datalevin.newClient`, `new_client`, and `newClient`; method names become `listDatabases`, `list_databases`, and `listDatabases`, respectively.

```
(require '[datalevin.client :as cl])

(def client
  (cl/new-client "dtlv://admin:secret@db-host:8898"))
```

The URI identifies the server and credentials. Database names are passed as separate arguments to the administrative functions.

1. Client Lifecycle

Function	Shape	Use
`new-client`	`(cl/new-client uri)`, `(cl/new-client uri opts)`	Create a pooled administrative client for a Datalevin server.
`disconnect`	`(cl/disconnect client)`	Close the client's pooled network connections.
`close-client`	`(cl/close-client client)`	Idempotent close helper. This is the resource cleanup function used by non-Clojure bindings.
`disconnected?`	`(cl/disconnected? client)`	Return true when the client pool is closed.
`get-id`	`(cl/get-id client)`	Return the server-assigned client UUID for this administrative client.

`new-client` performs authentication and maintains a connection pool. Creating a client includes secure password hashing work, so long-running tools should reuse one client instead of constructing a new one for every request.

Common options:

Option	Default	Meaning
`:pool-size`	`3`	Number of pooled connections held by the client.
`:time-out`	`60000`	Milliseconds to wait for an open connection or request before throwing.
`:ha-write-retry-timeout-ms`	Derived from HA timing and capped by `:time-out`	Extra retry budget for retryable HA write failover.
`:ha-write-retry-delay-ms`	`100`	Sleep between HA write retry rounds.

Credentials in the URI should be URL encoded when they contain special characters.

```
(def client
  (cl/new-client
    "dtlv://admin:secret@db-host:8898"
    {:pool-size 8
     :time-out 120000
     :ha-write-retry-timeout-ms 8000
     :ha-write-retry-delay-ms 150}))

(try
  (cl/list-databases client)
  (finally
    (cl/disconnect client)))
```

2. Database Administration

Function	Shape	Use
`create-database`	`(cl/create-database client db-name db-type)`	Create a server-side database. `db-type` is `:datalog` or `:key-value`.
`open-database`	`(cl/open-database client db-name db-type)`, plus schema/opts/info arities	Open a database on the server. `db-type` is `"datalog"`, `"kv"`, or `"engine"`.
`close-database`	`(cl/close-database client db-name)`	Force close a server-side database and disconnect clients using it.
`drop-database`	`(cl/drop-database client db-name)`	Delete a database. Close it first if it is in use.

Function	Shape	Use
`list-databases`	`(cl/list-databases client)`	List server-side databases.
`list-databases-in-use`	`(cl/list-databases-in-use client)`	List databases currently open on the server.

Database names supplied to `create-database` are normalized to kebab case by the server. Prefer stable, simple names and treat them as administrative identifiers, not user-facing display names.

`create-database` creates durable Datalog or KV database directories. The `"engine"` type for `open-database` is for server-side search engine activation, not for `create-database`.

Most application code does not call `open-database` directly. A remote `get-conn` or `create-conn` call opens the target database as needed. Use `open-database` from administrative code when you deliberately want to open or activate a server-side database without creating an application connection. The longest arity accepts `schema`, `opts`, and `return-db-info?`; `schema` applies to Datalog opens, and `return-db-info?` returns the server's database information map.

```
(cl/create-database client "app" :datalog)

(cl/open-database
  client
  "app"
  "datalog"
  {:user/email {:db/valueType :db.type/string
                :db/unique    :db.unique/identity}}
  {:wal? true})

(cl/list-databases client)
```

`close-database` is operationally stronger than closing one client connection: it asks the server to close the database and disconnect clients that are using it. Use it before maintenance operations such as dropping a database.

3. Users, Roles, and Permissions

Datalevin Server permissions are role based. Create users, create roles, grant permissions to roles, and assign roles to users.

3.1 Users

Function	Shape	Use
`create-user`	`(cl/create-user client username password)`	Create a login user. The username is normalized to kebab case.
`reset-password`	`(cl/reset-password client username password)`	Set a new password for a user.

Function	Shape	Use
`drop-user`	`(cl/drop-user client username)`	Delete a user.
`list-users`	`(cl/list-users client)`	List users.

Passwords are stored as salted hashes on the server. User creation and password reset intentionally perform expensive password hashing.

3.2 Roles

Function	Shape	Use
`create-role`	`(cl/create-role client role-key)`	Create a role. `role-key` is a keyword.
`drop-role`	`(cl/drop-role client role-key)`	Delete a role.
`list-roles`	`(cl/list-roles client)`	List roles.
`assign-role`	`(cl/assign-role client role-key username)`	Assign a role to a user.
`withdraw-role`	`(cl/withdraw-role client role-key username)`	Remove a role from a user.
`list-user-roles`	`(cl/list-user-roles client username)`	List roles assigned to a user.

3.3 Permissions

Function	Shape	Use
`grant-permission`	`(cl/grant-permission client role-key action object target)`	Grant a permission to a role.
`revoke-permission`	`(cl/revoke-permission client role-key action object target)`	Revoke a permission from a role.
`list-role-permissions`	`(cl/list-role-permissions client role-key)`	List permissions granted to a role.
`list-user-permissions`	`(cl/list-user-permissions client username)`	List effective permissions for a user through assigned roles.

Permission actions:

Action	Meaning
`:datalevin.server/view`	Read/view access.
`:datalevin.server/alter`	Includes view and allows modifying existing resources.
`:datalevin.server/create`	Includes lower actions and allows creation.
`:datalevin.server/control`	Administrative control; includes create, alter, and view.

Permission objects:

Object	Target examples
`:datalevin.server/database`	Database name, or `nil` for all databases.
`:datalevin.server/user`	Username, or `nil` for all users.
`:datalevin.server/role`	Role keyword, or `nil` for all roles.
`:datalevin.server/server`	Usually `nil`; subsumes database, user, and role objects.

Actions and objects are hierarchical. A granted action satisfies requests for lower actions in the same chain, so `:control` satisfies `:create`, `:alter`, and `:view`; `:create` satisfies `:alter` and `:view`. A server-object grant satisfies database, user, and role object requests when the target rule also matches.

Selected operation permissions:

Operation	Required permission
`list-databases`, `list-databases-in-use`	`:datalevin.server/create` on `:datalevin.server/database` with target `nil`.
`query-system`, `show-clients`	`:datalevin.server/view` on `:datalevin.server/server` with target `nil`.
`disconnect-client`	`:datalevin.server/control` on `:datalevin.server/server` with target `nil`.
`grant-permission`, `revoke-permission`	`:datalevin.server/alter` on `:datalevin.server/role` for the target role.
`ha-update-membership!`	Database `:datalevin.server/alter` plus server `:datalevin.server/control`.

Example: read-only access to one database:

```
(cl/create-user client "alice" "initial-password")
(cl/create-role client :app.role/reader)

(cl/grant-permission
  client
  :app.role/reader
  :datalevin.server/view
  :datalevin.server/database
  "app")

(cl/assign-role client :app.role/reader "alice")
```

Example: administrative server control:

```
(cl/create-role client :app.role/admin)

(cl/grant-permission
  client
  :app.role/admin
  :datalevin.server/control
  :datalevin.server/server
```

```
nil)
```

Use the narrowest permission that supports the task. Application users usually need database-scoped permissions, not server control.

4. Replication and HA

Function	Shape	Use
`replica-status`	`(cl/replica-status client db-name)`	Return async read-replica status for an open database.
`ha-update-membership!`	`(cl/ha-update-membership! client db-name spec)`	Commit an operator-driven consensus HA membership update.
`*ha-read-min-tx*`	`(binding [cl/*ha-read-min-tx* tx] ...)`	Dynamic Clojure var requiring HA read requests in the binding scope to observe at least transaction `tx`.

`replica-status` is for async read replicas. The returned map includes fields such as `:replica-applied-lsn`, `:replica-source-durable-lsn`, `:replica-source-committed-lsn`, `:replica-lag-lsn`, `:replica-degraded-reason`, and `:replica-last-error`.

```
(cl/replica-status client "app")
```

`ha-update-membership!` changes the authoritative membership for an open HA database. Chapter 22 explains the HA model, fencing, voters, leases, and the operational checks around this function.

The server accepts this call only outside `with-transaction`. The caller must have database alter permission for the target database and server control permission for the server. The target database must already be open and running consensus HA; the call is rejected for non-HA databases.

Useful `spec` keys:

Key	Meaning
`:ha-members`	Replacement data-node member list.
`:ha-control-plane {:voters [...]}`	Replacement control-plane voter config.
`:ha-control-plane-voters`	Alternative shape for replacement voters.
`:expected-membership-hash`	Optional compare-and-set guard.
`:clear-leases?`	Defaults to `true`; forces leadership to be reacquired.
`:replace-voters?`	Defaults to `true`; updates the Raft voter set.
`:timeout-ms`	Optional control-plane operation timeout.

```
(cl/ha-update-membership!
  client
  "app"
  {:ha-members
   [{:node-id 1 :endpoint "node-a:8898"}
    {:node-id 2 :endpoint "node-b:8898"}
    {:node-id 3 :endpoint "node-c:8898"}]
   :ha-control-plane
   {:voters [{:peer-id "node-a:9098" :promotable? true :ha-node-id 1}
             {:peer-id "node-b:9098" :promotable? true :ha-node-id 2}
             {:peer-id "node-c:9098" :promotable? true :ha-node-id 3}]}})
```

Membership updates are operator actions. Automating them without health checks, clock discipline, and rollback procedures can turn failover into a data-safety problem.

`*ha-read-min-tx*` is for read-after-write or monotonic-read requirements in remote Datalog HA clients. When it is bound to an integer transaction id, non-write Datalog read requests carry that minimum-tx requirement; stale read targets can reject the request, and the client will retry eligible HA endpoints.

5. System Database and Connected Clients

Function	Shape	Use
`query-system`	`(cl/query-system client query & args)`	Run a Datalog query against the server system database. Do not pass $; the server supplies it.
`show-clients`	`(cl/show-clients client)`	Return information about currently connected clients.
`disconnect-client`	`(cl/disconnect-client client client-id)`	Force disconnect a connected client by UUID.

`query-system` is for administrative inspection. The query is a normal Datalog query, but unlike `d/q`, arguments do not include a DB object:

```
(cl/query-system
  client
  '[:find ?user
    :where [?u :user/name ?user]])
```

Use `show-clients` before `disconnect-client`:

```
(def clients (cl/show-clients client))
```

```
;; client-id must be a java.util.UUID value returned by show-clients.
(cl/disconnect-client client some-client-id)
```

Do not confuse `disconnect-client` with `disconnect`. `disconnect-client` disconnects another server-side client session. `disconnect` closes the administrative client object in your process. `get-id` returns the UUID of the current administrative client when you need to compare it with `show-clients` output.

6. Operational Notes

- Keep one administrative client per operation scope and reuse it.
- Close clients with `disconnect` or `close-client` when the operation scope ends.
- Use server URIs such as `dtlv://admin:secret@host:8898`; pass database names separately.
- URL encode usernames and passwords in URIs when they contain special characters.
- Prefer least-privilege roles for applications and reserve server control for operators.
- Treat `close-database`, `drop-database`, `ha-update-membership!`, and `disconnect-client` as operational actions that should be logged.
- Use Chapter 22 for deployment guidance and this appendix for function lookup.

Summary

`datalevin.client` is the public API for administering a Datalevin server. It is not the normal query or transaction surface for application data; that remains the connection API covered in the main chapters. Use the client namespace when the unit of work is the server, a database lifecycle operation, RBAC, replica status, HA membership, or connected-client management.

Index

www.ingramcontent.com/pod-product-compliance
Lightning Source LLC
LaVergne TN
LVHW082027200926
848481LV00010B/2935